Picture Archiving and Communication Systems (PACS) in Medicine

NATO ASI Series

Advanced Science Institutes Series

A series presenting the results of activities sponsored by the NATO Science Committee, which aims at the dissemination of advanced scientific and technological knowledge, with a view to strengthening links between scientific communities.

The Series is published by an international board of publishers in conjunction with the NATO Scientific Affairs Division

A Life Sciences B Physics	Plenum Publishing Corporation London and New York
C Mathematical and Physical Sciences D Behavioural and Social Sciences E Applied Sciences	Kluwer Academic Publishers Dordrecht, Boston and London
F Computer and Systems Sciences G Ecological Sciences H Cell Biology I Global Environmental Change	Springer-Verlag Berlin Heidelberg New York London Paris Tokyo Hong Kong Barcelona Budapest

NATO-PCO DATABASE

The electronic index to the NATO ASI Series provides full bibliographical references (with keywords and/or abstracts) to more than 30 000 contributions from international scientists published in all sections of the NATO ASI Series. Access to the NATO-PCO DATABASE is possible in two ways:

- via online FILE 128 (NATO-PCO DATABASE) hosted by ESRIN, Via Galileo Galilei, I-00044 Frascati, Italy.
- via CD-ROM "NATO-PCO DATABASE" with user-friendly retrieval software in English, French and German (© WTV GmbH and DATAWARE Technologies Inc. 1989).

The CD-ROM can be ordered through any member of the Board of Publishers or through NATO-PCO, Overijse, Belgium.

Series F: Computer and Systems Sciences Vol. 74

Picture Archiving and Communication Systems (PACS) in Medicine

Edited by

H. K. Huang

Medical Imaging Division, Department of Radiological Sciences
University of California, Los Angeles
Los Angeles, CA 90024-1721, USA

O. Ratib

Digital Imaging Unit, Center of Medical Informatics
University Hospital of Geneva
CH-1211 Geneva-4, Switzerland

A. R. Bakker

BAZIS
Schipholweg 97, 2316 XA Leiden, The Netherlands

G. Witte

Radiological Clinic, Diagnostic Department
University Hospital Hamburg-Eppendorf
Martinistrasse 52, W-2000 Hamburg 20, FRG

With the assistance of

K. S. Chuang

Medical Imaging Division, Department of Radiological Sciences
University of California, Los Angeles
Los Angeles, CA 90024-1721, USA

Springer-Verlag
Berlin Heidelberg New York London Paris Tokyo
Hong Kong Barcelona Budapest
Published in cooperation with NATO Scientific Affairs Division

Proceedings of the NATO Advanced Study Institute on Picture Archiving and Communication Systems (PACS) in Medicine, held in Evian, France, October 14–26, 1990.

ISBN-13: 978-3-642-76568-1 e-ISBN-13: 978-3-642-76566-7
DOI: 10.1007/978-3-642-76566-7

Library of Congress Cataloging-in-Publication Data.
NATO Advanced Study Institute on Picture Archiving and Communication Systems (PACS) in Medicine (1990: Evian-les-Bains, France) Picture archiving and communication systems (PACS) in Medicine/edited by H.K. Huang ... [et al.]. p. cm.—(NATO ASI series. Series F, Computer and systems sciences; vol. 74) "Proceedings of the NATO Advanced Study Institute on Picture Archiving and Communication Systems (PACS) in Medicine, held in Evian, France, October 14–26, 1990"—T.p. verso.
"Published in cooperation with NATO Scientific Affairs Division." Includes indexes.

1. Radiography, Medical—Digital techniques—Congresses. 2. Imaging systems in medicine—Congresses. 3. Information storage and retrieval systems—Medicine—Congresses. I. Huang, H.K., 1939– . II. Title. III. Series: NATO ASI series. Series F, Computer and system sciences; vol. 74. RC78.7.D35N38 1990 616.07'54—dc20 91-11063

Softcover reprint of the hardcover 1st edition 1991
45/3140-543210 – Printed on acid-free-paper

Preface

This volume contains the proceedings of the NATO Advanced Study Institute on "Picture Archiving and Communication Systems (PACS) in Medicine" held in Evian, France, October 14 - 26, 1990. The program committee of the institute consisted of H.K. Huang (Director), Osman Ratib, Albert Bakker, and Gerd Witte. This institute brought together approximately 90 participants from 15 countries.

These proceedings are the accumulation of eight years of research and development results in PACS by various dedicated groups throughout the world. The purpose of this institute was to review the most recent technology available for PACS and some clinical results. The readers should notice the remarkable advances in this field by comparing the contents in these proceedings with those in a previous institute on "Pictorial Information Systems in Medicine" held August 27 - September 7, 1984 in Braunlage/Harz, Federal Republic of Germany, and published as Vol. 19 in this series.

The institute was organized according to four categories: PACS components and system integration, PACS and related research in various countries and manufacturing companies, clinical experience and research support, and participants' scientific communications.

In PACS components, we included image acquisition, workstations, data storage and networking. In system integration, topics on interfaces between Hospital Information System (HIS), Radiology Information System (RIS) and PACS, clinical reports, the ACR/NEMA standard, databases, reliability, and system integration were discussed. This lecture series emphasized the technical detail and "how to" aspects.

In PACS and related research in various countries and manufacturing companies, invited lecturers who are responsible for or participate in major PACS research and development in their respective countries summarize the current status. The summaries cover technical, administrative, and financial aspects.

The countries represented among the presentations include Germany, Italy, France, Belgium, Japan, Sweden, Turkey, The Netherlands, Austria, Switzerland, and the United Kingdom (in the order of the presentations). Researchers responsible for PACS research by major imaging manufacturers were also invited to present their current PACS status and its perspective. These manufacturers include General Electric, Philips/AT&T, and Hitachi.

In clinical experience and PACS research support series, researchers were invited to present their PACS clinical experience. The researchers are from centers including: University of Ottawa, Canada; University of Pennsylvania; Hokkaido University, Japan; Georgetown University; and UCLA. Two major PACS funding sources in the United States are the National Institutes of Health and the U.S. Armed Forces. The program directors of these two government agencies presented the current PACS funding situation.

Finally, selected topics related to PACS from the participants were organized into short presentations. The purpose of this series was to encourage information exchange among participants.

In addition, three workshops on data security, image quality and perception, and cost analysis, as well as two demonstrations on radiology literature search and receiver operating characteristics (ROC) software packages were organized in the evenings.

We are indebted to the authors for their excellent presentations and for delivering the manuscripts promptly, and to the participants for their enthusiastic discussions. The institute would not have been possible without the help of many people from the Medical Imaging Division, Department of Radiological Sciences, UCLA, especially Dr. K.S. Chuang and Ms. Betty Norton who administered the institute during the entire two-week period. We would like especially to thank the Unité d'Imagerie Numérique, University of Geneva Hospital for local arrangements. And last but not least we thank Dr. L.V. da Cunha, Director, ASI Programme for his guidance and NATO for providing the financial basis for this institute.

Los Angeles, Geneva,
Leiden, and Hamburg
January 1991

H.K. Huang
Osman Ratib
Albert R. Bakker
Gerd Witte

Table of Contents

European Collaborative PACS Projects

Participants' Scientific Communications

PACS Research and Development - A Review and Perspective

H.K. Huang

Department of Radiological Sciences, Medical Imaging Division,
University of California Los Angeles, Los Angeles, CA 90024, USA

Abstract: This paper describes the preparation, planning, and scientific program of the NATO Advanced Science Institute (ASI): Picture Archiving and Communication Systems in Medicine held in Evian, France from October 14-26, 1990. The three methods of PACS research and development based on the organization are described. The basic PACS components are reviewed along with some current designed concepts. The scientific program is organized in four categories: PACS components and system integration, PACS and related research in various countries and manufacturers, clinical experience and PACS research support, and participants scientific communication and workshop.

Keywords: Picture archiving and communication systems, digital radiology, image processing, communication network, computed radiography, PACS history

Introduction

The First International Conference and Workshop on Picture Archiving and Communication Systems (PACS) was held in Newport Beach, California, January, 1982 [1]. The meeting was organized by Andre Duerinckx and sponsored by the SPIE - The International Society for Optical Engineering. Thereafter the PACS conference combined with Medical Imaging and became an annual event, always held in late January and early February in Newport Beach, California. Sam Dwyer and Roger Schneider took the responsibility as co-chairmen of the conference as well as the editors of the proceedings [2,3]. The conference will move to San Jose, California in 1991.

In Japan, the Japan Association of PACS held its First PACS conference of Japan in 1982 and the First International symposium on PACS and PHD (Personal Health Data) was held in July, 1982 [4], sponsored by the Japan Association of Medical Imaging Technology (JAMIT). The

NATO ASI Series, Vol. F 74
Picture Archiving and Communication Systems (PACS)
in Medicine. Edited by H. K. Huang et al.

conference combined with the Medical Imaging Technology meeting, also became an annual event. In Europe, the EUROPACS (Picture Archiving and Communication systems in Europe) has held annual meetings since 1982 and is the driving force for European PACS standardization.

Although introduced in 1982, PACS has been very slowly implemented in the clinical environment for several reasons:

Technically, PACS consists of many components, including image acquisition, communications, archiving, display, and human-machine interfaces. It is difficult for a single manufacturer or a single department to assemble a large team with a sufficiently diverse background to cover all aspects of PACS. Economically, implementation of PACS is expensive, requiring substantial investment in research and development of all state-of-the-art PACS-related technologies. In addition, PACS maintenance is difficult. For example, when a vendor makes changes on the CT or MR image format, the software in the PACS acquisition node attached to the scanner must also be changed. Thus, the radiology department must have a close working relationship with the vendor as well as maintain a knowledgeable team to handle the software and hardware upgrade. Administratively, PACS is an image management system, and its implementation requires a reorganization of operational procedures in a radiology department. Psychologically, using digital video displays for radiologic images is not a traditional method in clinical practice, and as a result, it is not easily accepted by radiologists and clinicians.

The situation gradually changed starting two years ago. Virtually all the technical requirements of a PACS system had been met by industry and the cost of components became affordable. Administrators started to realize the importance of PACS in an integrated health care delivery system and some are ready for a longer term investment. The data derived from a few successful clinical PACS modules have provided solid data indicating that PACS is well accepted by physicians. Also, a new generation of radiologists, physicians, and clinical staff brought up in a personal computer and digital gadget era are no longer intimidated by the keyboard and dials. Therefore, we believe the time is ripe to request the NATO Advanced Study Institutes (ASI) to sponsor a workshop to review the PACS research and development in various countries. We initiated the contact with NATO Scientific Affairs Division in August, 1988 and submitted a proposal in April 1989. The NATO ASI requested some modifications in the proposal in July 1989. The revised proposal was resubmitted in September, 1989. The NATO ASI approved the proposal in November 1989 and authorized the program committee to start the workshop preparation. Dr. L.V. da Cunha, Director, ASI Programme has been providing us with guidance and encouragement throughout the preparation.

Methods of PACS Research and Development

Most PACS research and development efforts are initiated by university hospitals and academic departments and from research laboratories in major imaging manufacturers. There are generally three methods of approach. In the first approach, a multidisciplinary team, with technical know-how, is assembled by a department. The team becomes a system integrator, selecting PACS components from various manufacturers. The team writes the PACS software according to the clinical requirement. In the second approach, a team of experts is assembled to write a detailed specification of the PACS for a certain clinical environment. A manufacturer is contracted to implement the system. In the third approach, the manufacturer develops a turnkey PACS system and installs it in a department for clinical evaluation.

There are advantages and disadvantages in either approach. The advantages in the first approach is that the PACS system can be continuously upgraded by the research team using state-of-the-art components and therefore it will not become obsolete. The system so designed is geared to the clinical environment and the system is not at the mercy of the manufacturer. The disadvantages are that it requires a substantial commitment by the department to assemble a multi-disciplinary team. The system developed will be one of the kind, therefore service and maintenance will be difficult because it consists of components from different manufacturers.

The advantage of the second approach is that the PACS specification is written for a certain clinical environment and the responsibility of implementation is delegated to the manufacturer. The department acts as a purchase agent and does not have to concern with the installation. The disadvantage is that the specification tends to be over-ambitious because some outside experts may have underestimated the technical and operational difficulty. The designated manufacturer, lacking the learning experience of each component in a clinical environment, may tend to overestimate the performance of each component. As a result, the total PACS may not meet the complete specification. The costs of contracting the manufacturer to develop a specified PACS is also expensive due to the manufacturer's profit margin of building only one system.

The advantage of the third approach is that it is a generalized production system, the costs tend to be lower. In this approach, since it is a production, the manufacturer needs a few years to complete the production cycle. By the time it is commercially available, some components already become obsolete because of the fast pace in high technology development. Also, we are still doubtful whether a generalized PACS can be used for every specialty in a department and be used for every radiology department. We foresee that these three approaches will gradually merge as additional clinical PACS data become more available.

Due to various operating conditions, the methods of approach and the emphasis on PACS research and development in North America, European countries, and Japan is different. In the

United States, PACS research is mostly solicited from various departments to government agencies and manufacturers for support. All three methods of approach are used. In the European countries, PACS research is supported either through a national or a province resource. Because most of the PACS components are developed in the United States and Japan, they are not as readily available. The European research teams tend to work with a single major manufacturer. European research teams emphasize PACS modelling and simulation, as well as investigating the image processing component of PACS. In Japan, PACS research and development is a national project. The national resource is distributed to various manufacturers and university hospitals. A manufacturer integrates a PACS system and installs it in a hospital for clinical evaluation. The manufacturer's PACS specification tends to be rigid and leaves little room for the hospital research team to modify the technical specification.

PACS Components

In general, PACS research and development can be divided into three phases: Phase One involves system integration of various components required in a PACS, Phase Two is system installation in the clinical environment, and Phase Three is clinical evaluation.

PACS consists of three major components [5]: image acquisition, image storage, and image display stations. A digital network connects these components and a computer/image processing system as an integrated system, and transmits images throughout the department or the hospital. In addition, a radiology information system is needed for transferring patient information to the PACS, and a reporting system for generating and distribution of clinical findings to referral physicians.

Digital image acquisition consists of four units:

1. A digital network transmits images directly from CT, US, or MR to the archival computer system and to storage devices.
2. A magnetic tape drive reads images acquired by imaging modalities located either inside or outside the department which are not connected to the network (e.g., a CT study of a patient done in another hospital).
3. A high-resolution laser film digitizer converts conventional x-ray films to a digital image that maintains the diagnostic quality of the original film.
4. A CR system acquires a conventional x-ray image in a digital format.

Digital image storage relies on the fast magnetic disk arrays and optical disk library. Magnetic disks are more expensive in terms of bytes/dollar and are used in short-term image archiving, whereas the optic disk library has a very large capacity, costs less and is used for long-term archiving. These two types of storage devices are controlled through software to regulate the image flow between short-term and long-term archival.

Image display stations can be located in the radiology department, on patient floors, and in the offices of referring physicians. Each station consists of one or several video monitors, a computer, magnetic or small optical disk storage. The video monitor can have a resolution of 512, 1,024, or 2,048 lines. The 512-line monitor is inexpensive and can be used for CT, MR, US, and conventional x-ray images involving gross abnormalities, such as those often seen in the intensive care unit. The 2,048-line monitor is expensive and is used for displaying very high resolution radiologic images, most often used in the radiology department. At the intermediate level, the 1,024-line monitor, which is widely available and costs about one-fifth as much as a 2,048-line monitor, has adequate resolution for 90% of radiologic examinations, and its screen can be split to display 4 CT or US images of 16 MR images simultaneously.

Digital network development over the past 10 years has emphasized text information communication. A digitized x-ray image with 4MB has the equivalent of 50,000 lines of information and is very time-consuming to transmit using conventional communication protocols. Currently, using the Ethernet under optimal conditions, transmitting a 4 MB x-ray image within the same building takes about 20 seconds, which is too slow for routine clinical practice. The emerging FDDI (fiber optic distributive data interface) has the potential of improving the performance to five times better. Proprietary high speed imaging network is now available and can transmit a 4MB x-ray image in one second, satisfying most clinical applications. PACS digital network design needs a careful analysis of clinical requirement. A prevailing design concept is the distributive imaging network, which analyzes the operation in a radiology department to predict when a certain image will be required at a certain location (section) in the department. Images are then automatically distributed, from either the acquisition or the storage unit to the display station, before they are requested by radiologists or clinicians, allowing them to be reviewed instantaneously. The disadvantage of distributive image networking is that each display station needs a storage capacity to hold a certain number of images. This is expensive and will increase the cost of the display station, meaning that inexpensive stations for installation in physicians' offices will not be feasible.

We are very confident that PACS research and development has passed Phase One and is into Phases Two and Three, i.e., system installation and clinical evaluation. Although, we have some solid clinical data regarding these two latter phases it is not enough to be statistically significant. With several major PACS installations in the next couple of years, we anticipate that significant contributions from these centers will result.

Program Highlights

This NATO ASI scientific program is organized in four categories:

1. PACS components and system integration.
   In PACS components, we include acquisition, workstation, data storage and networking. In system integration, topics in HIS, RIS and PACS integration, clinical reports, ACR/NEMA standard, database, fault tolerance, and system integration will be discussed. This series of lectures will be emphasized in technical detail and "how to". They are scheduled in the morning.
2. PACS and related research in various countries and manufacturers.
   In this series, invited lecturers who are responsible or who are participating in the major PACS research and development in his/her respective country will summarize the current status. The summary includes both technical, administrative, and financial aspects. Also, researchers responsible for PACS research in various major imaging manufacturers are also invited to present the current PACS status and their perspective.
3. Clinical experience and PACS research support.
   In this series, clinical researchers who have clinical experience in PACS are invited to present their clinical experience. Two major PACS funding sources in the United States are the National Institutes of Health and the U.S. Armed Forces. The program directors of these two government agencies will present the current PACS funding situation in the United States.
4. Participants' scientific communication
   Selected topics related to PACS from the participants are organized in short presentations. The purpose of this series is to encourage information exchanges among participants.
   In addition, three workshops on data security, image quality and perception, and cost analysis will be organized by interested participants.

Summary

We believe that the development of PACS related technology is mature enough for system integration and clinical evaluation. The purpose of this NATO ASI workshop is for a small group of scientists, engineers, radiologists, and administrators to get together for an informal exchange of ideas and concepts on current technology, research and development opportunity, as well as sharing some clinical experience.

Acknowledgement

This work was supported in part by Public Health Service Grant No. R01 CA 39063, R01 CA 40456, and P01 CA 51198, awarded by the National Cancer Institute, Department of Health and Human Services, and by the UCLA Department of Radiological Sciences Research Fund.

References

1. Duerinckx, A ed. Picture archiving and communication systems (PACS) for medical applications. 1st International Conference and Workshop Proceedings SPIE - International Society for Optical Engineering, Vol. 318, 1982.

2. Dwyer SJ III, ed. Picture archiving and communication system (PACS) for medical applications. Second International conference workshop for picture archiving and communication systems (PACS) for medical applications. Proceedings SPIE - International Society for Optical Engineering, Bellingham, Washington, Vol. 418, 1983.

3. Schneider RH, Dwyer SJ III, Jost RG, ed. Medical Imaging III. Proceedings SPIE - International Society for Optical Engineering, Vol. 1090, 1091, 1092, and 1093; 1989.

4. Medical Imaging Technology. The 5th MIT and the 3rd PACS/PHD Symposia. Vol. 4, No. 2, July 1986.

5. Huang HK. Elements of digital radiology. Englewood Cliffs, New Jersey: Prentice-Hall, 1987:333, 364.

Principles of Computed Radiography

Gerd Witte

Radiological Clinic, Diagnostic Department, University Hospital Eppendorf,
Martinistraße 52, 2000 Hamburg, F.R.G.

Abstract: The introduction and use of computers has expanded the domain of diagnostic imaging, making possible such procedures as computer tomography (CT), magnetic resonance (MR), and digital subtraction angiography (DSA).
Although these new computer-aided techniques have considerably changed and improved the spectrum of diagnostic methods, nearly 70% of the medical examinations in diagnostic imaging are still performed with conventional X-ray procedures (3, 11). Computed radiography (CR) offers the chance of obtaining information from X-ray examinations in digital form so that the advantages of the digital techniques can be applied to "conventional X-ray" diagnostics.

Keywords: Computed Radiography, digital technique, postprocessing, dose reduction, electronic archiving (PACS).

Technique

In Computed Radiography (CR) a film cassette with a special storage film (image plate; IP) replaces the conventional screen-film X-ray cassette. The storage capacity of the IP is the result of anorganic crystals mounted on a polyester carrier-film. Since these crystals are activated by exposure to a normal X-ray tube. Only the screen/film system has to be replaced by the IP system, otherwise, the existing X-ray equipment can be used. After exposure, the digital image reading system scans the IP point by point with a laser beam. The resulting luminscent light value is transformed by a photo multiplier into an electric signal which is stored as a numerical value (fig. 1). The stored information can be computed as a picture and then shown either on a monitor

NATO ASI Series, Vol. F 74
Picture Archiving and Communication Systems (PACS)
in Medicine. Edited by H. K. Huang et al.

or, with the help of a laser camera as a "normal X-ray picture". After the scanning the remaining unused information on the IP is deleted by halogen light. The IP is then available again for further examinations. The lifetime of an IP is nearly the same as that of the usual screen/film system (5,7,9,11) (fig. 2). The principles of the image reading and Computed Radiography (CR) are shown in figures 1 and 2.

In the conventional X-ray examination, the X-ray film has different tasks, including recording, storage, transport and display of the examination information. In CR each phase of the examination can be performed separately and can therefore be processed independently of the other phases. Changing one step has no direct effect on other parts of the information chain. For example, a change of the dose of the exposure has no direct effect on the brightness in the final image (1,6,9).
Another advantage of the IP over the screen/film system is the directly proportional linear relation over a wide range between radiation dose and the brightness in the final image. The linearity of the relation results in a reduced sensitivity to incorrect exposure. In the conventional screen/film system on the other hand, the sensitivity is described by an s-shaped curve. A direct proportionality exists only in the small straight-line section in the middle of the curve. It is only in this small area, in which there is a direct proportionality between the dosis and the brightness in the final image, that a correct exposure is possible. All other areas below and above this small straight-line section are under- or overexposed and thus result in an incorrect exposure (fig. 3).

A pre-selection by means of a histogram determines the brightness distribution in the image and thus selects the optimal brightness range for the final picture. Thus, any changes that are seen in the final images, are not produced through different technical conditions (time, dose, exposure), but can be assumed to be due to a change in the patient's condition.

Clinical application and discussion

Our more than two years' experience with CR leads us to expect that in the future this technique can replace the conventional analog film technique in most examinations (1,

7, 9, 11). The advantages of luminiscent radiography over conventional X-ray technique are as follows:

- Possibility of dose reduction per examination unit.
- The application of digital postprocessing techniques.
- Electronic archiving.

Dose reduction

Since the IP has a greater sensitivity than the conventional screen/film system normally used in clinics, the radiation dose can be reduced without over- or under-exposure. It is, however, important to keep in mind that the dose reduction causes a loss of information. It is the task of the physician to define how much information is needed for each diagnostic procedure. For example, a control examination in skeletal diagnostics (scoliose, post-operative position controls), requires less information than a primary examination. The results of our examinations suggest that in thoracic diagnostics sufficient diagnostic information can be gained for certain specific clinical questions even with reduced doses (6, 10, 11).

Application of digital postprocessing techniques

Postprocessing provides us with better representations of the information gained from digital examinations. In clinical applications of CR, the most effective uses of postprocessing include changing contrast and brightness as well as frequency filtering and unsharp masking technique.

Frequency filtering allows a more distinct representation of structures with low density contrast, such as pulmonary nodules or discrete interstital changes. It has proved useful to define the different parameters for certain clinical questions and then establish them in programs, that run these parameters automatically. Since individual specifications and special changes of the postprocessing parameters are rarely called for, CR examinations are no more personell-intensive than the conventional examinations (2, 4, 8, 10).

Storing and Archiving

At present the digitally selected information in our system is stored on an optical laser disk (DOR) with a capacity of 2 gigabyte (2000 megabyte). This capacity is sufficient for about 800 to 1000 CR examinations, when a data compression is performed without any loss of information. The data set produced by a digital picture of a thorax consisting of 2000 x 2000 pixels requires a storage capacity of about 4 megabyte, whereas for a DSA picture with 512 x 512 pixels a storage capacity of 0,32 megabyte is necessary. This comparison shows that both the total number of examinations (nearly 70% of the diagnostic imaging procedures) as well as the amounts of data generated by CR examinations will determine the required capacity of storage and data transfer. Although not all components necessary for a complete "digital radiology department" are available today, the wide clinical introduction of CR is a necessary pre-condition for the further clinical application of PACS (1, 7, 8, 9).

Literature

1. Capp, M.P., H. Roehrig, G.W. Seeley et al: The radiological department of the future. Radiol. Clin. North Am. 23 (1985)349–355

2. Fraser, R.G., E. Breatnach, G.T. Barnes: Digital radiography of the chest: Clinical experience with a prototype unit. Radiology 148(1983)1–5

3. Hindel, R., W. Preger: Cost effectiveness prospects of picture archiving and communications systems. Health Policy 9(1988)91–101

4. Ishida, M., K. Doi, L.N. Loo, C.E. Metz, J.L. Lehr: Digital imaging processing: effect on detectability of simulated low contrast radiographic patterns. Radiology 150(1984)569–665

5. Johnson, G.A., C.E. Ravens: A survey of digital chest radiography. Radiol. Clin. North. Am. 21(1983)655–665

6. Merrit, C.R.B., C.C. Matthews, D. Scheinhorn, S. Balter: Digital imaging of the chest. J. Thorac. Imag. 1(1985)1–13

7. Merrit, C.R.B.: Computed radiographic imaging using photostimulable phosphors. Medicamundi 31[2](1986)45–56

8. Smathers, R.L., W.R. Brody: Digital radiography current and future trends. Brit. J. Radiol. 58(1985)285–307

9. Sonoda, M., M. Takano, J. Miyahara, H. Kato: Computed radiography utilizing scanning laser stimulated luminescence. Radiology 148(1983)833–838

10. Witte, G., H.W. Bause, W. Pothmann, B. Schwermer, M. Schofer: Die Thoraxaufnahme auf der Intensivstation als Digitale Lumineszenzradiographie (DLR) – Technik und klinische Anwendung. Fortschr. Röntgenstr. 149(1988)453–457

11. Witte, G., B. Schwermer, E. Bücheler: Digitale Lumineszenzradiographie (DLR) – eine Technik, die das konventionelle Röntgen ersetzt? Dtsch. Ärztebl. 37(1989)1591–1584

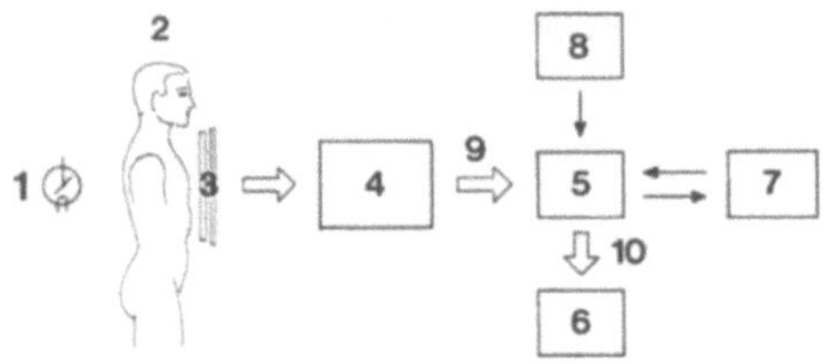

Fig. 1. Computed Radiography (CR)
1 = X-ray tube, 2 = patient, 3 = photostimulated luminescence image plate (I.P.), 4 = image reader
5 = image processor, 6 = image recorder (hardcopy monitor), 7 = Data recorder, 8 = console
9 = A/D converter, 10 = D/A converter

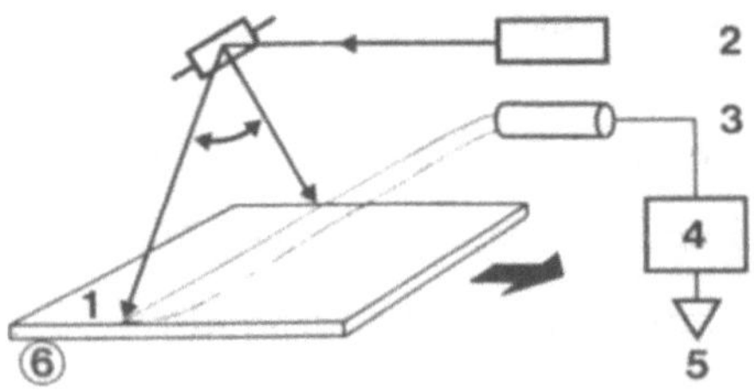

Fig. 2. Image reader
1 = photostimulated luminescence image plate (I.P.), 2 = laser beam, 3 = photo multiplier tube,
4 = analog/digital converter, 5 = image processor
6 = motor drive

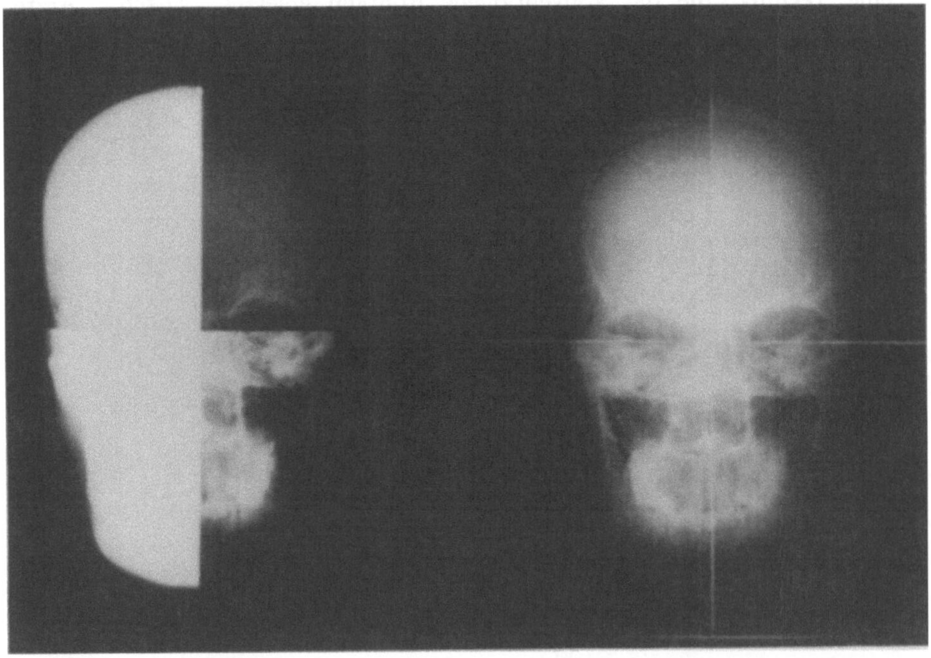

Fig. 3a. Scull phantom
analog film cassette examen (200 speed).
Exposure 75 KV and different m AS.
(right top 250 m AS, right bottom 25 m AS,
left top 10 m AS, left bottom 2,5 m AS

Fig. 3b. Same parameters as fig. 3a
digital computed radiography without
over- or under-exposure.

New Developments in Image Acquisition Systems: Advanced Multiple Beam Equalization Radiography

Leo J. Schultze Kool

Department of Diagnostic Radiology,
University hospital Leiden,
Bldg 1-C3-Q
PO box 9600
2300 RC Leiden
The Netherlands

Introduction

Chest radiography is one of the most frequently performed examinations, accounting for 30-40% of all radiographic examinations.

One of the main problems in chest radiology is the large absorption difference between lung and mediastinum, resulting in over- or underexposed areas on the X-ray. In order to solve this problem, a number of techniques have been developed including: unsharp mask, portal filters, and the digital beam attenuator. A different approach was developed by Plewes and Wandtke. Their system (SER) uses a small X-ray beam (1 x 1 cm) that sweeps over the patient in a raster like fashion. By measuring the patients absorption with a detector behind the film cassette, the x-ray tube output is adjusted to maintain a constant film exposure.

The herein described system (AMBER), developed by Oldelft, also uses a feed back mechanism to control the local exposure and can therefore be seen as a variant of the Plewes and Wandtke approach. The AMBER technique is, besides film, compatible with the digital detectors. Firstly, it offers the potential for acquiring images with an inherent decreased dynamic range, which proves to be important for the display of images either on film or on the viewing station and secondly, the technique renders an improved uniformity of the signal-to-noise ratio, which facilitates the application of imaging processing algorithms and the use of expert systems

This article will give a describtion of the system plus an overview of the results of some of the tests performed in the three year period during which the system has been installed, tested and further developed in the department of Radiology of the university hospital of Leiden.

NATO ASI Series, Vol. F 74
Picture Archiving and Communication Systems (PACS) in Medicine. Edited by H. K. Huang et al.

System description

The AMBER system basically is a two dimensional real time adaptive spatial filter. This is realised by modulation of the spatial X-ray intensity, as a function of the local patient absorption. Figure 1 shows a schematic drawing of the AMBER system. The system uses a horizontally oriented X-ray beam, that scans the patient vertically. This fan-shaped beam consists of 21 adjacent subsegments. The X-ray intensity of each subsegment is controlled by a corresponding element of the X-ray modulator, in front of the X-ray tube. Control means that the modulator can diminish the local beam intensity by diminishing the height of the fore slit locally. A continuous measurement of local patient absorption, by means of a multichannel (21) array detector is input for a control loop that adjusts each and every modulator element accordingly during a patient scan. In this way a real time spatial filter function is achieved. The maximum beam intensity (open position of the modulator), will always be determined by the mA selection on the X-ray generator. If the technologist for example selects an mA value, which is too low, this will result in underexposure of the mediastinum, while if the settings are extremely high, an unnecessarily high tube loading will be achieved Some technical data are presented in table 1.

Scatter reduction

In order to get an impression of the scatter reduction for different configurations of the AMBER system, and to compare it with conventional systems, scatter to primary measurements were obtained for six different situations. The method used, has been described by Plewes and Vogelstein. The S/P is derived from measurements taken with two small detectors placed behind, and just besides the shadow of a varying size lead-absorber stop. The former, behind the leadstop, measures scatter radiation only, while the latter measures scatter and primary radiation. When scanning, one can read the integrated detector currents and use them to calculate the S/P value for different sizes of the lead-stop. From these values the S/P for the zero lead stop can be extrapolated. Figure 2 shows the results of the S/P measurements for 6 different situations.

In the present AMBER unit, an after slit with a limited size of the lead blades (2.7 cm.) above and below the slit is combined with a bucky grid (ratio 12:1). The measurements show that the additional positive effect on scatter reduction due to this after slit construction is only small, because of this limited size of the lead blades. The measurements made with a construction with enlarged lead blades, (which offers the disadvantage of a more complex construction), however, demonstrates a considerable increase in scatter reduction.

Scatter to primary ratio measurements

S/P measurements were also obtained behind a chest phantom. The phantom consisted of blocks of lucite each 2.5 cm thick. 3 blocks for the "lung" areas and 7 blocks for the "mediastinum". Measurements were obtained both behind the "mediastinal" and "lung" part of the chest phantom. The AMBER system was used with a 12:1 ratio grid and the small after slit blades. The equalized and the non-equalized measurements were obtained with the same system, as it is possible to switch off the equalization device by the simple turning of a knob. Differences of the two measurements are thus only attributable to the influence of the equalization. The results (fig 3) of these measurements showed that with AMBER behind the mediastinal region of the phantom, a 37 % reduction in SPR was achieved. On the other hand, an increase of S/P was noted in the "lung" part of the phantom.

One has to realise that the measurements are valid for this phantom only. As absorption differences between lung and mediastinum in the population can differ considerably, both the S/P and the effect of equalization will vary.

Contrast detail study

The next step was to evaluate how this change in SPR would affect the image quality of the receptor. Two different types of receptors were used 1. film (TMG, Lanex regular) and 2. Stimulable phosphor plates (PCR). Although the phosphorplates are capable of digital enhancement, the quantum limitation and the relative low detector quantum efficiency indicate that a combination with AMBER could be quite useful.

The evaluation was done by means of a contrast detail phantom in combination with the above described chestphantom. Two contrast detailphantoms were used. One was placed behind the "mediastinum" and one behind the "lung". Radiographs of the phantom were obtained using three different exposures. Firstly, a conventional (non-equalised) exposure giving a density in the mediastinum and lung comparable to a standard chest radiograph. Secondly, a penetrated conventional (non-equalised) exposure giving a well exposed mediastinum and an over-exposed lung and thirdly, the AMBER exposure. The technique settings (mA) were chosen such that the film density in the lung region on the AMBER exposure equaled the filmdensity in the lung region on the conventional non-equalised exposure. The film density in the mediastinal region on the AMBER exposure was the same as that in the mediastinal area on the penetrated conventional exposure. The same technique settings were used for the exposures on the CR plates as well as on the film.

Thirty radiographs were obtained per technique. After randomization four observers were asked to indicate the number of holes they could detect in each row of the contrast detail phantom. The results of the study are visible in figure 4 They show that both for film and CR application of AMBER will significantly increase the detectability of low contrast details in the underpenetrated areas, while (for this phantom) the detectability of the details in the lung regions remains unaffected.

After contrast enhancement (unsharp masking, non-linear curve, kernel size 81 x 81, enhancement factor 5) improvement in detectability was found for the conventional exposure only. An explanation for the lack of improvement with the other two exposures could be that the SNR was sufficiently high with the penetrated and the AMBER exposure.

The result of the contrast detail study indicate that not only film but also phosphorplates will benefit from the use of AMBER. However, more studies will be needed to evaluate whether this combination offers a clinically significant image quality improvement.

Clinical studies

As this system is now commercially, it is to be expected that several clinical studies will soon appears in the literature. Already published papers concentrated on a nodule detection study and a study comparing the visibility of anatomical structures on both AMBER and comparitive conventional radiographs.

The nodule study showed that the detection of simulated nodules on a anthropomorphic chest phantom improved with 40% if the nodules were located in the mediastinal or subdiaphragmatic part of the phantom.

The visibility study showed an improved visibility of all anatomical structures positioned in either the under- or overpenetrated areas on the chest radiograph. It is to be expected that this improved visibility will lead to an increase in both sensitivity and specificity of the chest radiograph. A study assessing this issue is currently in progress, and is expected to be completed within the next two month's.

Literature

1. Plewes DB, Wandtke JC. A scanning equalization system for improved chest radiography. Radiology 1982; 142: 765-768.
2. Plewes DB. A scanning equalization system for chest radiography with regional exposure control: theoretical considerations. Med Phys 1983; 10: 646-654.
3. Vlasbloem H, Schultze Kool LJ. AMBER: a scanning multiple beam equalization system for chest radiology. Radiology 1988; 169: 29-34.
4. Elburg HJ, Boelens F, Schultze Kool LJ. Technical aspects of AMBER. Proc. Chest Imaging Conference 1987; 49-59.
5 Schultze Kool LJ, Busscher DLT et al. Advanced Multiple Beam Equalization Radiography (AMBER) in chest radiography: comparison with conventional chest images by means of a simulated nodule detection study. Radiology 1988; 169; 35-393
6. Schultze Kool LJ, Busscher DLT et al. AMBER: A promising technique for chest radiography. British Institute of Radiology; 1989 BIR report 20, 121-124.
7. Noordveld RB, Schultze Kool LJ, Kroon HM et al. The effect of AMBER on the image quality of film and stimulable phosphorplates. Proc. SPIE vol. 1090, 78-83.

table 1: Technical data:

tube voltage	: 141 kVp
total scantime	: 0.8 sec
local	
exposure time	: 0.25-0.50 msec.
FFD	: 1.80 m
film	: TMG (Eastman-kodak)
screen	: lanex regular (Eastman-Kodak)
Bucky grid ratio	: 12:1, 36 lines/cm
focal spot size	: 1.2 mm
Height after slit	: 4 cm
Detector	: Xenon
grid	: 12:1, 91 lines/inch

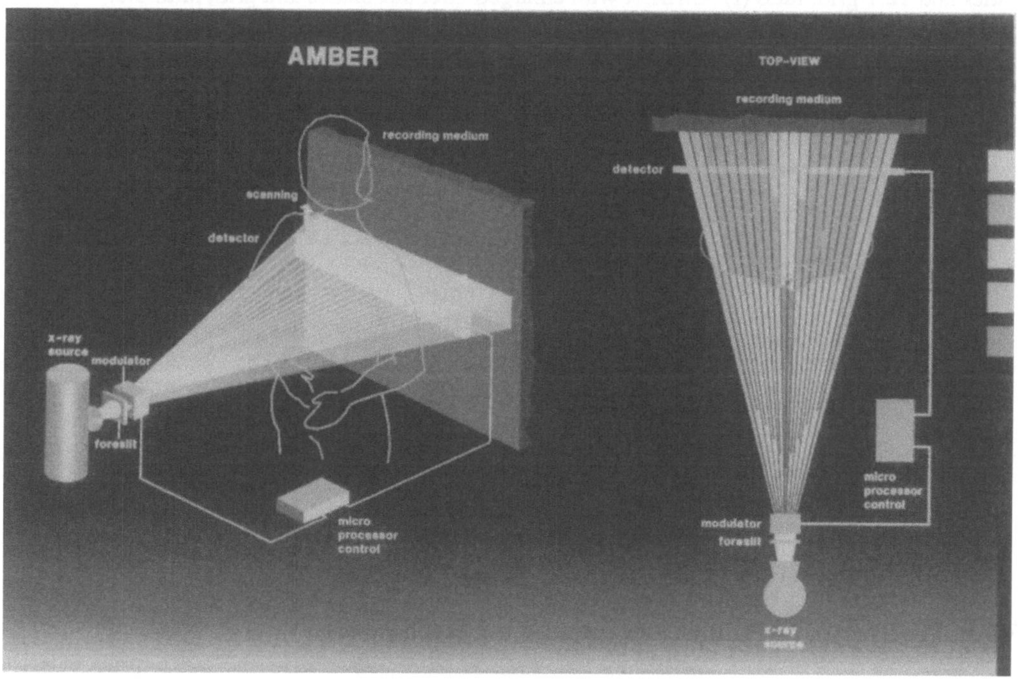

Fig. 1: Schematic drawing of the AMBER system.

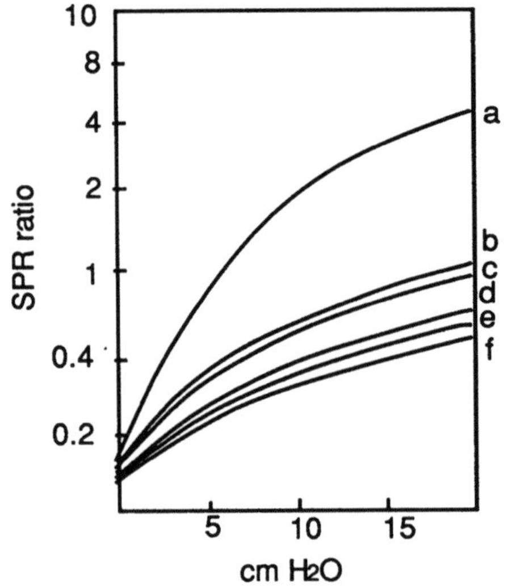

Fig. 2: SPR ratio versus scatter medium thickness.
This figure shows measurements for six different situations: (a) a conventional system without grid (b) a conventional system with grid, ratio 8:1 (c) AMBER with small after slit blades and 8:1 grid, ratio (d) a conventional system with grid, ratio 12:1 (e) AMBER with small after slit blades and 12:1 grid, ratio (f) AMBER with enlarged after slit blades and grid, ratio 8:1.

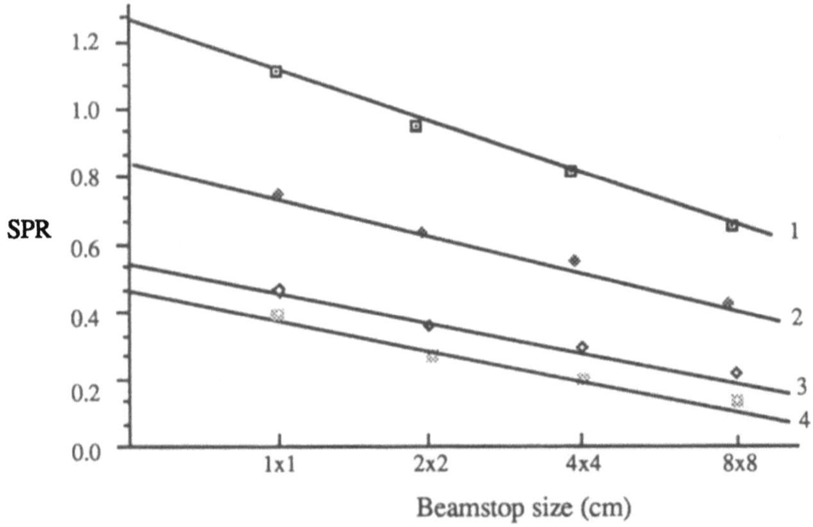

Fig 3: The affect of AMBER on SPR values of a chest phantom

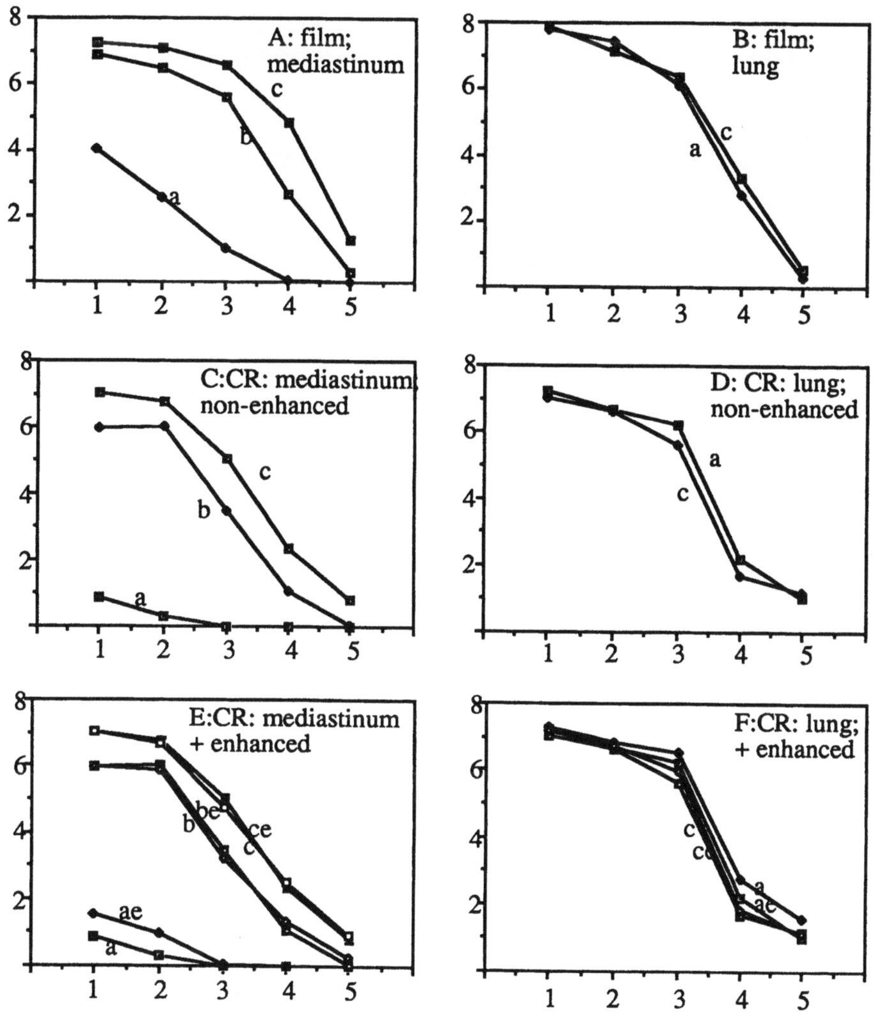

Fig. 4: graphs representing the average responses for the mediastinum and the lung of all the observers for film and CR. a–c the three exposures: a = conventional , b = penetrated conventional, c = AMBER, ae, be,ce results after enhancement

Horizontal 1-5 the 5 rows. 1 being the row with the highest contrast and 5 with the lowest. Vertical the number of detected holes.

Acquisition Interface for Computed Radiography

Bruce Ho, Craig Morioka, Lotfi Kaabi, Ricky K. Taira, Alek Hayrapetian, H.K.Huang

Department of Radiological Sciences, UCLA, Los Angeles, Ca. 90024

Abstract: Digital interface to the Philips Computed Radiography (PCR) units 701 and 9000 is not available as a standard component. Custom interfaces have been constructed using hardware and software techniques to achieve automatic data link without affecting the operational procedure or or slowing down the PCR. A complete acquisition interface design includes polling text patient information, image reformatting to the UCLA standard, and software interface to the network and database. Finally, an ethernet interface to the Luminsys computed radiograph plate reader is also reported.

Keywords: Archival, Communication, Acquisition, Computed Radiography

Introduction

Digital radiological images are acquired from three sources: the inherently digital modalities such as CT and MRI, laser film digitizers, and computed radiography (CR) units. Because up to 80% of radiological examinations involves radiographs, a direct integration of CR units to a Picture Archiving and Communication System (PACS) is very important operationally. The large CR image sizes (2K x 2K x 10 bits) and relatively high throughput (2 min. per image) makes the acquisition interface technically challenging. At UCLA, we have successfully interfaced to two CR units from Philips Medical Systems: PCR 901 and PCR 7000. In both cases, the image acquisition is automatic, and does not interfere with normal PCR operation in any way.

Hardware construction of a black box is used to intercept the data stream in the PCR 901, since the system does not have any provision for external computer interface. The black box also performs image reformatting before sending data to a VAX 11/750 via a DR11-W interface. The acquisition computer has since been upgraded to a Sun 4/370. The second generation PCR: PCR 7000 is equipped with a PACS Interface Processor (PIP) based on the OS-9 real-time operating system, which provides an ethernet connection and therefore eliminates the need for a custom hardware interface. Since the PIP is designed so that a PACS bottleneck can

NATO ASI Series, Vol. F 74
Picture Archiving and Communication Systems (PACS) in Medicine. Edited by H. K. Huang et al.

substantially slow down the PCR operation, our interface strategy includes a task control feature using Unix system commands, such that image acquisition through the ethernet take precedence over the subsequent image reformatting and transfer to the PACS storage.

Many clinical operations do not require a hardcopy print out. In those cases, it is much more economical to substitute an entire CR unit with a plate reader. Luminsys (San Jose, Ca) markets such a unit based on an IBM PC. Connectivity to this reader is implemented with a AT bus ethernet card and Excelan (San Jose, Ca) file transfer protocol (FTP) software.

PCR 901 Interface

The PCR 901 is based on photostimulable phosphor image plates. The plates come in sizes equivalent to the conventional screen-film jackets, and thus are adaptive to existing clinical x-ray equipments. Three major components make up the PCR system: the image reader, the image processor (see figure 1), and the film recorder which perform the functions of laser scanning/digitization, storage/image processing, and laser printing/film developing, respectively.

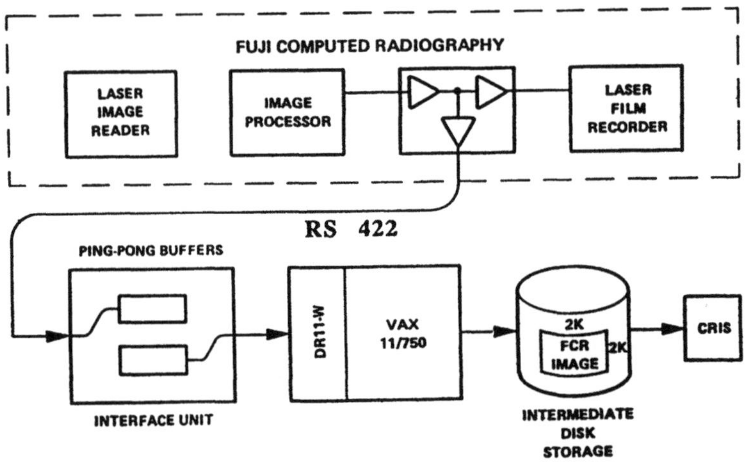

Figure 1. Architecture of the interface module as a gateway to PACS system.

The three sections are connected by cabling which carry both image data and timing signals. Our strategy for image data interception is to tap into the data stream between the image processor and film printer sections. This configuration allows us to extract processed images which are considered superior to the raw images generated directly by the laser reader. The interception circuit board is housed inside the processor cabinet and uses current loop receiver/driver components equivalent to those in the PCR. The signals are converted to differential pair for long distance (100 ft) transfer to the computer room. In additional to the interception circuitry, a black box is constructed to convert the signals to TTL (transistor-transistor-logic) level and generate the DR11-W handshake to interface with the VAX computer [1]. The configuration of the interface module, including the interception board, the black box, and the cabling is shown in figure 2.

The acquired image takes up 4 MByte of memory. To minimize the system memory requirement, we have decided to store each block of image data (32 MByte) as soon as it is acquired onto a magnetic disk. In slower disks, we have found that it is necessary to use 4 MBytes of contiguous disk space in order for disk writing operations to catch up with the rate blocks of image data are coming in. The acquired raw image can then be retrieved, postprocessed, and transferred to another area on the disk which is not contiguous. The time available between the end of one image acquisition and the beginning of the next one is 90 seconds minimum, which is sufficient to finish postprocessing and transfer.

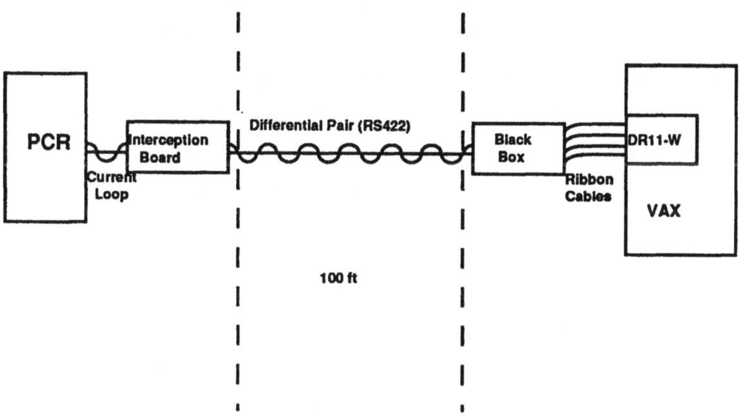

Figure 2. Cabling diagram for the interface module

Interception Circuitry

The interception circuitry acts in a eavesdropping manner, so that the PCR operation is not in any way disturbed. Each relevant signal is captured with a receiving buffer and driven out in two directions. One direction follows the original path to the printer, the other is sent to the black box through a 24 gauge twisted pair wire after a level conversion from current loop to differential pair. The differential pair driver and cabling follow the IEEE convention known as RS422. At the speed of the PCR signals (in the order of 100 KHz), this circuitry is capable of transmission over 3000 ft . This Y branch connection does not allow the PACS interface to control the PCR data flow. This means that data capture must occur at the same rate as plate digitization or film printing, or else parts of an image frame will be lost.

The relevant signals in this connection are 10 bits of image data, of which only the top 8 bits are extracted, and 3 control signals used to indicate the start of transmission, and clock image lines and pixels, respectively. Future upgrade to capture 10 bits per pixel can be made with very little effort.

Black Box

Because the PCR data rate and the VAX DR11-W throughput (1.5 MBytes/sec) are drastically different, some sort of buffering scheme is needed to synchronize the two input/output (I/O) processes so that they can each run at their own speed and still transfer data smoothly. This buffering is the main function of the black box which sits at the receiving end of the twisted pair cable. First, the signals are converted from RS422 to TTL levels (0 to 5 volts). The image data are then stored temporarily in one of two ping-pong buffers designed out of static random access memory (SRAM) [2]. The two buffers alternate in receiving from PCR and sending to VAX one block of data, each at the speed dictated by the I/O throughput of the processor involved. A block diagram of this black box is shown in figure 3. The read address generator is clock by a signal from the PCR, and the write address generator is clocked by the handshaking circuitry which exchanges REQ and BUSY signals with the DR11-W controller board in the VAX. The entire circuitry is designed out of discrete logic with no processor on board.

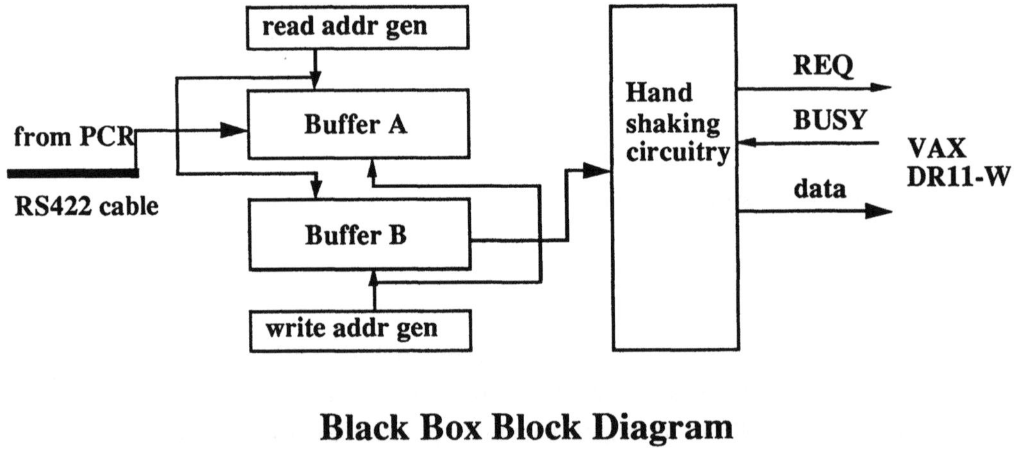

Figure 3. Functional block diagram of the black box with a ping-pong buffer design.

The DR11-W controller is an interface device originated by Digital Equipment Corporation (DEC, Maynard, Ma.). It has since then been established as an interface standard implemented by multiple vendors on the AT bus, VME bus, Multibus, as well as DEC's Q bus and Unibus on which it was first introduced. It features 16 bit, bidirectional, block and word mode I/O and has a simple protocol when used for processor/device interface (as supposed to processor/processor interface), in which the device activates one of two REQ signals to indicate its readiness for the next data exchange (either read or write) and the processor responds with a BUSY signal. The processor has an additional control over the device using a READY signal which allows the device to begin the first REQ for data transfer. No address bits are provided. The device appears to be a single I/O port to the processor. Also of interest to most designers are its three control and three status signals used for a more direct communication between the processor and the device without going through the data bus. In our design, one control line is used to send plate size information to the black box for the image reformatting function. The handshaking portion of the black box handles the DR11-W protocol for block mode data transfer. Each PCR image is reformatted into 2Kbyte image lines and grouped into 32 Kbyte (or 16 line) blocks for transfer.

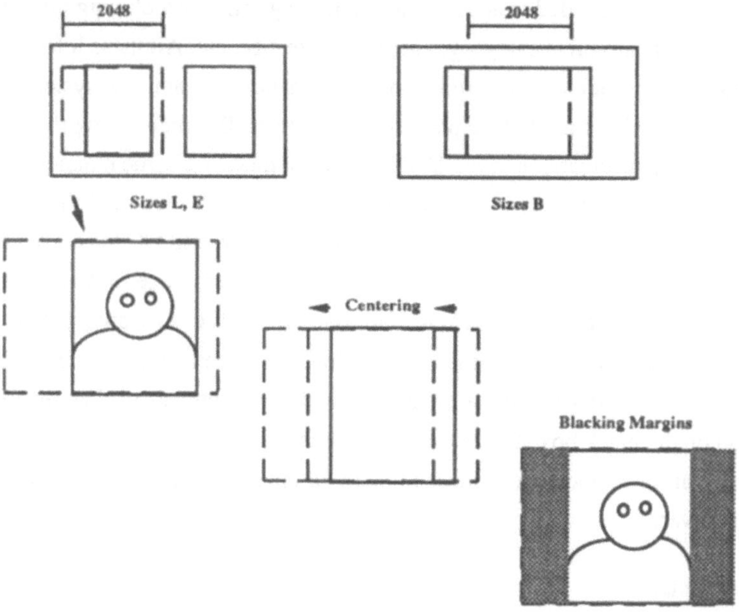

Figure 4. Each image is windowed into 2K pixels per line by hardware. Software postprocessing centers the acquired images and blackens the two sides for sizes L and E.

Image Reformatting

There are three plate sizes used: 14" x 17", 10" x 12", and 8" x 10", also known as sizes L, H, and B, each with a separate image format. The L and H sizes are printed two on one, which means two images are shown on one film, one original and one edge enhanced. Because of the duplication, each image ends up with a line size smaller than 2K pixels. The B size, which is printed with one on one format, does generate a line size greater than 2K pixels. To comply with the UCLA PACS standard, every image must be reformatted to a 2K x 2K square matrix. In the first two cases, pixel filling is used to extend the pixel count per line, while in the third case, cropping is done to reduce the pixel count to exactly 2K. This reformatting is done using hardware logic in the black box. The VAX computer reads the plate size through text information displayed on the PCR console, set a control bit on the DR11-W board which in turn controls the logic in the black box to either fill or crop each image line. Centering of these images after pixel filling and blacking of the two sides are done by postprocessing software as shown in Figure 4. Similarly, the number of lines in an image is either filled or cropped to 2K. This is controlled by software during the acquisition.

Text Interface

Some text informations are needed to accompany each image file for archiving purposes, namely, patient name, anatomical category, age, sex, plate size, etc. All these informations are entered by the technologist at the PCR terminal and automatically sent to the system console through a RS232 connection right before plate scanning starts. By intercepting the RS232 line, we can redirect the text informations to the VAX system in much the same manner we redirected image data. Also, we have introduced a new code to identify which section of the department a particular image originates from, i.e.. PD for pediatric radiology, and CC for coronary care unit. This code allows the acquisition computer to determine the proper image format (subsampling or not), select the location for permanent archive, and route image files to viewing stations where the responsible radiologist will perform the examination. All these text informations are read by the VAX computer in time to activate the image acquisition program, as well as set control bits for size selection on the black box. The text informations are linked into one ASCII string containing many control character sequences to mark all the data fields, so that it is easily interpreted by software.

Software Description

Three processes run continually on the capture computer for CR acquisition. The first listens though the RS232 text port for the creation of new studies on the CR system. Before the main read of the imaging plate (IP) is performed, the CR system sends patient demographics, study anatomy description, plate size, pre-read information (average and range of exposure values on IP), and image processing information to the first process on the capture computer. A descriptive image header is constructed from this information. The raw image data is then received via the black box. After the entire image has been transferred, the study is queued to the second processing stage (reformat). The reformat process standardizes the packaging of the image file, determines which archive to send the image, and scales the image data for the appropriate destination workstation type (e.g.. 2K x 2K primary viewing station or 1K x 1K review station). After reformatting, the image is queued to the final communication stage. The communication process uses a TCP/IP socket connection between the capture computer and the specified archive host to transfer the image file. Images which are successfully transferred are entered into a delete queue. This delete queue is serviced in a first-in-first-out fashion by the reformat process.

Four acquisition-related processes run continually on the archive machine: 1) receive communication process, 2) database update process, 3) archive manager process, and 4) communication out process. The database update process extracts the fields from the image header and inserts and modifies tables (patient description, study description, and image description) managed by a dedicated database machine. Other archive machines and workstations then have access to this information. The database update process uses the patient's age, anatomical region of study, referring clinical service, radiology service, and patient hospital id number to determine the destination workstation (pediatrics, chest, CCU, GU, etc.). The image is then simultaneously queued for image archival onto optical storage and for image communication to the destination workstation.

Upgrade to Sun 4/370

Our original Pediatric Radiology PACS module uses a centralized configuration, in which every acquisition, archive, and display task is processed by one VAX computer. When it becomes evident that the host computer is too overloaded to provide efficient services, we move the PCR interface and a laser digitizer interface to a Sun 4/370 system. After the completion of acquisition and reformatting, image files are then transferred to the VAX when it is not busy. A DR11-W simulator board from Ikon corporation(Seattle, Washington) is incorporated into the Sun's VME backplane. A generic DR11-W device driver from Sun Microsystems is used for software integration into the system. The Fortran programs used for acquisition were rewritten in C and ported from the VAX's VMS operating system to Sun's Unix. The text interface software is also rewritten to make use of the Sun system's serial I/O port. We configured the serial input format, i.e.. raw mode, baud rate, etc, by making an ioctl call to the termios structure.

PCR 7000 Interface

The interface strategy for the latest model PCR 7000 differs substantially form that of the PCR 901. A real-time system called the PIP has been inserted between the image processor section and the film printer to intercept and simulate the communications between them. The PIP provides an ethernet outlet and FTP capability to link to an external acquisition computer. We have dedicated a Sun SPARC station 1 for this purpose. Because the communication signals are no longer exchanged directly, it is possible to slow down the PCR's operation to accommodate a slower PACS interface if necessary. However, fast film processing rate is given priority for our clinical operation. It is crucial, therefore, that the PACS interface does not produce noticeable delay in PCR film production. To avoid unnecessary network congestion, we installed an independent ethernet backbone between the PIP and the SPARC station, separate from the main trunk which runs throughout the department. The overall interface configuration is shown in Figure 5.

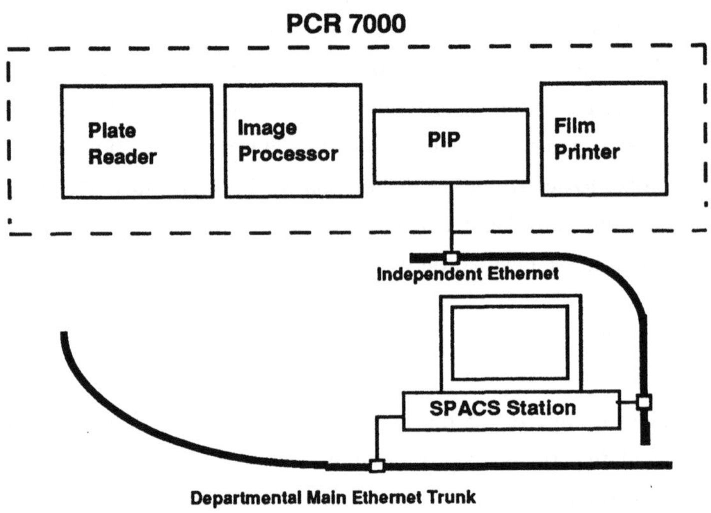

Figure 5. Configuration of ethernet interface to the PCR 7000.

Image Formats

The same three plate sizes are used in the two PCR models. Our service contract with Philips Medical Systems allows us to use the two systems interchangeably for any one plate. The PIP captures and transfers all 10 bit grey value for each pixel generated by the PCR. After reformatting, each image file is 8 MByte in size (2K x 2K x 10 bits extended to 2 bytes). Plate size and Text patient information are all sent in a header file along with the image data file through the ethernet. The SPARC station reformats the image files according the plate size information in the header file. The programs for acquisition, reformatting, and communication are broken up into three software modules linked by queue structures similiarly to the software design in the Sun implementation of PCR 901 interface. Because of the large image size, both reformatting and network transfer are time consuming tasks which can burden the SPARC station and cause a delay in its response in acquiring new images. We have installed a total of 800 MByte worth of magnetic disk space to temporarily store up to 100 image files. This buffering space allows us to postpone reformatting and transfer tasks for in the interest of yielding high priority to acquisition.

Task Control

Task control is implemented with signal and kill Unix system calls. The reformat and communication processes are both forked from the acquisition program, so that the two child process ID are known to the mother program for use in sending kill calls. A kill call has the effect of change a flag value in the child process. Each of the child process is implemented in loops, i.e. the communication program will transfer one block of image data per loop and restart at the beginning of the loop for the next block. At the end of each loop the child process checks the flag value to determine whether the acquisition program has requested that it suspends itself with a pause system call. After the latest PCR image is completely acquired, the acquisition program send another two kill calls to turn the child processes back on, and the cycle repeats as the next PCR image becomes available. The task control process is depicted in figure 6.

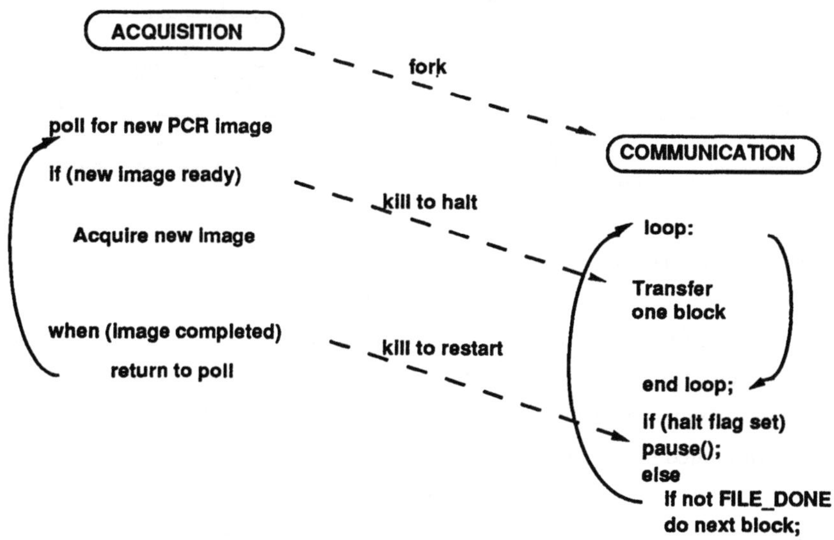

Figure 6. Flow chart for task control algorithm

With this task mechanism implemented, we do not observe any slowing down of the PCR operation even when processing films at maximum speed.

Luminsys CR Plate Reader

The CR system from LUMISYS (San Jose, CA) is a standalone, compact unit, based on a PC/AT host computer. The system consists of a plate reader and printer connected through a SCSI interface (see figure 7). Plates of the size 14" x 17" are are digitized into a 2048 x 2487 x 12 matrix, with a spot size of 174 micron. The upgraded system to be released soon, will digitize five different image sizes with the same spot size. For the 8" x 10" plates, the spot size will be 100 microns. The system has a 300 MB magnetic disk for temporary images storage and an optional 1024 x 1240 monitor to allows the operator to view images immediately after digitization.

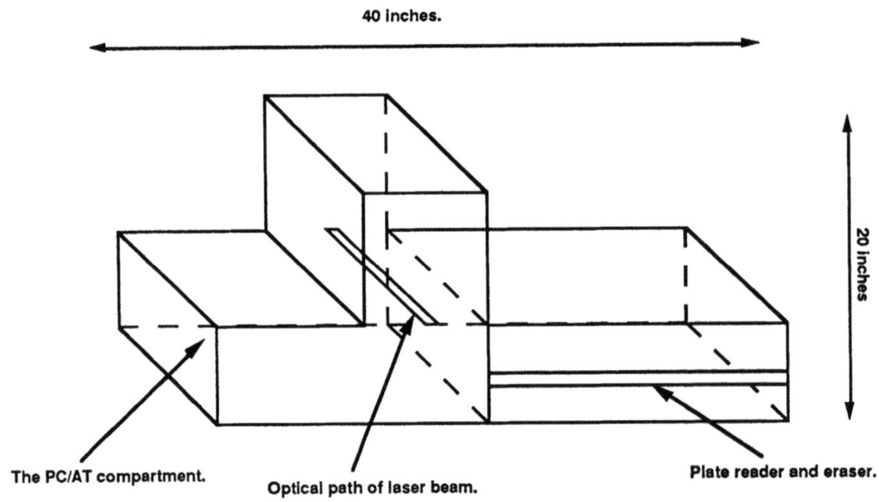

Figure 7. Block diagram of the Luminsys plate reader.

The AT bus allows for easy installation of an ethernet card and which we use to connect to the department's ethernet backbone. During plate scanning operation, the scanning program prompts the operator to enter the patient data (patient name, ID number, examination date, etc.), which are then incorporated into the image header file. After digitization, a send queue is stuffed with the name and storage path of the image. The operator is asked if there are more plates to be scanned. A send routine is activated once all plates have been scanned. This routine transfers all scanned images over the ethernet to the cluster controller computer according to the text information in the queue. The cluster controller computer is then responsible for routing the images into the database.

References

1. Ho, B.K.T, Morioka, C., Mankovich, N.J., Stewart, B., and Huang, H.K.: ImageAcquisition for the Pediatric Radiology PACS Module. In: International Symposium on Pattern Recognition and Acoustical Imaging (Roger H.Schneider and Samuel J.Dwyer III, eds.). Proc. SPIE, 767, 554-557, Feb. 1987, Newport Beach, California.

2. Ho, B.K.T.: Digital to Analog Converter and the Hardware Interface. In: Computers in Medical Physics (Anthony R. Benedetto, H.K.Huang, Don P. Ragan, eds.). Medical Physics Monograph No. 17, pp.59-71. Woodbury. New York.: American Institue of Physics, Inc 1990

PACS Image Acquisition: Film Scanners

Ricky K. Taira, Mansur Loloyan, and Alek Hayrapetian

Department of Radiological Sciences, UCLA School of Medicine, Los Angeles, CA 90024-1721 U.S.A.

Abstract: The digitization of films for a PACS operation is described. This paper describes their operation, performance, PACS integration, and clinical operational concerns.

Keywords: Film Digitizers, Aliasing

1. Introduction

Film will continue to be common in an "all-digital" radiology department for the following reasons: (1) need for historical films taken before installation of digital radiography (DR) units, (2) inclusion of film studies performed by outside departments (emergency, operating rooms, etc), (3) inclusion of films brought in from outside hospitals, and (4) studies performed on film when DR system was unavailable. It therefore is necessary for PACS systems to include film digitization nodes.

Candidates for film digitizers include: (1) video digitizers, (2) optical drum scanners, (3) solid state photo-sensor digitizers (e.g CCD cameras), and (4) optical laser scanners. Video camera digitizers are not acceptable for primary diagnosis since they lack sufficient spatial and density resolution (typically < 1024 x 1024 x 8 bits). Optical drum scanners have high spatial and density resolution but they suffer from slow digitization speed. The film alignment procedure furthermore is rather tedious. They are thus not practical for high volume film digitization. Solid state CCD cameras have the potential to become excellent digitizers for radiographic film although further development is still required. Optical laser scanners have excellent spatial and density resolution and have reasonable scanning time. Currently, they are the best devices for digitizing films for a PACS operation.

Laser film scanners utilize a narrow, intense beam of light that is easily deflected. Many of the limitations inherent in video electron beam scanning (poor MTF, veiling glare, phosphor grain

NATO ASI Series, Vol. F 74
Picture Archiving and Communication Systems (PACS) in Medicine. Edited by H. K. Huang et al.

noise, inadequate optical power) are overcome with laser technology [7]. The high energy incident per unit area of the scanning spot enables these devices to employ low numerical aperture beams which have a large depth of field. As in video scanning, the image is read out in a rectilinear raster scanning motion with rapid retrace. The image to be read is continuously sensored in the scanning direction and discretely sampled perpendicular to the beam sweep.

Figure 1 shows a block diagram of a typical laser scanner system. A low power laser (typically HeNe) and a rotating multi-facet mirror are used to scan the laser light across a single line of the radiograph. The light transmitted through the radiograph is detected and converted to an optical density signal using an optical fiber bundle, a photomultiplier tube (or photodiode), and a negative logarithmic amplifier. The analog optical density signal is then quantized to 10 to 12 bits using an A/D converter. The digitized optical density value is addressed by a pixel and

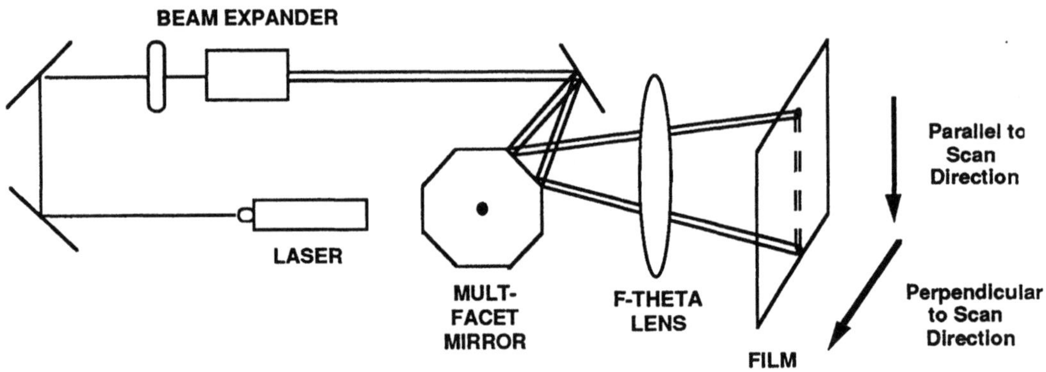

Fig. 1a - *Laser scanner optics*

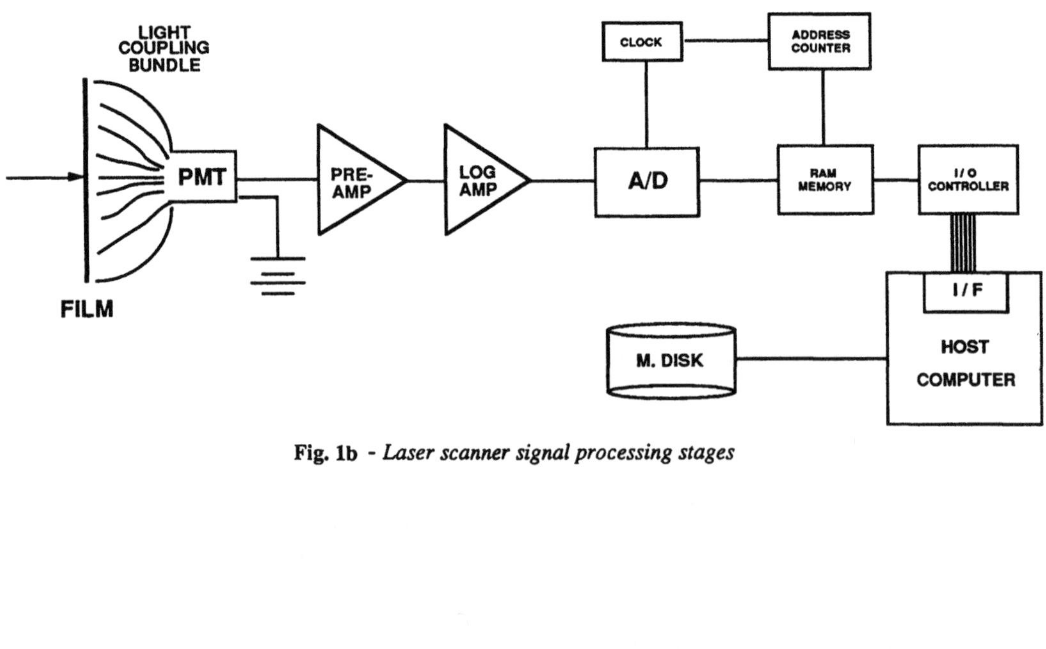

Fig. 1b - *Laser scanner signal processing stages*

line counter then stored in RAM memory which may range in size from one line to the entire image. Once the image buffer is full, an I/O controller sends the digital data to a host computer where the image is stored onto a magnetic disk. Common computer interfaces include DR-11W (parallel DMA interface), GPIB (IEEE-488), SCSI, or proprietary to the manufacturer. A mechanical transport system controlled by a finely calibrated step motor then moves the film to the next scan line. A microprocessor inside the scanner (not shown) coordinates the film transport step motor, data transfers to host, and polygonal mirror scanning (pixel addressing).

2. Performance

Spatial Resolution: The spatial resolution of a laser scanner depends on the following factors: (1) laser spot irradiance profile, (2) pre-amplifier and log amplifier frequency transform characteristics, (3) sampling rate of A/D, and (4) scan line spacing. Most laser spots can be models by a two-dimensional Gaussian function. The distance at which the profile falls to $1/e^2$ is referred to as the laser spot width. The spot will vary slightly with respect to film position. At the periphery of the scanning field, the spot will be larger than at the center due to the oblique angle of the beam and the film. The spatial resolution will be slightly worse at the periphery if the spot size is not adjusted for scan position.

Laser film scanners are not shift invariant imaging systems due to the finite size of the digitized pixels. The frequency transform characteristics can vary significantly for a fixed sampling distance and spot profile depending upon the position and orientation of the film signal relative to the center of the beam profile. Figure 2 shows the extreme cases of in-phase and out-of-phase sampling. Contrast frequency response curves for in-phase and out-of-phase sampling have been measured by Ishimitsu *et. al.* [5]. Sampling artifacts are discussed in Section 4.

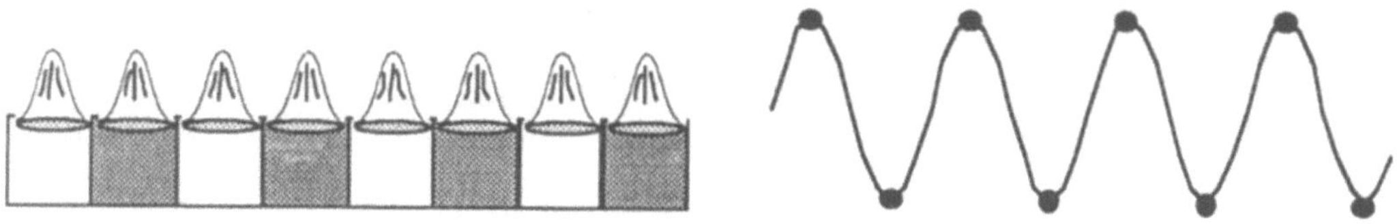

Fig. 2a - *In-phase sampling*

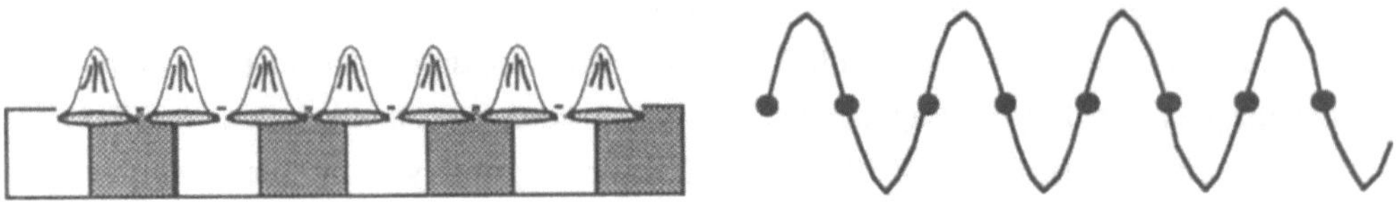

Fig. 2b - *Out-of-phase sampling*

Sensitometric Curve: Ideally, the conversion of optical density (OD) to a digital value should be linear for a wide dynamic range (0-4 OD). All laser scanners today have extremely linear optical density to gray level transforms over the range of diagnostic film densities. Most have switches allowing the user to select the range of optical densities (0-2 OD, 0-3 OD, 0-4 OD, etc) to transform into a full 10 to 12 bits of digital data. Most allow the user to subtract the baseline fog levels from the digital data as well. Note that the sensitometric curve is highly linear for optical density and not necessarily for X-ray exposure. For the digitization to be linear with respect to exposure requires that the scanner be calibrated for each film type (i.e. store the H-D curve for each film).

Signal-to-Noise: The introduction of noise limits the number of significant bits in the image. Noise may arise from a variety of sources including: 1) fluctuations in the laser power supply, 2) quantum noise in the laser and light collection system, 3) microphonic noise (vibrations which are transferred to the resonator structure causing amplitude modulation), 4) dust in the imaging optics, and 5) electrical noise in the signal conditioning electronics [1]. Noise levels are a function of optical density since quantum noise is greater for higher (darker) optical densities.

Spatial Linearity: The high intensity of a laser allows the use of low numeric aperture beams and which provide a large depth of field. Thus image quality remains essentially constant within the scan field. The use of a f-theta lens provides a constant linear displacement of the laser spot with respect to a constant change in the scanning angle[2]. Spatial distortions are very small and negligible in general.

Speed: The bottleneck in scanning speed is typically the transfer of data from the scanner's image memory to the host computer's magnetic disk (Fig. 1b). Scanners which have image buffers of only one line and/or slow interfaces have relatively long scanning times (60 - 120 seconds for a 2048 x 2048 x 10 bit matrix). The speed at which the light detection system can convert light photons to an electrical signal also limits scan speed (minimum integration time for adequate photon detection statistics). A photomultiplier typically takes 10 to 30 nsec. Photodiodes are slightly slower. The type of A/D converter also effects scan speed. The flash type A/D's are very fast (a few hundred nanoseconds). They, however, are accurate to only 8 bits. The successive approximation type A/D's are more accurate (10 to 16 bits) but require about a microsecond per conversion.

3. Configuration and Operational Issues

Configuration: Laser film scanners should be viewed as an image acquisition device in a similiar manner as CT, MR, and CR systems. Therefore, they must be interfaced to a general purpose PACS capture computer which is connected to the PACS network. A multi-tasking computer is desirable. The interface from the laser scanner to the capture computer is typically GPIB, SCSI, DR-11, Ethernet via a vendor supplied host, or proprietary. The film digitization software should be integrated with the following subsystems: 1) PACS and RIS database management systems: this will minimize the amount of information that the operator would have to type (see below). 2) Image display software: digitized images should be displayed on CRT's to verify image quality, image centering, and image orientation. 3) Network communication software: digitized images must be sent to archive as well as workstation nodes. Film digitization source code and/or object libraries must be obtained from the vendor in order to integrate heterogeneous PACS software subsystems.

Which Films: Upon deciding to proceed with an all digital radiology department, a decision as to which films to digitize is necessary. For large hospitals, it is not practical to digitize all films stored in the library. Possible scenarios include: 1) digitize all films from the past n-months, 2) digitize only the most recent films in each patient's jacket, and 3) digitize necessary current and comparison films for new and/or existing PACS patients only. Films which have been digitized should have an identifying mark to indicate that the image has been included into the PACS database. This will facilitate the identification of films which have already been digitized.

When to Digitize: A key logistic issue is when to scan a film. For current films, two methods are described: 1) polling for films from a fixed site (e.g. a light alternator), and 2) event driven as radiographs are generated. The first method, polling, was implemented at the UCLA Medical Center in 1987. A PACS technologist screened all films placed on a designated alternator and digitized all films which were not present in the PACS database. The alternator was checked in the morning, afternoon, as well as in the evening. Long delays however were seen between film generation and film digitization. These delays were mainly attributed to films not being present at the alternator site at the time of polling (film were in the wards and scattered about the department). The average digitization delay was about one day. In the event-driven method, films are digitized immediately after being processed. The ideal situation is to have a laser scanner system in line with the film processor. Historical comparison films may be scanned during off-hours prior to the patient visit if scheduling information is available (for example via a RIS system).

Required Text Information: Text information for digitized film images is similiar to other acquisition systems. The following are recommended information.

1. Patient Demographics: For each patient entered into a PACS, hospital ID, name, sex, and birthday are required. The operator should be able to provide a patient ID to search RIS/PACS/HIS databases for this information. If no records are found, screens should be provided for the user to enter the demographic information.

2. Study Information: A study reference number is needed to identify date and time of study, why the study was requested, study description, the radiology service, and the referring clinic/physician. A RIS bar-code system would allow this information to be obtained if entered previously at a registration area. If the study has not been previously entered, screens for data entry are required. Internal study reference numbers must be generated for PACS/RIS searches.

3. Image Information: An image reference number is needed to identify X-ray imaging technique used (kVp, mAs, grid type, X-ray unit), technologists ID, patient orientation, film parameters (film size and type), and digitization parameters (number of pixels in x and y directions, pixel depth, calibrated pixel size in x and y directions, mm offset from left edge of film, mm offset from top of film, and digitization optical density range).

4. Network Routing Information: A workstation node ID is needed to identify which PACS node(s) to sent the digitized image. This information also may be determined via referring clincial service, referred radiology service, patient age, and anatomical region of study.

Performing the Scan: User parameters often include pixel size, sampling distance, optical density range to quantize the available range of digitization, film fog level, and field of interest. Different models of course have different amounts of parameter flexibility. Many fix the field of view according to size of film (8" x 10", 10" x 12", 14" x 17" etc.). This is not always desirable especially in neonatal films. Spatial resolution parameters (spot size and sampling distance) should be selected based on type of radiographic exam, intended viewers, and/or matrix size of output display device. ROC studies indicate that the required pixel sizes for primary diagnosis are about 200 micrometers for general radiography (abdomen and chest), 150 - 100 micrometers for pediatric exams, and 75 - 50 micrometers for mammographic and bone exams. Depending upon the OD range of the film to be scanned, the user should appropriately set the optical density range. A desirable feature for a scanner would be to perform this automatically via sampling a course histogram of the image and performing an autorange of the light detection system and log amplifier settings in a fashion similar to computed radiography systems[6]. The feature is important if auto stack feeders of films becomes popular.

Visual Image Quality Check: After the completion of each scan, a means for viewing the digitized image is necessary to confirm region of interest, orientation, and overall image quality. A subsampled low resolution video display of the image is usually acceptable (except in situations where vertical scan line artifacts are present due to dust in the imaging optics).

4. Clinical Concerns

Caution should be practiced when digitizing radiographs of bone fractures, mammograms, and/or radiographs obtained with high frequency anti-scatter grids. These high spatial bandwidth films often result in aliasing artifacts in the digitized images[4]. They arise due to undersampling of the film image[3]. Sampling at the Nyquist frequency (i.e. two times the maximum spatial frequency in the film image) is not always practical because of the need for large matrices (large disk storage), slow scanning time, more precise motor control movements, and degraded S/N at higher sampling frequencies. To prevent aliasing artifacts, a means of attenuating the high frequency components of the image above one-half the sampling frequency is necessary prior to sampling. This can be performed by: 1) **increasing the spot size**: the presampled image is the result of the film image convolved with the laser spot profile. Thus a larger laser spot will result in a lower frequency pre-sampled image spectrum. 2) **implementation of analog low-pass filter**: in the direction parallel to the scanning direction, the frequency response depends not only on the sampling rate and laser spot profile but also on the the temporal frequency response of the light detection system and signal conditioning electronics. The frequency transform characteristics of the analog circuitry should behave as a low pass filter so that the pre-sampled image data satisfies the condition previously stated. Note that in the direction perpendicular to the scan direction, method two above does not apply. (In the perpendicular direction, the temporal response of the analog circuitry is not important). Thus when scanning radiographs with perpendicular frequency components greater that the Nyquist frequency aliasing artifacts may be observed depending upon the size and shape of the laser spot. For this reason, most laser scanner systems employ spots which are not circularly symmetric (typically longer in the perpendicular direction). In summary, in the parallel direction the spot motion is continuous and the system resolution is determined from spot size, spot shape, and response time of the detection electronics. In the perpendicular direction, the spot motion is discrete and resolution is determined by line spacing distance and spot size and shape only. One should note the tradeoff between good high frequency response and the avoidance of aliasing artifacts in the digitized image.

Acknowledgements

This work was supported in part by Public Health Service Grant No. R01 CA 39063, R01 CA 40456, and P01 CA 51198, awarded by the National Cancer Institute, Department of Health and Human Services, and by the UCLA Department of Radiological Sciences Research Fund.

References

1. A.D. Berg, J.P. Wheeler, "Laser Noise in Writing Applications," *Optical Engineering*, **15**:84-89, 1976.

2. P.M. Emmel, "System Design Considerations for Laser Scanning," *SPIE: Laser Scanning and Recording for Advanced Image and Data Handling*," **222**:2-14, 1980.

3. M.L. Giger and K. Doi, "Investigation of Basic Imaging Properties in Digital Radiography. 1. Modulation Transfer Function," *Medical Physics*, **11**(3): 287-295, 1984.

4. H.K.Huang, *Elements of Digital Radiography*, Prentice-Hall, Inc., Englewood Cliffs, N.J., 78-81, 1987.

5. Y. Ishimitsu, K. Arai, R.K. Taira, and H.K. Huang, "Radiological Laser Film Scanner Sampling Artifact," *Computerized Medical Imaging and Graphics*, **14**(1): 25-33, 1990.

6. M. Sonoda, M. Takano, J. Miyahara, and H. Kato, "Computed Radiography Utilizing Laser Stimulated Luminescence," *Radiology*, **148**: 883-838, 1983.

7. J.C. Urbach, S.F. Tibor, and G.K. Starkweather, "Laser Scanning for Electronic Printing," *Proc IEEE*, **70**:597-618, 1982.

third party Ethernet interface for the Data General computer of the scanner and a GE proprietary interface to ACR-NEMA standard.

The Ethernet interface allows network services only while the scanner is not in clinical operation. This requires the clinical operator to initialize the communication server process. In routine operation this allows data transfer only during longer periods with no scheduled clinical procedures. The off-line data transfer is performed using FTP protocol in pull-mode in which the capture process first transfers the scanner's log file and then extracts the image files for each scan. This operation proved not useful for a PACS that is expected to distribute the images immediately for primary review since the data transfer is only possible during night hours. Also scans are frequently deleted due to limited storage capacity of the scanner and are not available for transfer when the network service is finally initiated.

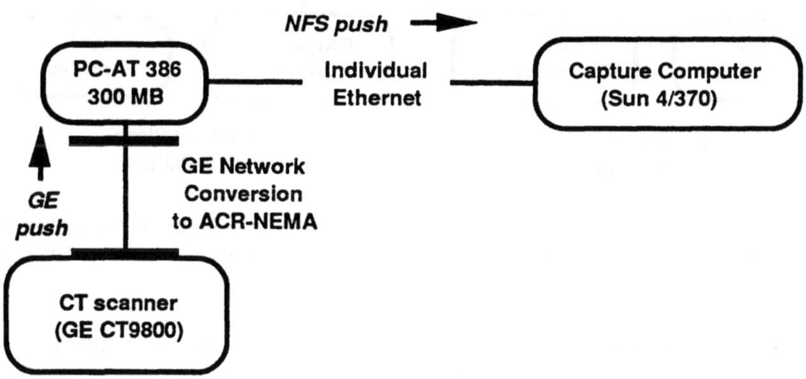

Figure 3. Interface configuration between CT scanner and capture computer

An improved scheme utilizes the proprietary communication features of the scanner and a GE interface to ACR-NEMA standard [2]. Due to a lack of an ACR-NEMA interface for our UNIX based PACS computers we use a PC-AT computer as a gateway to our capture computer. Figure 3 shows this complex chain of components for the connection of the CT scanner. It enables an on-line data transfer in push mode with only little affect on the clinical operation. Here the capture process continuously receives the image files sent from the scanner and identifies the scans from the individual images. Since the scanner does not send a specific message that indicates the completion of a scan the capture process has to wait for the occurrence of an image with a different scan number in order to determine a scan as complete. This algorithm causes a significant delay in the processing of the data. Another disadvantage is that the data transfer process has to be disabled for certain procedures because it compromises the scanning speed. Honeyman et al.describe a similar interface configuration for the GE CT scanner with an ACR-NEMA interface for a Sun workstation that became recently available [3].

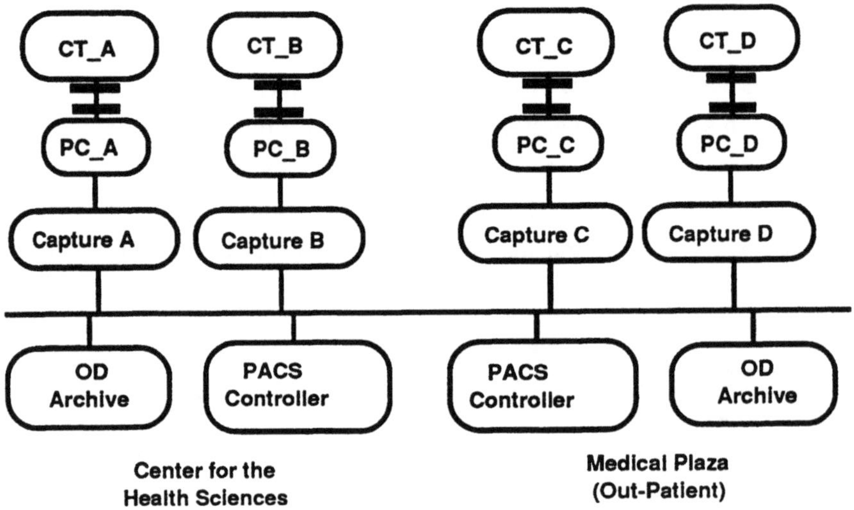

Figure 4. CT scanner network configuration at UCLA with one PACS controller and optical disk archive for each the in-patient area in the Center for the Health Sciences and the out-patient area at the Medical Plaza.

Figure 4 presents an overview of the four CT scanners and their connections to the PACS network at UCLA. Each connection requires individual Ethernet links between the scanner interface and the ACR-NEMA interface and the capture computer. We currently use one capture computer for each CT scanner. The capture computers have a storage capacity of 500 Megabytes that allows temporary storage of images for up to ten hours of scanning time. The transfer time is fifteen seconds per image and under routine operation the acquisition time averages ninety minutes to two hours per scan depending on the number of images per scan.

Acquisition of Ultrasound images

In contrast to the CT and MR scans the images available from current ultrasound devices are provided in the format of video frames. The descriptive of a frame are encoded in the the video data and can not easily be retrieved. Only a specialized additional module enables a transfer of image data along with sufficient descriptive information through a network interface. Most ultrasound systems are equipped with lower-cost computers which are not designed to handle the formatting of the data file and the file transfer concurrently with the clinical procedure. Thus the data can only be transferred in between clinical procedures.

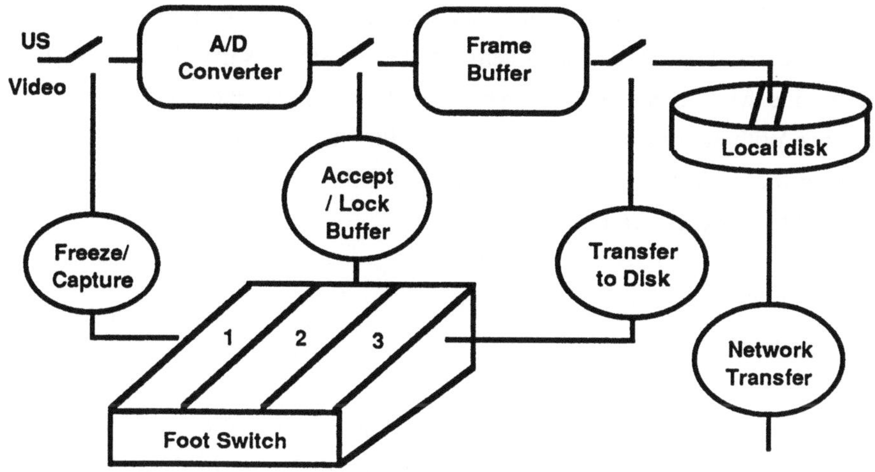

Figure 5. Operator initiated transmission of digital ultrasound image data to capture computer instead of the conventional analog film exposure.

Figure 5. shows a diagram for the scheme for the capture of Ultrasound images from an ATL UM4 system for a PACS for Ultrasound images [4]. A special module replaces the conventional multi-format camera. This module allows to digitize and store (lock) an digitized image in the frame buffer, to further save the up to sixteen images in the frame buffer on a local disk, and finally to transmit completed scan series across an Ethernet network to a capture computer. The only task of the capture process is to continuously screen the scan directory for new data files and schedule the jobs to the reformat queue. Even though the data transfer is executed by the clinical operator this scheme does not necessarily compromise the clinical operation since the data transfer feature replaces the exposure of the analog film on the Ultrasound device and the PACS system is capable to produce the scans on the display stations as quickly as the conventional process. The transfer times average one to two minutes per scan depending on the number of images per scan.

Discussion

The integration of existing imaging equipment of multiple modalities and from different manufacturers requires significant development effort, depends on the technical support of the manufacturer and is not always feasible.

In our implementation of acquisition schemes we found the on-line pull mode most favorable for an immediate data acquisition because it does not require any action or attention of the clinical operator. This scheme, however, requires that the imaging device is capable to service the data transfer concurrently with the clinical operation without affecting its performance Without these capabilities the reliability of the image acquisition depends on the operator of the imaging device and compromises the performance of both the clinical procedures and the PACS operation.

Acknowledgement

This work was supported in part by Public Health service Grant No. R01 CA 39063, R01 CA 40456, and P01 CA 51198, awarded by the National Cancer Institute, Department of health and Human Services, and by the UCLA Department of Radiological Sciences Research Fund.

References

1. K.K.Chan, R.K.Taira, H.K.Huang, *Integration Issues in PACS*, Proc. NATO ASI: PACS in Medicine 1990, Evian, France, (this volume).
2. ACR-NEMA Digital Imaging and Comm. Stds. Ctee., *ACR-NEMA 300-1985: Digital Imaging and Communications*, November 1985.
3. J.C.Honeyman, et al., *Clinical and Research Radiology Network at the University of Florida*, Computer Applications to Asist Radiology, SCAR 90, 390-395 (1990).
4. W.S.Weinberg, F.N.Tessler, E.G.Grant, H.Kangarloo, H.K.Huang, *Development and Implementation of Ultrasound Picture Archiving and Communication System*, Proc. SPIE Medical Imaging IV 4-9 Feb. 1990 Newport Beach, CA in print.

Display Workstation: Hardware Architecture

Paul S. Cho and Bruce K.T. Ho

Department of Radiological Sciences, UCLA School of Medicine, Los Angeles, CA 90024-1721, USA

Abstract: In radiological applications two classes of display workstation are usually considered. These are the high-definition (2K) workstation for primary diagnosis and the medium-definition (1K) workstation for uses primarily outside the radiology department. The hardware architecture of each class of workstation is discussed at the level of system integration for the 2K workstation and at the board-level design for the 1K workstation.

Keywords: Display workstations; diagnostic, high-resolution, gray-scale.

Introduction

There are two classes of display workstation usually considered in radiological PACS. The first class is a high-definition, high-performance workstation to serve radiologist for primary diagnosis[4,6]. This class of workstation should have the maximum attainable resolution (currently 2K) and must be able to load large image data at a high speed to keep up with the normal pace of the routine image reading. The second class is a medium-definition (1K) workstation for remote viewing applications[1,3] such as viewing in intensive care units, operating rooms, conferences, class rooms, etc.

The important characteristics of these workstations are the ability to display large number of images simultaneously, fast image updating, easy-to-use patient directory in the form of an on screen menu, and some basic image processing functions which the radiologist and physician can use in real-time or near real-time. These functions include zoom-and-scroll, window-and-level, rotation, contrast inversion, etc. In addition, the workstations should cover the wide range of radiological images including CT, MR, angiograms, and high resolution radiographs ranging in size from 256x256 to 2048x2048 in spatial resolution and from 8-bits to 12-bits in density resolution.

High-Definition(2K) Display Workstation

In recent years, the 2K display technology has been developed and improved to meet the increasing demands of the high resolution market in such applications as large PC board design,

NATO ASI Series, Vol. F 74
Picture Archiving and Communication Systems (PACS) in Medicine. Edited by H. K. Huang et al.

detailed CAD/CAM, and computer graphics. There are at least five companies that design and market 2K display controllers for these types of applications. These include Metheus Corp. (Hillsboro, OR), CalComp (Anaheim, CA), Raytheon (Sudbury, MA), Univision (Burlington, MA), and benchMark Technologies (Surrey, U.K.). These controllers drive 2K color monitors at a typical video bandwidth of 350-MHz[7].

For radiological applications, gray-scale monitors are normally required. Presently, a full 2048-line gray-scale monitor is commercially available from MegaScan Technology Inc. (Hopkinton, MA). The 19-inch, 72-Hz, non-interlaced monitor provides 2560x2048 pixel resolution (200 dots/inch) at the video bandwidth of 500-MHz. First to adopt the MegaScan monitor for radiological application was PIXAR (San Rafael, CA) which developed a video serializer for the PIXAR-II image computer. The serializer receives 8-bits of red, green, blue, and alpha data every 8-nsec from the associated framestore processor of the PIXAR-II. These pixel data are registered by the serializer which, in turn, outputs the data to a DAC every 2-nsec. The DAC then generates a 500-MHz composite video signal on a 50-ohm BNC output.

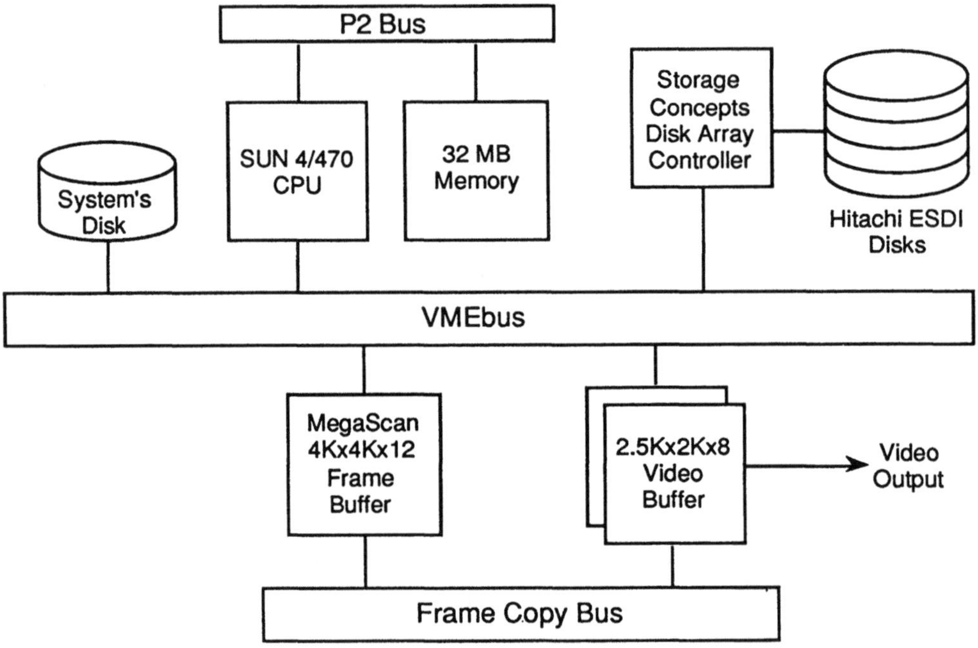

Figure 1. An example of a high-definition (2K) display workstation based on a SUN 4 host, MegaScan 2K display subsystem, and Storage Concepts disk array storage system.

More recently, MegaScan designed its own gray-scale 2K display controller which consists of two 9u-based boards for the VME bus. The 4096x4096x12-bit DRAM-based frame buffer board is connected to a 2560x2048x8-bit VRAM-based video buffer board through a proprietary bus which can transfer data at 50 MPixels/sec. Image data can be copied from an arbitrary rectangular window in the frame-buffer to an arbitrary location on the video buffer. Pixel-replicated zoom, image panning, and 12-bit-to-8-bit LUT conversion can be applied independently to each window. The video signal is generated by a Triquint GaAs DAC to drive the monitor at 500-MHz bandwidth.

An example of a high-definition display workstation is shown in Figure 1. The workstation has three functional components, namely, minicomputer host, display subsystem, and storage device. It is important that the host computer has high processing power to handle multiple requests simultaneously. In addition to the commands input by the user, there may be various jobs running in the background such as communication programs, display window manager, database manager, etc. The CPU must not become the bottleneck of the workstation operation. The general purpose bus should provide sufficient address space, number of physical slots, and electrical power to support the workstation peripherals. In addition, the I/O hardware and software architecture should allow fast data throughput. The SUN 4/470 minicomputer features an improved I/O architecture and memory management and utilizes a 22-MIPS, 3.8-MFLOPS processor.

The display subsystem consists of one MegaScan frame buffer and two video buffers. More of these boards can be added to increase the number of display monitors. Image loading speed is limited by the disk speed. The frame buffer is specified at 20 Mbytes/sec transfer rate. In order to achieve the display speed which satisfies the clinical requirement, a disk array has been configured. The Storage Concepts C-51 disk array storage system consists of four to eight Hitachi ESDI drives (2.6-Gbytes to 5.2-GBytes formatted) and a disk array controller[2]. With four disks, data transfer rate (disk to frame buffer) of 8 Mbytes/sec can be realized. That is, sixteen CT images, each slice being 512x512x12-bit, can be displayed in one second.

Medium-Definition(1K) Display Workstation

An example of a medium-definition (1K) display workstation is given from our own 1K display hardware which has been designed specifically for radiological application.

In order to achieve near-real-time zoom and scroll speed, a data transfer period of 132-ns is required. The actual speed as limited by the graphics controller chip, frame buffer, and DMA (direct memory access) controller is 200 ns/cycle. This yields an image update time of 50-ms or 1.5 frame time. Because of the smaller image sizes of CT and MR, those two types of images can easily be manipulated in real time. However, when dealing with 2K radiographs, pixel sampling reduces data width to 2 pixels per transfer, and a worst case response time of 3 frames is obtained.

A large DRAM-based frame buffer (16-32 Mbytes) is used in our design instead of the VRAM-based buffer found in most high end graphics systems. A significant price advantage is gained this way. In particular, we chose a frame buffer marketed by MMI (Micro Memory Inc., Chatsworth, CA), which features a minimum transfer cycle time of 125-ns in block mode. A separate display controller board holds exactly one megabyte of VRAM-based display buffer, pixel mapped to the screen (see Figure 2). During image manipulation, new image data are driven by our custom designed DMA controller over the VSB bus from the frame buffer to the display buffer (see Figure 3). A transfer time of 200-ns can be reached by the DMA controller, as required by our specification[5].

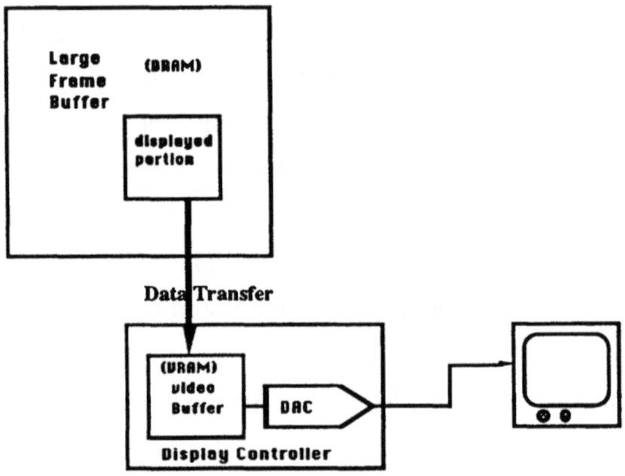

Figure 2. Display design with a large DRAM frame buffer. Image data must be transferred to a video buffer on the display controller board before displayed on the screen.

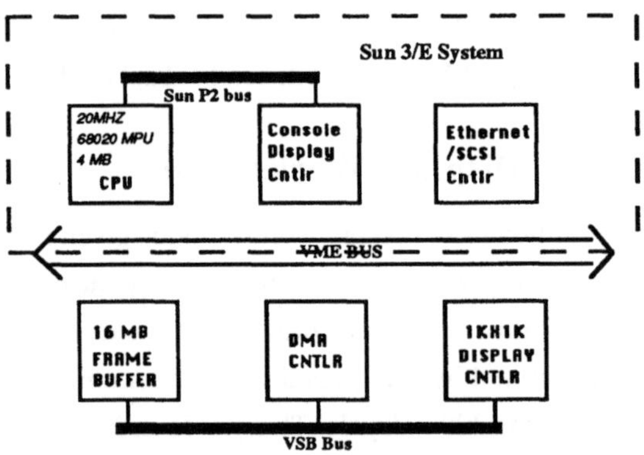

Figure 3. Display station based on a Sun 3/E host, a frame buffer, and custom designed DMA and display controllers.

Because the large frame buffer rides on the VSB bus, multiple display boards can retrieve raw image data from the same frame buffer, thus allowing expandability of the number of screens. Multiple frame buffer boards can also be incorporated if needed at an affordable cost.

Figure 4 shows a block diagram of the display controller. The GSP (graphics system processor) TMS 34020 manufactured by Texas Instruments (Dallas, TX) handles screen refresh, DRAM refresh, and image data transferred from either VME or VSB bus. The GSP is a 32-bit programmable processor capable of advanced graphics instructions such as PixBlt (Pixel Block Transfer), plane masking, transparency, Boolean processing, clipping, etc. It is chosen in this design because of its 20 MBytes/sec transfer speed from external bus to the video buffer. Subsampling and 12-bit-to-8-bit conversion are done by discrete logic in the path of the data flow. The data width of 32-bit coming off the bus is reduced to 16-bit during this conversion step.

The DMA reads from the frame buffer and writes into the video buffer simultaneously over the VSB bus. This is made possible by the VSB protocol which allows a participating slave (video buffer) to listen to conversation between the bus master (DMA) and the responding slave (frame buffer). The coordination of the three boards is done through CPU commands send over the VME bus.

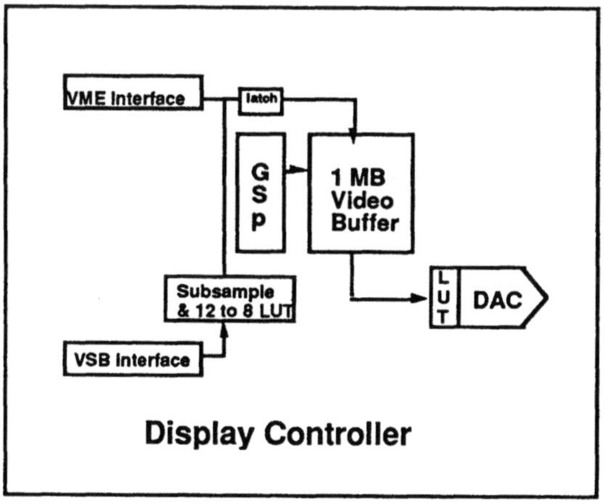

Figure 4. Display controller board with TMS34020 as graphic system processor. Both VME and VSB interfaces can be used for image data transfer.

Acknowledgements

This work was supported in part by Public Health Service Grant No. R01 CA 39063, R01 CA 40456, and P01 CA 51198, awarded by the National Cancer Institute, Department of Health and Human Services, and by the UCLA Department of Radiological Sciences.

References

1. Arenson R.L., Seshadri, S.B., Kundel, H.L., DeSimone, D., Van der Voorde, F., Gefter, W.B., Epstein, D.M., Miller, W.T., Aronchick, J.M., Simon, M.B., Lanken, P.N., Khalsa, S., Brikman, I., Davey, M., and Brisbon, N.: Clinical evaluation of a medical image management system for chest images. American Journal of Roentgenology. 150, 55-59 (1988).
2. Bock, M.: Parallel disk systems multiply disk level performance. Computer Technology Review. Winter (1989)
3. Cho, P.S., Huang, H.K., Tillisch, J., and Kangarloo, H.: Clinical evaluation of a radiologic Picture Archiving and Communication System for a coronary care unit. American Journal of Roentgenology. 151, 823-827 (1988).
4. Dwyer, S.J., Cox, G.G., Cook, L.T., McMillan, J.H., and Templeton, A.W.: Experience with high resolution digital gray scale display systems. Proc. SPIE 1234, 132-138 (1990).
5. Ho, B.K.T., Chan, K.K., and Huang, H.K.: Integrated system design for high-speed image decompression and display. Proc. SPIE 1091, 332-336 (1989).
6. Taira, R.K., Simons, M., Razavi, M., Kangarloo, H., Boechat, M.I., Hall, T., Chuang, K.S., Huang, H.K., and Eldredge, S.: High resolution workstations for primary and secondary radiology readings. Proc. SPIE 1234, 18-25 (1990).
7. Wilson, A.C.: Where are the 2Kx2K display controllers? ESD: The Electronic System Design Magazine. August, 36-43 (1988).

PACS Workstation: User Interface Design

Osman Ratib, Yves Ligier, Matthieu Funk, Christian Girard, René Perrier, Denis Hochstrasser

Digital Imaging Unit, Center of Medical Informatics, University Hospital of Geneva
1211 Geneva-4, Switzerland

Abstract
This paper covers some of the recent concepts in user interface design for digital imaging workstation. In a multimodality PACS network, digital workstations should allow to access, display and analyze digital images from different imaging modalities. In the design of such workstations a special emphasis must be placed on the design of the user interface to allow clinicians with minimal or no computer manipulation skills to use complex analysis tools. General guidelines of a graphic oriented user interface, based on windows and icons, are proposed. A prototype software was designed and developed at the University Hospital of Geneva (OSIRIS software) based on these guidelines. This software was designed to be easily implemented on a variety of hardware platforms providing the users with an identical and consistent user interface.

Keywords: Workstation, PACS, User Interface, Digital Imaging

Introduction

Recent developments of Picture Archiving and Communication Systems (PACS) in clinical environment allow physicians and radiologists to assess radiographic images directly through imaging workstations [1]. The development of medical workstations was primarily oriented toward the development of a convenient tool for rapid display of images. There is, however, an increasing need for quantitative analysis of the images. Therefore PACS workstations should be divided into two categories depending on their specific usage: 1) high resolution viewing stations: their function is to replace the conventional "viewbox" commonly used by physicians to review film-based images. These stations must provide a resolution sufficient to match the diagnostic quality of conventional films. The number of interactions and manipulations required from the users are usually reduced to their minimum to allow for a more efficient utilization in a busy clinical environment. 2) image analysis workstations usually designed for the evaluation of specialized imaging modalities that require more quantitative evaluation of the images. This type of workstation must therefore provide a large number of image processing and analysis tools. The design and performance of these two categories of workstations are quite different due to their differences in utilization.

Principles of an interactive user interface

In designing graphic user interfaces for computers, where the functions are activated through graphic icons and selectable menus and triggered through mechanical devices such as a mouse, the goal is to create a tool that rapidly becomes transparent to the user allowing him or her to achieve complex tasks with the minimal amount of technical training.

Most interactive computer systems rely nowadays on underlying graphical tools used for display and often referred to as Graphical User Interfaces (GUI). Although there are some hybrids, most GUIs consist of three major components: a windowing system, an imaging model, an application program interface (API). Most window-based user interfaces provide a rich environment designed to facilitate display and manipulation of text and graphic documents in windows that can be stacked one on top of another like papers on a desk or tools on a workbench. A typical window environment may also have several applications in operation simultaneously on the screen. Each application

NATO ASI Series, Vol. F 74
Picture Archiving and Communication Systems (PACS) in Medicine. Edited by H. K. Huang et al.

typically has a main or primary window that displays data and in which users carry on their primary interaction with the application. Additionally, applications usually have one or more secondary windows that carry different documents or data.

Workstation design

Image display in windows: In a software designed to display medical images it is necessary to define the way images should be displayed and handled in a window. Because medical images often come in sets (sequential images or set of different views or sections) the window must provide a way of handling more than one image at a time. Such a software should usually support two modes of presentation of a set of images:
• Tiled images
• Stacked images
The user should be able to switch from one mode to another at anytime.

In the Tile mode the images are displayed side by side in a grid that is optimized to be as square as possible based on the number and the size of images to be displayed. The scroll controls of the window should allow the user to move across the set of images. The whole set behaves as a single large image. Therefore zooming or reducing the images will automatically change the magnification factor of the whole set.

In the Stack mode the images are virtually displayed in a stack. The users should then be able to "browse" through the stack up and down using a special control. In a movie mode the user should also be able to dynamically display the images in sequence.

Image contrast and intensity adjustments: the most essential tool in medical image manipulation is the adjustment of contrast and intensity. The user should be able to adjust the contrast and intensity of an image interactively in real time. Selection of contrast through selected numerical values or on intensity histograms should be avoided and should be reserved for advanced users who want to use specific image processing features. Commonly, physicians need to be able to visually adjust the image contrast and intensity through some sort of interactive control tools. A set of linear and non linear intensity transformation functions (ITF) should also be selectable. Finally the user should have the choice to display the images in black and white or in pseudocolor scales. Sets of standard as well as custom color lookup tables (CLUT) should be available.

Because the dynamic range of some images may largely exceed the range of intensity levels that can be displayed at one time on a screen, it is important to allow for optimal mapping of the image values into the subset of intensity (or color) values available for display (typically medical images will have intrinsic dynamic range of up to 12 or 16 bits deep while most of the common display systems support 8 bit displays that will allow only 256 levels to be displayed at one time). The interactive adjustment should be done in real time on the available dynamic range of the display (typically 256 levels) and when done, the image in its full dynamic range should be remapped to the screen.

Also because images displayed in different windows could be used for comparative evaluations, each window should be adjustable separately. Due to hardware constraint, it is often not possible to perform this adjustment in each window individually in real time. It is therefore usually accepted that in some configurations the real time adjustment will affect the whole screen (including all the windows displayed on it) and when done, only the active window (the top most in the stack of windows) will be readjusted to match the setting that was selected interactively.

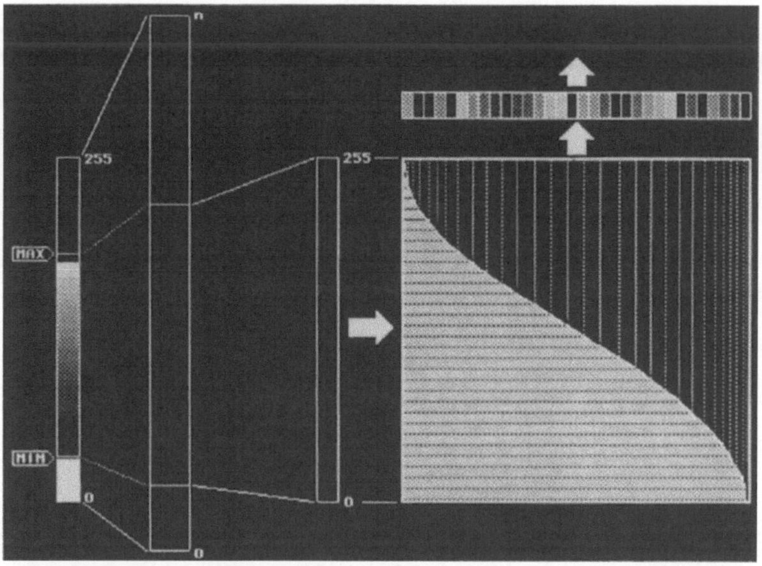

Fig. 1: Schematic illustration of the mapping process used for intensity and contrast adjustment of an image that has a wider dynamic range than the one supported by the display hardware.

Annotations and image related information: Medical images are usually accompanied with clinical information and related data. Three types of annotations and data presentations should be supported and accessible from the workstation.

1) Each image window should be able to display a secondary window containing the image interpretation report as well as any clinical information related to the images.

2) Annotations should be added to the image when needed in form of short text that can be positioned anywhere in the image. All the annotations are always handled as "objects" separate from the image itself that can be moved, altered and deleted at will. An arrow or a line should usually connect an annotation to its point of "anchor".

3) In order to allow for longer text to be attached to an annotation, a "Post-It" window should be accessible. Such a window should act as a pop-up window that is activated by clicking on a selected annotation on the image. When a Post-It is available the corresponding annotation should have a visual clue that allows the user to know that there is more information available.

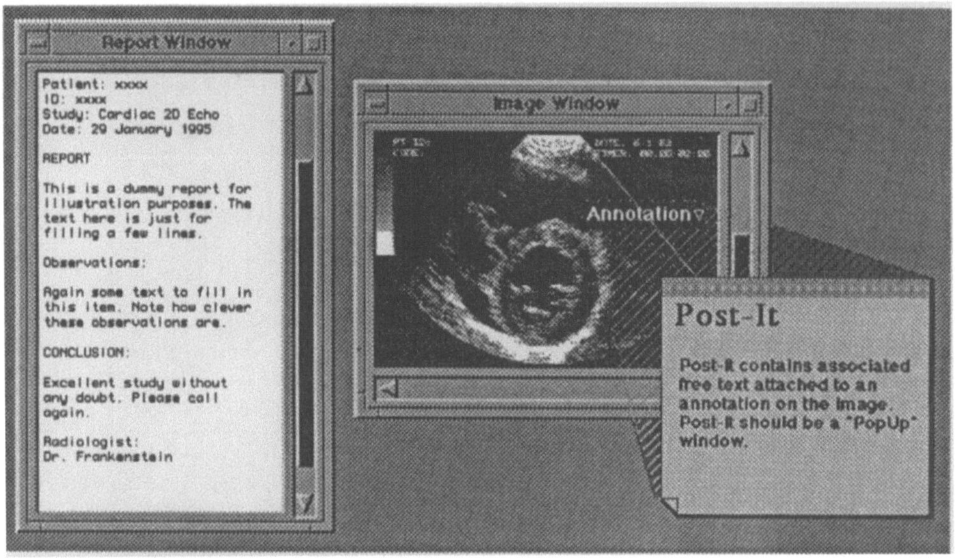

Fig. 2: Illustration of the window-based user interface supporting three types of display for annotation and image-related information (see text for details).

Processing and analysis tools: They can be divided into the following categories:

Image manipulation tools
- Zooming and panning
- Contrast and intensity adjustment
- Magnifying glass
- Rotation and flipping
- Reordering images

Image processing tools
- Filters
- Adaptive Histogram Equalization
- Isocontours
- Image combinations
- Display of global and regional histogram
- Cross section histogram

Annotation and drawing tools
- Annotations
- Arrows and lines

Standard analysis tools
- Coordinates, angles and local density measurement
- Regions of interest (ROI)

  - Copy and paste an ROI between images.
  - Clear or delete an ROI.
  - Reshape and resize ROIs.
  - Merge of several ROIs in one single ROI.
  - Generate an ROI from the intersection of two ROIs.
  - Shrink and grow of an ROI.

In addition to this list of general purpose tools, clinical tools can also be implemented. Usually such tools are more specific to certain types of images. Among the most commonly used clinical tools one can mention ejection fraction calculation, vascular stenosis evaluation, flow measurement, dynamic tracer washout analysis etc...

Discussion and conclusions

The display of images in separate windows offers the advantage that the user has full control of the way the images are presented on the screen by moving and resizing the individual windows. However, the interactive window management required to move images from one location of the screen to another may not be suitable for reviewing a large number of images and may interfere with the radiologist's routine which relies mainly on visual interpretation of the images. The window environment remains particularly convenient however when direct manipulation of the images is required either for image processing or for quantitative analysis.

Part of an ongoing PACS project at the University Hospital of Geneva, a portable software called OSIRIS [2] for the display and manipulation of images is being developed based on the specifications described here. This project is based on preliminary exploratory work in this area at UCLA where a prototype of such a software called CALIPSO [3] was developed on a Macintosh computer. During the development and evaluation phase of the project, the Macintosh was found to be an excellent training platform for clinicians and residents to familiarize themselves with different image processing and image analysis techniques and to explore their effects on different imaging modalities. During this first phase a clinical survey showed that none of the users seemed to be confused by the large variety of tools that the program offers. On the contrary most of the users made suggestions of additional tools that they would like to have. Thus, there is a general demand for more sophisticated analysis tools if they are made available in a user-friendly interface which does not require extensive computer manipulation skills. This confirm our hypothesis that if complex analysis tools are made available in simple systems one can expect a wide acceptability by the radiologists and clinicians. Another important observation is that when physicians are exposed to such analysis tools, their imaginations are simulated and almost all the users that were interviewed had envisioned specific applications of some of these tools in their daily work.

In the design of the OSIRIS program special care was taken to provide a very consistent approach with the different image processing tools. The way the different tasks are performed are always very similar allowing the user to rapidly become familiar with the operation of all the features of the program. This consistency in the design and operation of a large variety of analysis tools applicable to different imaging modalities is unique and non existent in currently available commercial systems. Most manufacturers implement image processing and analysis programs that are manufacturer and modality specific. Recent developments of PACS have lead the manufacturers to provide systems that can handle and display images from different modalities. Very little effort was made, however, to unify the different image analysis techniques in a unique set of standard tools. The OSIRIS software is developed under X11 and OSF/Motif windowing environment to ensure its portability to any UNIX platform.

References

1. Huang HK., Mankovich N., Taira R., et al.: Picture Archiving and Communication Systems (PACS) for Radiological Images: State of the Art. CRC Critical Reviews in Diagnostic Imaging; 28 (4): 383-428, 1988.

2. Ligier Y., Funk M., Ratib O., Perrier R., Girard C.: The OSIRIS User Interface for Manipulating Medical Images. Proceedings of NATO ASI meeting on "Picture Archiving and Communication System (PACS) in Medicine". Evian, 1990.

3. Ratib O., Huang HK. CALIPSO, an Interactive Software Package for Multimodality Medical Image Analysis on a Personal Computer. J. of Medical Imaging, 3: 205-216, 1989.

Reading Room Design for PACS

Steven C. Horii [1], Howard N. Horii [2]

[1] Georgetown University Hospital, Washington, D.C. 20007, U.S.A.
[2] Grad Associates, P.A., Gateway One, Newark, NJ 07102, U.S.A.

ABSTRACT Despite the rapid progress made in the electronic design of imaging workstations for medicine, much less effort has gone into the design of environments in which such systems will be used. Based on studies of radiologist film reading sessions, considerable time will be spent working at such viewing systems. If the rooms in which the workstations are placed are not conducive to comfortable work, it will certainly not favor electronic viewing over film reading.

In examining existing reading environments, it is also apparent that they are not optimum, even for film. Since some of the problems for film and electronic viewing overlap, such as heat generation and glare from light sources, it should be possible to develop solutions which are applicable to both environments or to rooms which will feature both viewing systems.

This paper will discuss some of the approaches to designing environments in which viewing of images is supported by the room architecture and engineering and not degraded by it.

KEYWORDS: radiology, design, PACS, ergonomics

INTRODUCTION

Extensive work has been done on the design of workstations for medical imaging to satisfy the requirements of medical imaging specialists, and to make their performance more efficient from human factors and productivity viewpoints [2,3,7,8,10,11,14,20,21]. In particular, analyses of the work patterns of radiologists [3,8,11] have been combined with advances in person-computer interaction (user interfaces) resulting in viewing and reading systems which are far ahead of early approaches [3,7,14].

In addition to the workstation itself, we also examined the environment in which such equipment will be used. Unfortunately, aside from the papers by Bart ter Haar Romeny *et al.* [22] and Balter and Janower [1], little attention has been given in the medical imaging literature to such environmental design. The problem has both architectural and engineering concerns, and requires not inconsiderable use of human factors design principles

NATO ASI Series, Vol. F 74
Picture Archiving and Communication Systems (PACS) in Medicine. Edited by H. K. Huang et al.

FACTORS IN WORKSPACE DESIGN

There are two excellent books on designing radiology departments [6,19], and both discuss some elements of reading room design. Both volumes were written before electronic PACS became a serious alternative to film-based operation. However, some of the basic design considerations are not altered by switching to softcopy reading.

SIZE OF SPACE

Often a first consideration is to determine the amount of available space to house the reading area. Finding space is a problem in any health care facility we have seen. Even if a new structure is being built, space may be at a premium because of high construction costs or available lot size. More commonly, one will be faced with finding space in an existing facility, or modifying the facility available. Scott [19] recommends about 8 by 10 feet (2.6 by 3 meters) for each person reading. Since a four-panel viewbox array, an alternator, and a four-monitor PACS workstation have about the same width (5 to 7 feet or 1.6 to 2.3 meters) this estimate is a reasonable one. It certainly allows for the one or two other people usually present in a teaching hospital setting, and can accommodate more for shorter periods of time.

USE OF SPACE

The use to which the room will be put has a great impact on the space required as well as a number of other considerations. For a reading room, the main activity is interpreting examinations. Often, however, such rooms may be used for short conferences, consultation with referring physicians, and teaching. Unless there is a requirement to provide services other than for those associated with film reading, it is simpler to design based on the predominant use. Combination use rooms often, but do not always, involve some compromise of one or the other function. For example, in a reading room also used for lengthy conferences, slide projection requirements may preclude optimum placement of viewboxes or light-baffling walls.

ACCESS TO SPACE

Despite the common picture of radiologists relatively fixed at their reading stations, in fact, there is a lot of movement of people in and out of the reading area. The radiologist may move to another reading area for consultation, leave temporarily to attend to a patient having a procedure done, or move to a different place to make or answer a phone call. Similar actions are done by the housestaff, and different ones by referring physicians and fileroom personnel. All of this points to a high-traffic area, so that in these ways, the use defines the required access.

Fire and building codes will fix the number and placement of exits from a space. In addition to emergency egress, though, access is also involved with movement of people to other areas as they do their work. A facility such as a main or large satellite reading room needs to be accessible to virtually anyone in the department. It should not, however, be located so as to serve as a general traffic corridor, nor should such use be encouraged, as it is disruptive to the work of the radiologist. Aside from the radiologists working in the room and referring physicians coming in to look at examinations, there are technologists, clerks, transcription personnel, secretaries, administrators, and housekeepers (this list being based on access frequency and not on job importance, though cynics might assume otherwise) who regularly need to go to the reading area. It is impossible to make such a facility easily accessible to all of these people. Rather, since the radiologists are most dependent on the space, some thought should be given to optimizing the location of the room for access to the areas they need to reach often.

SUPPORTING SERVICES

In support of the uses of the reading room, we need to consider supporting services. Heating, ventilation, and air conditioning (HVAC), electrical power, communications, illumination, and acoustic control are typical factors which we will include with services. All of these will be discussed in more detail in subsequent sections of this paper. While all of these are part of the environment, electrical power requirements and communications systems have less of an immediate impact on bodily comfort than the other listed items. Providing these services for a workspace requires the coordination of several types of designers and contractors. A good architect will work with the client and the client's contractors so that the construction proceeds as smoothly as possible.

BUDGET CONSIDERATIONS

At the base of any workspace design, be it new or renovation, is the budget allotted for the project. This may determine what kinds of solutions to problems can be realistically implemented. Again, a good architect serves the client by being able to provide a number of possible solutions to any problem, so that an optimum design can be achieved for any budget level.

TASKS AND ENVIRONMENT

The major task to be supported in a reading room is the interpretation of medical images. However, this consists of a number of smaller tasks. A number of these were described in a very detailed fashion in the paper by Rogers *et al.* [15] and in a more general fashion by Mun *et al.* [12]. Aside from looking at images on alternators or workstations, a large amount of writing is done, results are dictated, and teaching is often incorporated into these processes. This means that facilities to support these activities must be provided, and features of the room should not hinder any task to a significant degree, or preferably not hinder any task at all. Some of these actions have requirements which are contrary to others. For example, image reading is best done with as little ambient light as possible, but some light is needed for writing and for searching through film folders.

The devices which need to be located in the room are also based on the work to be done. Film alternators and workstations are the equipment typically designed for. However, report generation requires some form of dictation system, and consultation requirements force heavy dependence on telephones. Because we have found that telephone use may be quite frequent, it would benefit users if the phones were located on the workstation or alternator tables, or were at least within reach without standing up.

ENVIRONMENTAL FACTORS

The environment can be thought of as a combination of the physical space or structure, the services provided, and the equipment and people within. In addition to the size and layout of the room we are working with, there are ceiling constraints and flooring systems to be considered. Fixed ducting, fire sprinklers, emergency lighting, and structural beams all affect placement of lighting fixtures. Flooring may be flat, flat with cable troughs, or raised. Raised flooring is an advantage in that it allows for very easy re-routing of cables, but does require more maintenance than plain flooring (the supports may get out of alignment, resulting in panels which rock when walked on).

Structural requirements may include floor loading and vibration studies, and some local building codes will affect the impact of these issues on any design. Addition of heat sources or expanding a facility for a larger number of people will have to take these into account when considering HVAC requirements.

COMFORT FACTORS

Most important to the users of any given space is how comfortable they will be when using it. Quite separate from the user interface issues raised by electronic workstations are those things which directly affect comfort. These factors include seating, heat, noise, and light. Though people can operate within wide ranges of these variables, adverse values will impair function either temporarily or permanently (chronic exposure to high noise levels, for example, can do both). In some instances, the degree of impairment may not be noticed by the person working, but various measurements will show distinct performance losses [16].

SEATING

We have found that reading sessions may last from fractions of an hour to four hours [13], so that comfortable seating is a must. The "ergonomic" design chair has become quite popular in offices in which video display terminals (VDTs) are used. The major features of these chairs are adjustability, good low back (lumbar) support, arm rests (for keyboard users), and good padding. There are apparently a set of ANSI standards for such chairs, though the authors have not seen them. We have found this type of chair to be acceptable to the radiologists, and are presently in use in some reading rooms and offices.

HEATING AND AIR CONDITIONING

The HVAC design should, having taken into account the equipment and personnel heat load, maintain the room temperature in a comfortable range. For most people, this range changes with seasons, but is approximately 63-71°F (17.2-21.7° C) in winter, and 65-75° F (18.3-23.9° C) in summer [4]. The relative humidity of the air affects perceived temperature, and the most comfortable range is from about 20 to 60 per cent at 75° F [18]. Electronic equipment is also constrained by temperature and humidity, but will generally operate well in the same ranges as those for which people are comfortable.

ACOUSTICS

Sound is one of the environmental factors which is highly subjective. Most of us know that sounds may or may not be irritating depending on the context. A newborn's first cry is probably welcome to most ears, but a child's cry which wakes one in the middle of the night is much less so. It is sound quality as well as loudness and context which alter its perception. Noise is thought of as sound which is unwanted. Despite the problems defining noise, most designers think that an office-type workspace has a noise threshold of about 58dBA [5]. There are specific OSHA limits for sound exposures, but these are for sound levels far above what should be encountered in any reading room environment. Standard methods for noise control are in a selection of materials for ceiling, floor, and walls. Acoustical ceiling and carpet are excellent sound absorbers, and matte wall finishes or even acoustic absorption panels on the walls are better than plain gloss or semi-gloss wall treatments.

LIGHTING

Lighting is, we feel, one of the most critical issues to be faced in reading room design. This is particularly true for rooms in which CRT displays will be read since their brightness is usually less than that of film on a viewbox. As described previously, the tasks of reading images and writing notes will have opposite lighting requirements in a radiology reading room. Aside from general room illumination and task lights, the image displays themselves (whether CRT or alternator) are a source of additional light. The problem is that the lighting may result in glare from CRT screens or from film, and this may impair reading by reducing contrast sensitivity of the eye or by fatiguing the iris [17]. The curved screens of CRTs are an additional complication in that they reflect light sources from a wider angle than a flat screen or film. In our reading room at present, most task lighting is stray light from the alternators. All the ceiling lights are usually turned off. This poses some problem for the PACS workstation or for writing notes in log books as some illumination is needed to see the keyboard and control panel or book page.

Glare is a problem for at least two reasons. The first is that the glare light will cause the pupil to constrict more than it would if there were no glare. This results in less of the information illumination getting to the retina. Secondly, the glare source can result in veiling glare within the eye itself. This results from the scattering of light by the vitreous portion of the eye, and is increased with extraneous light entering the eye (and also increases with age).

Lighting design is often aimed at achieving some brightness on the work surface and is adjusted for the task to be performed. This rarely takes into account the problems posed by glare, but with increasing use of VDTs, some lighting manufacturers are designing systems which control the output light beam.

Other methods of reducing the reflection from CRT screens include etching the envelope glass, coating the glass, and placing filters or screens over the tube. Etching the glass reduces specular reflection, but also blurs the image on the phosphor, reducing the apparent spatial resolution. Antireflection coatings work well, but suffer from sensitivity to fingerprints and smudges. Frequent cleaning of such screens is a must. Filters work by attenuating the light reflected off the tube more than the light coming from the tube, since the reflected light passes through the filter twice. Filters themselves are subject to reflection from their surfaces. Screens act as collimators, but are subject to collecting dirt. For high-resolution displays, the mesh of the screen may be visible.

One consistent problem we have seen in reading room design is in the placement of viewboxes and alternators. Placement on opposing walls with the reading surfaces facing each other should be avoided if at all possible. Even placement on orthogonal walls can result in glare problems for anyone not viewing normal to the illuminator surface. Unfortunately, even some texts [23] violate this principle in example designs. This is not a problem if the two sets of viewboxes or alternators are never used simultaneously, for the unused one is simply turned off.

Local fire codes may restrict very low illumination levels, and some consideration must be given to special lighting for exits in this regard. Also, good solutions for task lighting at a workstation are difficult to find. One problem is that some task lighting, when bright enough to read or write by, illuminates a user's face or clothing enough for them to become a glare source.

CONCLUSION

We have tried to provide an outline of room design based on conventional architectural principles applied to the problems posed by electronic image displays. In addition to the usual concerns about the equipment (powering it, cooling it, and paying for it) the most successful environments for the equipment will be those in which human factors have been given equal or more consideration than equipment requirements alone.

REFERENCES

1. Balter, S., Janower, M.L.: Radiographic viewing conditions. Proc. SPIE, v. 43, Appl. Opt. Inst. Med. II, pp 225-227, 1973.

2. Beard, D., Pizer, S.M., Rogers, D., Cromartie, R., Desirazu, S., Ramanathan, S., Rubin, R.: A prototype single-screen PACS console development using human computer interaction techniques. Proc. SPIE v. 767, Medical Imaging I, pp 646-653, 1987.

3. Beard, D., Creasy, J.L., Symon, J., Cromartie, R.: Experiment comparing film and the FilmPlane radiology workstation. Proc. SPIE v. 914, Medical Imaging II, pp 933-937, 1988.

4. Diffrient, Niels, Tilley, Alvin R., Harman, David: *Humanscale 1/2/3, 4/5/6, and 7/8/9.* M.I.T. Press, Cambridge, MA, 1981.

5. Diffrient, N., Tilley, A.R., Harman, D.: *op. cit.*, 4/5/6 p 13.

6. Fischer, Harry W.: *Radiology Departtnents: Planning, Operation, and Management.* Edwards Brothers, Inc., Ann Arbor, MI, 1982.

7. Fuchs, H., Pizer, S.M., Creasy, J.L., Renner, J.B., Rosenman, J.G.: Interactive, richly cued shaded display of multiple 3D objects in medical images. Proc. SPIE v. 914, Medical Imaging II, pp 842-849, 1988.

8. Good, B.C., Cooperstein, L.A., Gur, D.: Radiologists' perceptions of and preferences for a concise history form during radiologic interpretation. Proc. SPIE v. 914, Medical Imaging II, pp 970-975, 1988.

9. Horii, S.C., Horii, H.N., Kowalski, P.: An eclectic look at viewing station design. Proc. SPIE v. 914, Medical Imaging II, pp 920-928, 1988.

10. McNeill, K.M., Fisher III, H.D.: A model for radiologic workstation user interface design. Proc. SPIE v. 767, Medical Imaging I, pp 713-716, 1987.

11. McNeill, K.M., Seeley, G.W., Maloney, K., Fajardo, L., Kozik, M.: Comparison of digital workstation and conventional reading for evaluation of user interfaces in radiology. Proc. SPIE v. 914, Medical Imaging II, pp 872-876, 1988.

12. Mun. S.K., Benson, H., Welsh, C., Elliott, L.P., Davros, W.: Baseline study of radiology services for the purpose of PACS evaluation. Proc. SPIE v. 914, Medical Imaging II, pp 978-987, 1988.

13. Mun, S.K. *et al .*: *op. cit.*, p 985.

14. Ratib, 0., Huang, H.K.: Integration of a desktop image analysis work station based on a Macintosh II computer in a clinical PACS. Scientific Exhibit # 915, RSNA 74th Sci. Assy. and Ann. Meeting, Chicago, IL, 1988.

15. Rogers, D.C., Johnston, R.E., Brenton, B.C., Staab, E.V., Thompson, B.G., Perry, J.R.: Predicting PACS console requirements from radiologists' reading habits. Proc. SPIE v. 536, Medicine XIII/PACS III, pp 88-96, 1985.

16. Sanders, Mark S. and McCormick, Ernest J.: *Hurnan Factors in Engineering and Design, Sixth Edition.* McGraw-Hill, Inc., New York, NY, 1987.

17. Sanders, M.S., McCormick: *op. cit.*, pp 414-415.

18. Sanders, M.S., McCormick: *op. cit.*, p 432.

19. Scott, Wendell G. (Ed.): *Planning Guide for Radiologic Installations, Second Edition.* Williams and Wilkins, Baltimore, MD, 1966.

20. Schuttenbeld, H.H.W., ter Haar Romeny, B.M.: Design of a user interface for a PACS viewing station. Proc. SPIE v. 767, Medical Imaging I, pp 844-848, 1987.

21. Taira, R.K.: Operational characteristics of pediatric radiology image display stations. Proc. SPIE v. 767, Medical Imaging I, pp 571-576, 1987.

22. ter Haar Romeny, B.M., Raymakers, J., van Waes, P.F.G.M., Helder, J.C., de Graaf, C.N., van Rijk, P.P., Schuttenbeld, H., Zuiderveld, K.J., Tiemann, J., Scharnberg, B.: The Dutch PACS project: philosophy, design of a digital reading room and first observations in the Utrecht University Hospital in the Netherlands. Proc. SPIE v. 767, Medical Imaging, pp 787-792, 1987.

23. Tuddenham, W.J.: Rooms for Film Processing and Display, in: Scott, W.G., *op. cit.*, p 79.

In addition to these specific references, a related paper was published as:

Horii, S.C., Horii, H.N., Mun, S.K., Benson, H.R., Zeman, R.K.: Environmental designs for reading from imaging workstations: ergonomic and architectural features.
Proc. SPIE, v. 1091, Med. Imaging III, pp 172-183, 1989.
J. Digital Imaging 2(3), pp 156-162, 1989.

Data Storage and Compression

Paul S. Cho, Kelby K. Chan, and Bruce K.T. Ho

Department of Radiological Sciences, UCLA School of Medicine, Los Angeles, CA 90024-1721, USA

Abstract: The PACS data files migrate and reside among a variety of storage devices during their life-cycle. High-capacity Winchester disks are used ubiquitously. High-speed disks such as parallel transfer disks and disk arrays are often configured in the diagnostic display workstations. Optical disk jukeboxes provide the central archive. This paper reviews the current storage technologies and also discusses the data compression techniques applicable to the radiological images.

Keywords: Images; storage, retrieval, and compression.

Data Storage Devices

In PACS database, image files are likely to be distributed among various storage devices. One most commonly employed distribution strategy is the hierarchical storage architecture in which the images at the peak demand are kept in fast storage devices while those less accessed are stored in slower mass storage. These storage devices include Winchester disks, parallel transfer disks, and optical disks.

High-Capacity Winchester Disk Drives

The past decade has seen a great increase in the storage capacity of the Winchester hard disk drives. For example, the capacity of the 5.25-inch Winchester, which was first introduced in 1980 (Seagate ST 506), has increased from the initial 5-Mbytes to 800-Mbytes. A typical high-capacity 5.25-inch drive contains as many as eight platters, providing 16 surfaces for data storage. To house more platters, the drive motor is now designed to fit inside the disk platter spindle rather than underneath the stack of platters. Also, the bit density (specified in bits per inch, or bpi) has been increased by the use of high-density magnetic media such as alloys of cobalt or chromium. Thin films of these alloys coated on the platter can provide more than 28,000 flux transitions per inch compared with 16,000 for the thicker iron oxide media. How closely the tracks can be spaced together (track density, tpi) depends on how precisely the head can be positioned. High-capacity drives rely on the track-following servo system to guide the

NATO ASI Series, Vol. F 74
Picture Archiving and Communication Systems (PACS) in Medicine. Edited by H. K. Huang et al.

heads to the exact position above the platter. The servo-reference tracks are normally prerecorded on one of the platter surfaces. Other factors that are important in the design of high-capacity Winchester are size of the read/write head, the gap between the head and the platter surface, the gap between the poles of the head, and the bit encoding technique.

With the increase in bit density, higher data transfer rates became possible. Data transfer rates are determined by the disk-to-computer interface specifications (Table 1). Four widely implemented disk interfaces are the enhanced small device interface (ESDI), the small computer system interface (SCSI), storage module device (SMD) interface, and the intelligent peripheral interface (IPI).

	ESDI	SCSI-1	SMD-E	IPI
ANSI Standard	X3.170-1990	X3.131-1986	none	X3.129-1986 X3.130-1986
Transfer Rate (MB/sec)	3	1.5 (async), 4(sync)	3	10
Width of Data Path	serial	8-bits	serial	16-bits
Disk Size (inch diameter)	5.25	5.25	8	8, 5.25
Max. Cable Length (feet)	10	20, 80 (differential)	50	400

Table 1. Comparison of various disk interfaces.

ESDI, which was officially adopted as an ANSI standard X3.170 in early 1990, permits maximum transfer rates of 3 Mbytes/sec and some drives have realized 2.5 Mbytes/sec [1]. The SMD interface has never been formally adopted as a standard. The original SMD interface was specified at 1.8 Mbytes/sec, which was later increased to 3 Mbytes/sec. The enhanced version is called SMD-E (SMD-Extended). The drives are typically 8-inch Winchester with the current maximum capacity of over 1,200-Mbytes [16].

The SCSI interface is widely accepted for its versatility. While ESDI and SMD are device-level interface connecting directly to the hard disk drive, SCSI is a system-level interface. It is, in fact, a bus that can connect a variety of other peripheral devices such as tape drives, optical disk drives, printers, and scanners. SCSI-1, which became an ANSI standard X3.131 in 1986, allows data transfer at a maximum 1.5 Mbytes/sec asynchronous rate and at a maximum 4 Mbytes/sec synchronous rate. The width of data path is 8-bits, compared with the serial transfer of ESDI and SMD-E. Furthermore, SCSI can perform concurrent I/O operations efficiently. When a peripheral device receives a command, it can disconnect itself from the host computer while it performs the operation, thus freeing up the system bus to handle other tasks. When the peripheral is finished with the command, it connects to the SCSI to transfer data or status information. With this scheme, multiple peripherals can process commands simultaneously.

Further increase in data transfer rate at a maximum 10 Mbytes/sec is expected in the second generation SCSI or SCSI-2 [6,23].

IPI supports a theoretical maximum transfer rate of 10 Mbytes/sec through its 16-bit wide data path. It has been mainly used to interface to 8-inch Winchester drives, although 5.25-inch IPI drives are appearing. Also available are 2-head parallel transfer IPI-compatible drives, with data transfer rate of 6 Mbytes/sec [17].

Much of the performance improvement for these disk interfaces is based on the cache design. For example, in the *segmentable cache* design, the cache is subdivided into several track buffers. With each new I/O request, the controller determines if the needed data is already in one of the track buffers. If so, the controller does not have to read from the disk, thus saving the total seek and rotational latency times. If the needed data is not in any track buffer the controller reads the track that contains the requested data and overwrites the least recently used track buffer with this data.

Parallel Transfer Disks and Disk Arrays

Data transfer rates above 10 Mbytes/sec are achievable through parallel transfer disk (PTD) or disk array. The PTD allows multiple read/write heads to simultaneously transfer data. Disk array, on the other hand, configures multiple conventional Winchester disks in parallel. Two common approaches are software striping and hardware paralleling. In software striping, the disks are connected to the system bus through traditional controllers. Blocks of data are segmented and moved to and from the disk drives in parallel. The data segmentation and recombination are handled by the software. In the hardware approach, a parallel drive array controller simultaneously manages multiple disk drives. For example, eight ESDI disk drives, each capable of 2.5 Mbytes/sec data transfer rate and having 800-Mbytes can be configured in parallel to deliver 20 Mbytes/sec transfer rate with a total storage capacity of 6.4-Gbytes. If the application does not require large storage capacity but needs a high transfer rate, one 900-Mbyte PTD can provide 20 Mbytes/sec.

Optical Disk Drives

WORM (write once, read many) optical disks have been widely used as the archival media in radiological PACS. These disks are available in different sizes, 3.5-inch, 5.25-inch, 12-inch, and 14-inch, ranging in capacity from 180-Mbytes to 10-Gbytes. More recently, erasable MO (magneto-optical) disks have been introduced (1987). Currently, the manufacturers are concentrating on the 5.25-inch MO disks for which the ISO (International Organization for Standard) is finalizing the standard. Also, the Japanese are working towards standardization of the 5.25-inch MO disk file format for medical images in an effort to ensure media interchangeability [12]. Since the 5.25-inch MO disks are highly portable and reusable, they may be used as communication media. In one scheme, after the images are written to the MO disks at a central location, the disks are hand-carried to the diagnostic workstations for reading. Once all the cases are reviewed, the disks are returned to the central copying station for recycling[12].

Today, there are over a dozen optical disk manufacturers and about 40 other vendors who integrate the WORM and MO products. The optical disk suppliers include SONY, Canon, Ricoh, Maxoptics, Hitachi-Maxell, Panasonic, Mitsubishi, Sharp, NEC, Pioneer, Kodak, Toshiba, etc. By far, the most popular optical disk-to-host interface is the SCSI. The data transfer rates for the MO drives are typically around 500 KBytes/sec for write and 650 KBytes/sec for read operation. In comparison, the WORM drives are slower on the average, 250 KBytes/sec for write and 500 KBytes/sec for read.

Data Compression

Data compression techniques [14,19] play a key role as a leveraging technology in image management systems. Despite the ever increasing performance of communications and storage technologies, the demands have always exceeded the capacities. Even as the technology advances, the requirements for spatial resolution, dynamic range, and image volume continue to increase. Thus image compression can contribute either by lowering the threshold of technological feasibility, or by increasing the functionality of future systems.

For radiological images, existing lossless or reversible methods achieve [15,21,22] compression ratios with an upper bound in the neighborhood of 3:1. More typically the ratio is in the range 1.5-2.0, but probably closer to 1.5.

Irreversible methods can achieve higher compression ratios at the expense of image fidelity. Among the irreversible methods, vector quantization [7] and transform encoding [20] show the greatest promise. The discrete cosine transform (DCT) has been the transform of choice within the field because of its superior mathematical properties [13,20]. The DCT is, however, computation intensive. Thus, in the past, simpler and less desirable transforms have sometimes been used. More recently, developments [5] in both fast cosine transform (FCT) algorithms and digital electronics have rendered this consideration somewhat obsolete.

Full-Frame DCT Compression Method

The full-frame DCT algorithm is diagrammed in Figure 1. Starting with the original image, a discrete cosine transform is performed on the entire frame. This differs from block quantization [18,24] methods which divide the image into a number of small matrices before performing transforms on each sub-matrix. Subdividing the image before transform has the potential for introducing block artifacts in the reconstructed image. Although methods exist for reducing the visibility of such artifacts, they may be enhanced again if post-processing such as windowing or histogram equalization is performed on the image.

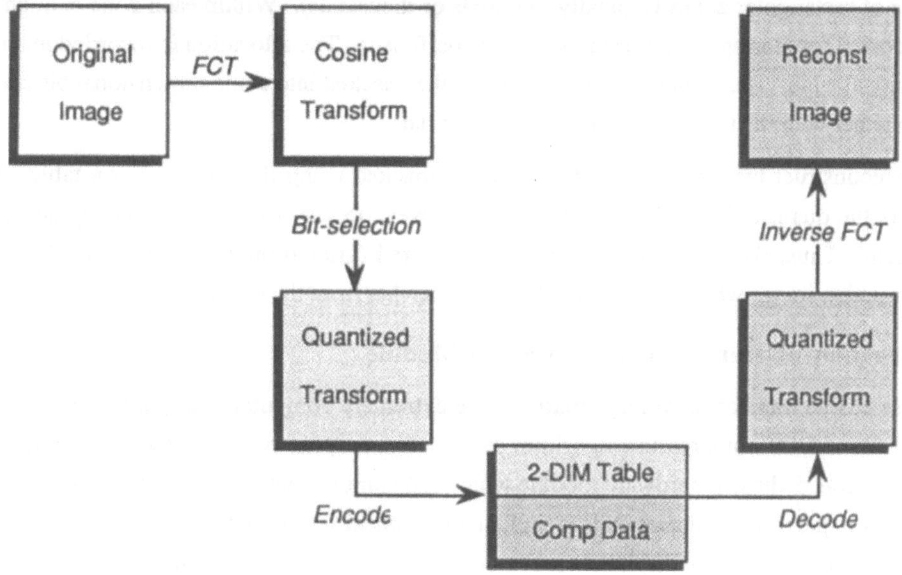

Figure 1. Flow diagram for compression and decompression process. During decompression, the reconstructed image is obtained via an inverse FCT of the quantized transform. Thus the original image is not recovered exactly.

The dynamic range of the transform coefficients is typically much larger than that of the original image[5]. Thus transform by itself introduces data expansion. In order to achieve compression, a quantization or bit-selection step is used to obtain the quantized transform. This step determines how many bits of precision are used for storing each transform coefficient. The quantization method described earlier[4] follows a form of thresholding. Two parameters (p,k) are used. The quantized transform $Q(u,v)$ $(0 \leq u,v \leq N\text{-}1$; NxN is the image size) is obtained from the cosine transform $F(u,v)$ via the relation:

$$Q(u,v) \;=\; \mathbf{nint}\,[\,2^{p'} F(u,v)\,], \qquad (1)$$

$$\text{where } p' \;=\; p+1 \;;u,v < k$$
$$\;=\; p \qquad ;u \geq k \text{ or } v \geq k \qquad (2)$$

and **nint** denotes the nearest integer function. The parameter p determines (via p') the accuracy maintained for the transform coefficients. In effect, $F(u,v)$ is rounded off to the p'-th binary place. Larger values of p result in better image fidelity but lower compression ratio. The parameter k serves as a fine adjustment. Transform coefficients with either frequency coordinate less than k ($0 \leq k \leq N$-1) are kept with one extra bit of accuracy.

The next step is the encoding process. The quantized transform is divided into a large number of rectangular zones (typically hundreds or thousands). Within each zone enough bits are allocated for storing the largest magnitude coefficient. The allocation is recorded in a two-dimensional table. The quantized coefficients are then packed into a one-dimensional bit-stream. This together with the table forms the compression data.

To reconstruct the image, the bit-stream is unpacked using the bit-allocation table. This recovers the quantized transform. There is, however, no way to recover the original cosine transform. Thus, the quantized transform is inverted (after proper rescaling) to obtain the reconstruction image which, in general, differs somewhat from the original image.

Construction of Hardware Compression Module

Because cosine transforms of large matrices are extremely computationally intensive, running the full-frame DCT compression algorithm on a general purpose computer requires many hours to complete. A dedicated hardware module which can compress and reconstruct an image within seconds is required to implement this technique in a clinical setting [8,9].

An important goal in the hardware module design is to implement high speed image compression for the various types of clinical modalities including CR, angiography, CT, and MR, with matrix sizes of 2K, 1K, 512, and 256, respectively. Pixel depth of 12-bits and 8-bits must all be processed by one module. A compression speed of 4-sec for 2K images and 1-sec. for 1K images are set as the criteria for the highest speed applications. Finally, the module must be mass produceable, low cost, and portable to each archiving and display station any where in the hospital.

Our compression module is based on a quad digital signal processor (QDSP) board. This board is integrated into a host computer system through a VME (Versa Module European) backplane. Our current development system uses a Sun workstation model 3/E, although any computer system equiped with a VME backplane may be substituted. A block diagram of the host computer with the compression module is shown in Figure 2. A large (16-Mbytes) frame buffer is included mainly for the enormous scratch space needed during the cosine transform computation. Enclosed in dashed lines are the essential workstation components: a CPU (central processing units), a video display controller for the console, and a third board for connecting to ethernet and SCSI devices, e.g. a hard disk or a tape drive. We chose a gray-scale monitor for the console in order to show both text and images.

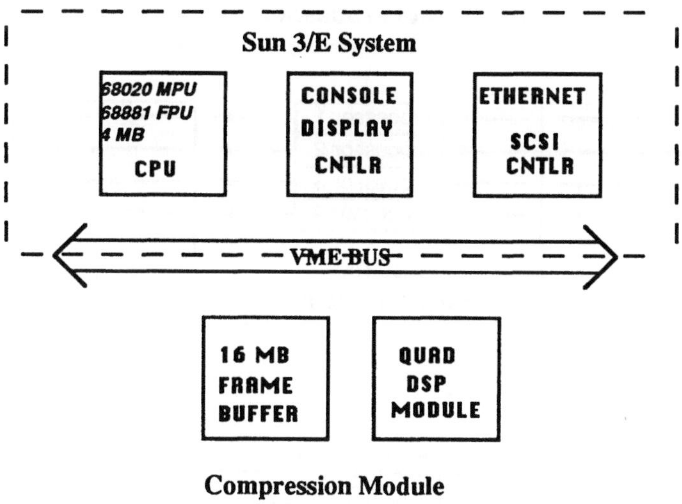

Figure 2. Block diagram of the compression module integrated into a Sun 3/E host computer.

Parallel Processing and the Quad Digital Signal Processor Board

The speed of the compression module comes from parallel processing and the use of high performance digital signal processors (DSP)[10,11]. The DSP's, manufactured by Motorola (Phoenix, AZ), are inherently optimized for linear fast Fourier transform (FFT). Our algorithm very effectively took advantage of this optimization by exploiting the close analogy between cosine transform and Fourier transform. More importantly, the efficiency obtained through parallel processing is possible because each row and column of an image is independently transformed and quantized. Multiple processors simply work on separate image lines during the two passes (horizontal and vertical) of the two dimensional transform (see Figure 3). Mathematically, the two passes are simply summations of the pixel values multiplied by the transformation kernel, carried out first in the x-direction, then in the y-direction. Quantization (trimming of lower bits) is performed on the cosine coefficients obtained at the end of the second pass, and results in the drastically reduced data volume.

The QDSP board groups together four signal processors to achieve the intended parallelism. Each of the four processor is coupled with enough cache memory to perform computations on one complete line pair of image data up to 2K in size. The layout of this parallel processor board is show in Figure 4. The four prominent square components are the DSP's. Each is accompanied by associated cache memory chips constructed from static random access memory (SRAM). On top are the two 96-pin connectors for attaching to the VME backplane. Near the front plate (on bottom of the figure) are four LED's (light emitting diodes) which flash continuously during the processors' computation cycles.

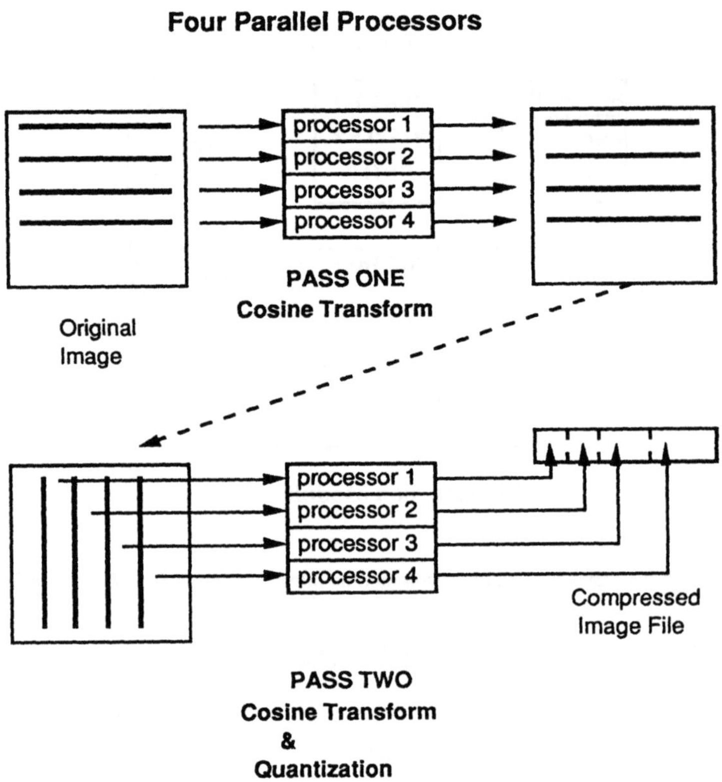

Figure 3. Parallel processing hardware makes high speed compression possible. Four separate rows of image data are cosine transformed during the first pass, and four columns during the second pass.

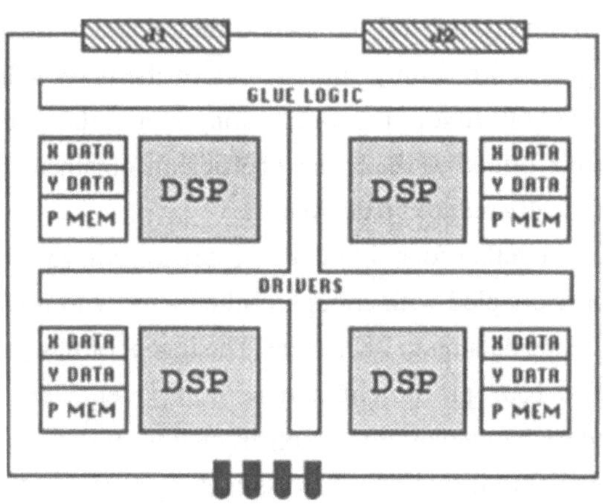

Figure 4. The layout of the parallel processor board with four processors chips and their associated cache memories.

Frame Buffer and Back Plane

Between the two passes of the transform, the direction of data flow changes from horizontal to vertical (or from x- to y- direction) within the data matrix. The halfway transformed data at this stage actually grows in pixel depth (a property of digital computation) and must be stored in a very large intermediate buffer. For a 2K x 2K image (regardless of original pixel depth), 16-MByte of image memory space is required. In our development system, we implemented this required memory with an external frame buffer attached to the common backplane as shown in Figure 2. This frame buffer is a commercially available product manufactured by Micro Memory Inc. (Chatsworth, CA). On systems such as the Sun 4/490, with a minimum of 32-Mbytes of system memory, the frame buffer may be eliminated. Without the external frame buffer, there is no noticeable loss of speed except when compressing very large image sizes such as 2K x 2K x 12-bit, in which case, the speed is found to be up to 80% slower.

The electronic backplane, also called a "bus", is responsible for linking multiple boards of a single system in a coherent fashion. In particular, the VME bus is chosen because of its acceptance as the preeminent industry standard. It allows us many more options, for example, when we look for a suitable frame buffer, or a host CPU. The adoption of the VME bus also makes the board portable to any Sun Microsystems (Sunnyvale, CA) workstations, which make up all the major nodes of our PACS system as well as the display stations of many imaging manufacturers [2].

Data Transporter

In our Sun 3/E development system, a compression speed of four seconds for 1K image was achieved. We found that data transfer between the processor board and the frame buffer is the limiting factor in the compression speed, not the actual processing. In our continued development, a custom designed data transporter was built to circumvent this bottleneck. This data transporter is fast enough to drive the data traffic between the frame buffer and two QDSP boards (running 8 processors) simultaneously. Figure 5 illustrates the configuration of this four board combination. A secondary bus, the VSB (VME Subsystem Bus), links the data transporter to the frame buffer in this data transfer strategy. This configuration reaches one second compression for 1K images. Because of the capability of the data transporter, we are confident that speed will not be a constraint in the clinical implementation of image compression.

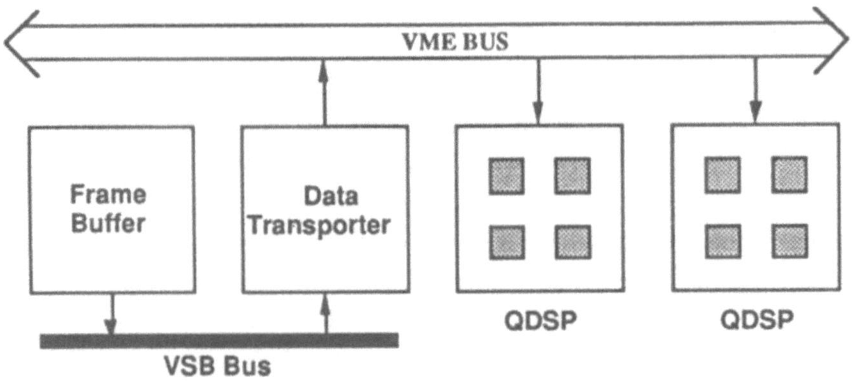

Figure 5. High speed compression is achieved by this four board configuration with one frame buffer, 2 processor boards, and a data transporter.

Module Portability

As was emphasized previously, the compression module is portable to all Sun 3 and 4 computers as well as to any VME based computer system. Currently, software compatibility has only been demonstrated on the many Sun systems in our laboratory and a Motorola System 1131, which we also used during development. The necessary software modification for switching between these two types of system is very minimal. It is expected that our software package can be modified to work on any other computer running the Unix operating system with a reasonable amount of effort. The software package consists of "compress", "decompress", and "image_subtract" programs intended for processing individual images manually. Automatic operations of the compression module can be implemented in the future at various PACS stations such as a viewing station, an archiving node, a laser film printer, etc. by incorporating these programs into a master PACS program.

	Sun 4/370 (32 MB)	Sun 3/E (16 MB frame buffer)
4 processors	4 sec/16 sec	4 sec/16 sec
8 processors with Data Transporter	N/A	1 sec/4 sec

Table 2. Time to compress an 1024/2048 image.

Results

We have fabricated the QDSP module in a printed circuit board form which is suitable for mass production. Our laboratory and staff are set up to do small-quantity manufacturing of this module for product distribution. The completed modules have been tested in four types of Sun workstations: Sun 3/E, 4/330, 4/370, and 4/490. We now have the capability to install the compression modules both in our hospital as well as in other institutions as needed for the purpose of quality evaluation.

A table of performance measurements is given in Table 2 for the Sun 3/E and Sun 4/370 systems on image sizes of 1K x 1K x 8-bit and 2K x 2K x 8-bit. In case of the Sun 3/E, boost performance provided by the data transporter is also given. There is no noticeable difference between the two system configurations for these image sizes. However, as noted previously, systems without an external frame buffer will underperfom when processing 2K x 2K x 12-bit images because the system must use virtual memory, thus slowing down overall performance. This may be improved with future Sun operating systems which allow user to dedicate required memory for one particular process.

Acknowledgements

This work was supported in part by Public Health Service Grant No. R01 CA 39063, R01 CA 40456, and P01 CA 51198, awarded by the National Cancer Institute, Department of Health and Human Services, and by the UCLA Department of Radiological Sciences.

References

1. Alford R.C.: The evolution of ESDI. BYTE. 15(6), 297-306 (1990).
2. Arenson R.L., Chakraborty D.P., Seshadri S.B., and Kundel H.L.: The digital imaging workstation. Radiology. 176, 303-315 (1990).
3. Bramble, J.M., Huang, H.K., and Murphy, M.D.: Image data compression. Invest. Radiology. 23(10), 707-712 (1989).
4. Chan, K.K., Lou, S.L., and Huang, H.K.: Radiological image compression using full-frame cosine transform with adaptive bit-allocation. Comput. Med. Imag. and Graph. 13, 153-159 (1989).
5. Chan, K.K: Implementation of fast cosine transforms with digital signal processors for image compression. Proc. SPIE 914, 782-785 (1988).
6. Glass, B.L.: The SCSI bus. BYTE. 15(2), 267-298 (1990).
7. Haskell, B.G. and Hang, H.M.: Comparison of discrete cosine transform and vector quantization of medical imagery. Proc SPIE 626, 399-408 (1986).
8. Ho B.K.T, Chan K, Ishimitsu Y, Lo S.C., and Huang H.K.: Expandable image compression system: A Modular Approach. Proc. SPIE 767, 286-289 (1987).
9. Ho B.K.T., Chan K.K., Ishimitsu Y., Stewart B.K., Lo S.C., and Huang H.K.: High speed image compression system: prototype and final configuration. Proc. SPIE 914, 786-791 (1988).
10. Ho B.K.T. and Huang H.K.: Specialized module for full frame radiological image compression. Optical Engineering 1989 (to be published).
11. Ho B.K.T., Chao J., Wu C.S., and Huang H.K.: Full frame cosine transform image compression for medical and industrial applications. Machine Vision and Applications. 1990, (to be published).
12. Inamura, K., Umeda, T., Takigawa, A., Inamoto, K., Kozuka, T., and Ohyama, N.: A trial of PACS employing magneto-optical disks. Proc. SPIE 1234, 50-59 (1990).
13. Jain, A.K.: A sinusoidal family of unitary transforms. IEEE Trans. on Patt. Anal. and Mach. Intell. PAMI-1, 356-365 (1979).

14. Jain, A.K.: Image data compression: a review. Proc. IEEE 69, 349-389 (1981).

15. Lo, S.C. and Huang, H.K.: Error-free and irreversible radiographic image compression. Proc. SPIE 536, 170-177 (1985).

16. Lyons, H.N.: Innovative technology breathes extended life into 8-inch Winchester. Computer Technology Review. 9(10), 43-45 (1989).

17. Mooney, J.M.: The IPI standard: the peripheral answer to RISC. VMEbus Systems. August, 51-60 (1989).

18. Nakagawa , Y., Shinoda, M., Gotoh, T., and Ohta, R.: Hybrid image compression method for medical images. Proc. SPIE 914, 815-822 (1988).

19. Netravali, A.N. and Limb, J.O.: Picture coding: a review. Proc. IEEE 68, 366-406 (1980).

20. Pratt, W.K.: Digital image processing, pp. 242-247, John Wiley & Sons, 1978.

21. Rhodes, M.L.,Quinn, J.F., and Silvester, J.: Locally optimal run-length compression applied to CT images. IEEE Trans. Med. Imag. MI-4, 84-90 (1985).

22. Roos, P., Viergever, M.A., van Dijke, M.C.A., and Peters, J.H.: Reversible intraframe compression of medical images. IEEE Trans. Med. Imag. 7, 328-336 (1988).

23. Wright, M.: SCSI host-adapter boards meet the needs of Multibus and VME bus computer systems. EDN. 31(21), 81-90 (1986).

24. Yonekawa, H., Ishimitsu, Y., and Akune, J.: Adaptive cosine transform coding using a new quantization algorithm. Proc SPIE 1093, 474-485 (1989).

Database Architecture and Design for PACS

Brent K. Stewart and Ricky K. Taira

Medical Imaging Division, Dept. of Radiological Sciences, UCLA School of Medicine, Los Angeles, CA 90024

Abstract: Relational database management systems using Structured Query Language (SQL) are well-suited to PACS development due to their inherent administration, fault tolerance, portability and networking capabilities. Several important aspects of database system development for PACS are database design, SQL programming tools and RIS-PACS coupling.

Keywords: Database Administration, Database Design, Fault Tolerance, Relational Database

Introduction

Previous PACS modules developed at UCLA (including Pediatric Radiology, the Coronary Care Unit, and Neuro Radiology) operated independent of one another. Each module had its own database management system and as such, could not easily access the data stored and managed by the other modules. We have, in the last year, integrated a database management system (DBMS) for PACS. The requirements of a PACS database management system must include: (1) Integration - PACS modules must be able to share information. Heterogeneous hardware and software systems may exist; (2) Performance - Efficient access to user and program requested data; (3) Scalable - Easy to accommodate growth as the application scope increases; (4) Reliability - Hardware redundancy and automatic database recovery capabilities; (5) Open Architecture - interface to other DBMS; (6) Advanced Development Tools - 3GL, 4GL, object oriented; (7) Flexibility - Easy to make database schema changes.

Database Management Systems

The database management system must properly address integrity, security, concurrency, and recovery issues, improve the power/ease-of-use ratio, minimize maintenance burdens, and maximize performance, especially over networks. Traditional database systems addressed some of these concerns, but required a highly competent degree of programming skill, direct management of hardware procedural details, and lacked a theoretical foundation. Without the systematic functional guidelines that theory could have provided, products were developed empirically.

NATO ASI Series, Vol. F 74
Picture Archiving and Communication Systems (PACS) in Medicine. Edited by H. K. Huang et al.

Relational Systems

In 1969, mathematician E. F. Codd, developed a relational theory of data, which he proposed as a universal foundation for database systems.[1] His relational model, based on the set of mathematics of relations and first-order predicate logic, covers the three aspects of data that any DBMS must address: structure, integrity and manipulation. The relational data model has emerged as the sine qua non of database design technology.

A relational DBMS presents the database to the user as collections of tables, which obey a certain discipline. The DBMS ensures that all database tables comply with these requirements, so that mathematical operations and strict logic can be applied to them. The DBMS transparently translates these logical requests into an optimized internal-access strategy. Thus, a relational DBMS relieves you of developing or maintaining integrity code in your applications and offers a level of productivity and reliability superior to that of traditional systems.

SQL

The only concrete expression of the relational model that has gained industry acceptance is Structured Query Language (SQL). Subsequently, SQL has been adopted as a standard by ANSI, the International Standards Organization, the Open Software Foundation and Federal Information Processing Standards. However, SQL is a language for interacting with relational databases, not a full application development language. It is the relational nature of SQL that has propelled it as the language of choice for cooperative processing (as in the client/server approach) and distributed databases. Distributed database processing among networked heterogeneous computers are possible as workstations can communicate in a standard language with different remote servers.

Database Design

The first step in database design is to establish an external data model. This model is a database representation of the real "micro-world" whose static and dynamic behavior is being mimicked. The most difficult part of the entire database design is gaining an extremely good understanding of the particular micro-world: the functioning of the present film-based radiology department, data requirements for all application transactions and reports, etc. Examples of the information necessary for the external data model include: patient demographics, image and study information, active workstation cases, diagnostic reports, optical disk archive volume directory and performance statistics. The extensive data gathering activity involves hard thinking, negotiation and much iteration. We have identified six classes of database user queries: (1) workstation user requests, (2) PACS technologist requests, (3) resource management queries, (4) workload and performance queries, (5) acquisition task requests, and (6) research queries.

Architecture

When creating the architecture and design of a database, several decisions must be made: centralized versus distributed data access, what method will be used to generate database structures, and whether or how the data will be partitioned. The question of whether to use a centralized or distributed database architecture depends on the data access patterns of the PACS by the users, on network speed and loading considerations, and on the ease of design and data integrity issues.

By storing a single copy of each piece of information, a centralized database server[6] cuts down on data redundancy and inconsistency. When an application requires information from the database, it uses a local client library to create a database query which is sent across the network to the server. Distributed database systems,[3] on the other hand, store and manage the integrity of a database across many different systems. Distributed database technology evolved due to relational database advances and because networks connecting the nodes (e.g. Ethernet and 56 kbps modem) were extremely slow. However, as higher-speed network links become commonplace (e.g. FDDI, UltraNet and T3 lines) and database servers become increasingly powerful and sophisticated, the performance differences between centralized and distributed database architectures fade.

Partitioning, Replication and Allocation

Three important concepts in database design are partitioning, replication and allocation. A horizontal partition of a table contains all of its columns and a subset of its rows, created by performing a relational selection operation. Two concepts determine how tables are horizontally partitioned: (1) the activity of a row (frequency of access) and (2) location access patterns (which nodes typically access a particular row). A vertical partition of a table contains a subset of the columns of the table and all of its rows, created by performing a relational projection operation. The grouping of attributes is based on the access activity of individual attributes and their affinity for one another in transaction processing. Tables are replicated in a database network for two reasons: to maximize local availability of data, thus improving performance and to provide backup copies of tables in the event of network or server failure. While replication provides some security it also introduces integrity problems.

Allocation is simply a combination of partitioning and replication. The allocation process decides which sites store what partitions and is a key element in distributed database design. The guiding principle is to allocate partitions and tables to maximize local processing. However, for frequently updated tables, replication degrades database performance, because all copies of a table must be updated to maintain the integrity of a distributed database. The network speed and architecture are important considerations here. Transaction management, concurrency control, and recovery from failure present major challenges when you develop a distributed DBMS because they entail considerable software complexity.

Synthetic and Analytic Database Design

Two complimentary approaches exist to the overall database design problem: the synthetic approach[2] and the analytic approach.[4] In practice, one certainly employs a combination of both. The synthetic approach is applicable when you are designing a brand new database from scratch. Also known as entity-relationship modeling, the synthetic approach involves determining what entities (objects or things) are required, what relationships exist among these entities and deciding what properties the entities should have. The analytic approach, by contrast, applied when you already have an existing design and it is necessary to analyze that design in order to understand it properly before recasting it into a more desirable form. The analytic approach to database design uses normalization theory (a set of five normalization rules) to group data elements into relational tables. The five rules are applied successively. Casting the database tables in third normal form is usually sufficient.

Database programming tools

SQL statements are embedded in programming languages, where they retrieve sets of rows from the database and pass them to host-language variables for further processing. SQL statements can be either pre-compiled or generated at run-time Embedded SQL statements can be translated into optimized database queries called stored procedures. These stored procedures can execute quickly because the access strategy pre-determined. However, for insert/update/delete SQL commands with nulls or select commands using various keys or columns in a database table, the number of permutations involved becomes unmanageable for the use of stored procedures. It is then better to use bit-codes and some kind of SQL statement assembler to generate the desired SQL statements at run-time.

SQL is also incorporated in front-end tools other than programming languages, where SQL may be hidden altogether. Easy-to-use tools and applications with forms, iconic, or object orientations can better exploit the relational features underlying SQL. These can directly manipulate relational data as sets, rather than one row at a time, as with procedural tools. This is the direction of the future.

Database Primitives

The function calls to the UCLA PACS database are categorized into PACS primitives, PACS database primitives, and SQL primitives. There are four cardinal database operations using SQL: 1) INSERT data, 2) SELECT table rows which have particular characteristics, 3) UPDATE a table row, and 4) DELETE a table row. Each database table has four associated function calls that act only on that table, one for each cardinal operation. The primitives we have developed use an Application Programming Interface (API) to the SQL server for the C programming language. The host language passes SQL statements as character arrays to the SQL server for execution, and the receiving program loops over the resulting sets in the traditional way. The main attraction of this approach is that we use SQL within a familiar, standard (and thus portable) language.

Database Administration

Database servers are complex programs that require a trained administrator. While even inexperienced microcomputer users can often manage local databases, a database server administrator must understand database design, data integrity and security, backup and recovery performance tuning and the particular quirks of the SQL server implemented at their site. The database system must be able to handle multiple users while providing reasonable levels of performance, security and integrity. In addition, due to the necessary up-time constraints of a PACS, the database architecture should have sufficient fault tolerance and redundancy mechanisms.

Integrity

Data integrity refers to the accuracy and consistency of the database. Four types of database integrity are datatype checking, domain constraints, entity integrity and referential integrity. Datatype checking is simply verifying a values datatype to be correct and within the range of values supported by the system. Domain constraints, generally called rules, are generalized mechanisms for specifying the set of logically related values from which the value in a particular column can be drawn. Entity integrity requires that no component of a primary key be allowed to have a null value. Referential integrity guarantees the logical consistency of the database by ensuring that the value of a primary key and the foreign keys that refer to it match. A trigger is a referential integrity mechanism consisting of a collection of SQL statements describing an action to be executed when a specified data modification operation is attempted on a given column or table.

Fault Tolerance

SQL server vendors design a certain amount of fault tolerance into their systems. Common ones being atomic transactions, transaction logging and disk mirroring. A transaction is a sequence of related operations that the database system guarantees is atomic; all of the operations in a particular transaction either execute successfully or abort. A transaction can terminate normally (a commit operation) or abnormally (a rollback operation). Abnormal termination, as in a system crash, means aborting every database operation in the transaction.

The database server must provide facilities for backing up the data and recovering it when the system crashes or the database is somehow corrupted. When a catastrophe occurs, the server can use a recent complete backup of the database and its transaction log to bring the database up-to-date (roll-forward operation).

Disk mirroring provides non-stop recovery in the event of media failure. All writes to a device are copied to a separate physical device. In the event that one devices fails, the other contains an up-to-date copy of the database and associated transaction log. Disk mirroring has the added advantage that either device is available for select operations. The database server internal access strategy algorithm reads from the disk where the last I/O was closest to the current read request.

Redundancy: Hot Backup and Multiple Network Paths

To insure maximum system up time, two master database SQL servers have been implemented. Each server machine maintains identical data. This configuration is known as "hot" backup. In the normal operating mode, all master database inserts, updates and deletes are mirrored on both database servers. If one server is disabled, all transactions will continue to be processed by the other. In addition, a log of all database operations are written into a database recovery table on the surviving server. The table contains a "time stamp" column and a column storing the SQL statements executed while the one server is down. When the disabled server is brought up again, a client process selects from the database recovery table on the other server and executes the SQL commands in the order of the time stamp. Using this scheme, the PACS can continue to function as long as one of the two database servers is operational.

In addition to the above hot backup scheme, the above two SQL servers are both connected to our global Ethernet and high-speed UltraNet[5] backbones. The UltraNet software contains a sockets compatibility library that uses the TCP/IP/UDP stack via the UltraNet network interface to communicate over UltraNet. This allows the data from the servers to be transferred transparently over UltraNet. The data path for the servers can be switched from UltraNet to Ethernet (and vice-versa) by simply changing the server node names in the network interfaces file. Then the servers must be brought down and re-initialized so that the alternate network is used by the servers. This switch can be affected though a C function, shell script or manually.

Security

Security is an important consideration in PACS text and image database management for both patient confidentiality and medico-legal issues. Three major security mechanisms are account control, privilege control and the use of views. Most sophisticated database management systems have identification and authorization mechanisms using accounts and passwords. Application programs may also supply additional layers of protection. Privilege control refers to the granting to and/or revoking from database user access to specific tables, columns and views. Views are created using projection and selection on more than one table and/or other views. Permission to access the data in a view is explicitly granted, regardless of any permissions in force on the underlying tables.

RIS-PACS Coupling

A RIS-PACS interface is necessary for two important reasons: to provide PACS with auxiliary text data which further describes patient images and to trigger PACS related actions at the occurrence of various RIS events. The RIS information needed includes: a) study requisition information b) study results (radiology reports). There are four types of events which a RIS system needs to communicate to PACS: patient scheduling, patient admission, patient discharge and patient transfer to a different clinical service.

In phase one of our RIS-PACS integration, only one-way communication is supported required from RIS-PACS. The RIS system must initiate all data transfers. That is, all appropriate data will be "pushed" from RIS to the PACS database. From the RIS computer, appropriate data to be inserted into the PACS database can be made using SQL remote procedure calls. In the second phase, PACS will request (pull) information from the RIS. Goals for a third phase of integration include PACS initiating a query to RIS, and PACS sending image location information to RIS.

Acknowledgement

This work was supported in part by Public Health Service Grant No. R01 CA 39063, R01 CA 40456, and R01 CA 51198, awarded by the National Cancer Institute, Department of Health and Human Services, and by the UCLA Department of Radiological Sciences Research Fund.

References

1. Codd, E.F.: A Relational Model of Data for Large Shared Data Banks. Comm of the ACM, June 1970.
2. Date, C.J.: A Practical Approach to Database Design. In: Relational Database: Selected Writings, pp. . Reading, Mass.:Addison-Wesley 1986.
3. Davis, R.: Sharing the Wealth. BYTE 14(9), 267-274 (1989).
4. Emerson, S.L.: The Practical SQL Handbook. Reading, Mass: Addison-Wesley 1989.
5. Stewart, B.K.: Three Tiered Network Architecture for PACS Clusters. In: NATO ASI: PACS in Medicine. Berlin: Springer-Verlag 1990.
6. Van Name, M.L.: Serving Up Data. BYTE 14(9), 259-264 (1989).

PAPYRUS: a portable image file format

Osman Ratib, Yves Ligier, Ron Appel, Jean Raoul Scherrer

Digital Imaging Unit, Center of Medical Informatics, University Hospital of Geneva
1211 Geneva-4, Switzerland

Abstract

As part of a hospital-integrated PACS at the University Hospital of Geneva, a standard file format (PAPYRUS) was developed based on the ACR-NEMA specifications on image communication ([1]). The purpose was to define a unique file structure that can handle images as well as the related clinical information to be used in multimodality image archiving and communication systems. The PAPYRUS format can be referred to as an "encapsulated" format for ACR-NEMA messages. The first version of the PAPYRUS format has been extensively modified after its evaluation by a technical working group of the European teleradiology project ("TELEMED" project). After several technical meetings the different partners of TELEMED came to an agreement that required a few functionalities to be added to the original proposal. Among the modifications two features supported by the SPI (Standard Product Interconnect for compatibility of Digital Imaging) were discussed: 1) the unique image identification code and 2) the image folder concept that contains references to image data sets. Related images can be distributed in separate files or directly in the folder file.

Display and image manipulation programs are being developed on different hardware platforms based on the PAPYRUS image file format as part of the Geneva PACS project as well as the TELEMED project for teleradiology and remote expert consultation in Europe.

Keywords: PAPYRUS, PACS, Image file format, Teleradiology

Introduction

In a multimodality image archiving and communication environment one of the major problems is the handling of images from different sources having different formats and structures. If the load is reasonably small it is always possible to have programs handling different file formats depending on the source of the images, but such programs are difficult to update and every time a new format is added all the programs will have to be modified.

The best solution is therefore to convert all the different file formats into one unique format before archiving so that all the programs that perform image manipulation and display will only have to deal with this standard format. Unfortunately there is no consensus on the standard image format. The major problem is the question of handling the data that are related to the images (patient information, study information, clinical information, etc.). One easy way is to separate the images from all other data in different files. However, in a distributed system where data as well as images will be transferred and stored in different locations over a network it is quite dan-

NATO ASI Series, Vol. F 74
Picture Archiving and Communication Systems (PACS) in Medicine. Edited by H. K. Huang et al.

gerous to separate images from the related information. If problems occur one may end up with orphan images that are then difficult to match with patient and study information. It is also often more convenient in data communication to "package" the images with the related data in one file.

Therefore the best choice is a unique file structure that can handle images as well as all the related information. The study and patient information can be stored in "headers" at the beginning of the file or in special sections of the file. The structure of these headers must be general enough to handle all kinds of information related to the patient and the study as well as the different image formats. Several formats have been proposed and used to store medical images. The simplest format contains a single header with a single image per file containing the minimum amount of information. Alternatively, the file structure can be very complex for the storage of any number of images and any size of information.

Another problem that is encountered when designing a file structure is how the data are stored in different fields in the headers. The most common way is to decide on a unique format with fixed fields that can contain anything from the patient name to the size of the image. However in an effort to come up with a more standardized approach some groups have proposed the use of a format where each field is preceded by a label (also called opcode or tag). One way is to use the standard ACR/NEMA (see next section) codes to label each field. The advantage is that if a foreign image is read by a program that is unaware of the real header format, it is still able to read the information. Each field label contains the code of the following field as well as its length in bytes.

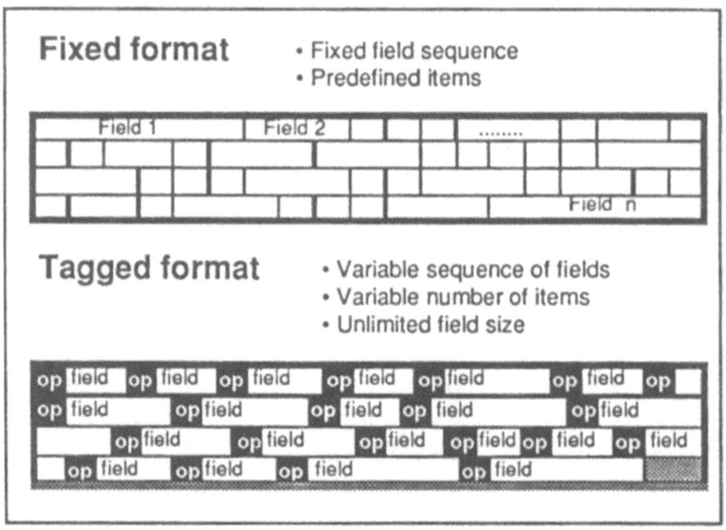

Fig. 1: Two different types of possible data structures in file headers.

At the University Hospital of Geneva, we selected a tagged file format based on the ACR-NEMA specifications (2) for data communication until a final agreement about the ACR-NEMA file format is reached. The fields in the headers are of variable length and are always preceded by a label and the length of the field in bytes. The file resembles an ACR-NEMA message and could be directly read as such. The only difference is that in order to allow for multiple images to be stored in a single file or in separate files, an additional group of data called the "FOLDER INFORMATION" was added. This additional group carries the group number 0041 (odd number reserved for users by the ACR-NEMA), and contains the number of images actually in the file and a set of pointers to the location of these images.

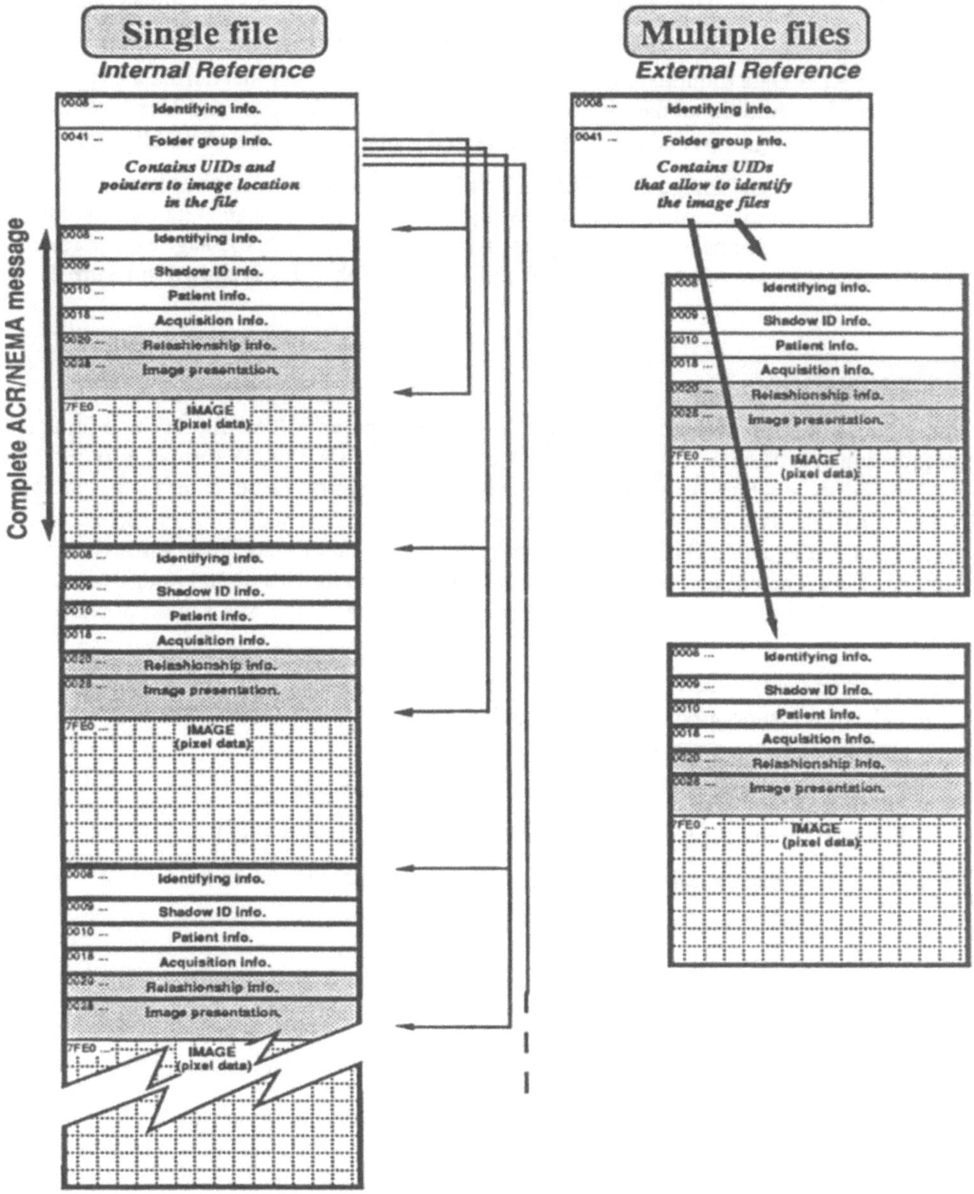

Fig. 2: Illustration of the "Folder by reference" principle. The shadow group 0041 is used to store information about the location of the images.

File structure

Each image of a given radiological procedure and its associated information are contained within the same file (like in an ACR/NEMA message), and are organized into groups representing the general categories of information. Each group is further subdivided into related data elements. Within each group there are data elements which are mandatory and some which are optional. The mandatory items should provide data necessary to display an image and to interpret it. Optional data elements may be included to provide supplemental information.

The PAPYRUS Data dictionary defines the groups and elements that define a Papyrus-conforming Data Set. Additional data items can be kept in Manufacturer- and User-defined Data Elements in shadow groups and are to be transferred transparently in PAPYRUS file.

Images are stored as "Image Data Sets". Each Image Data Set contains only one image and corresponds to a complete ACR/NEMA message. As such, each Image Data Set consists of a list of ACR-NEMA groups of elements necessary to form a complete message except for group 0000 (Command Information). The general architecture of the files is shown in figure 2.

Folder structure

Individual Image Files can be stored in sequence in a single "Folder" file. A folder is a means of associating sets of related Data Sets. In the context of the PAPYRUS format, a folder is simply a Data Set type which incorporates, either by reference or by value, other Data Sets. As defined here, a folder is not a display metaphor, a query mechanism, a directory structure or a data base structure. It is only a Data Set that permits the definition of association information about other Data Sets. The Folder concept is directly adapted from the SPI digital image communication standard. A request to incorporate this concept in the ACR-NEMA standard is also pending.

In the current version the PAPYRUS format only supports the folder "by reference". Folders "by value" are not supported. However in order to allow for multiple images to be stored in a single physical file, two types of references are supported:

- Internal references (pointers to Data Sets in the folder file)
- External references (references to Data Sets stored in other files)

A schematic representation of these two types of references is shown in figure 2.

References

1. Ratib O.: PAPYRUS, Image file format, version 2.2. Int. Rep. HCUG-CIH Reg. 900016, 1990.

2. ACR-NEMA standard publication. "Digital Imaging and Communications". #300-1988.

Teleradiology and Network Strategies

Janice C. Honeyman [1] and Samuel J. Dwyer, III [2]

[1] Department of Radiology, University of Florida Medical Center, Gainesville, Fl 32610
[2] Department of Radiology, University of Kansas Medical Center, Kansas City, KS,66103

Abstract: A teleradiology system is a unique combination of digital data networks and computer systems with the purpose of electronic transmission of radiographic images from one site to another. These systems can be either wide area networks (WAN) designed to transmit radiographic images over large distances (usually greater than 2 KM) or local area networks (LAN) designed to communicate with images over shorter distances (usually less than or equal to 2 KM). This paper discusses the strategies used to implement teleradiology systems at two University Medical Centers.
Keywords: Teleradiology, Communication Networks, PACS

Introduction

Teleradiology systems electronically transmit radiographic images and consultative text from one site to another. They are networks designed to provide the prompt interpretation of radiologic images for patients in under-served rural areas as well as other medical facilities having no full time radiologist. A Wide Area Network (WAN) is a communication system extending over large distances (covering more than a metropolitan area), often employing multiple communication link technologies such as copper wire cables, coaxial cable, fiber optics links, digital switched circuits, and microwave and satellite links. They can also integrate multihospital/clinic health care provider consortiums (HMOs). A Local Area Network (LAN) communicates over shorter distances (usually within a campus or a city), and employs many of the same technologies as a WAN. A WAN typically uses leased lines or communications owned by a public carrier or communication provider while the owner of a LAN typically owns and administers the transmission media. Teleradiology systems can improve emergency service coverage. With the transmission of images to radiologists homes during evenings or weekends, radiologists are able to conveniently provide prompt interpretation of radiologic images to physicians in the emergency room, intensive care unit, or

NATO ASI Series, Vol. F 74
Picture Archiving and Communication Systems (PACS) in Medicine. Edited by H. K. Huang et al.

elsewhere in the hospital. The immediate availability of images for emergency room and intensive care unit physicians leads to earlier intervention and better patient management on the basis of image data provided. These systems offer consulting-at-a-distance with subspecialty radiologists. Radiologists in the community or in rural areas could gain immediate access to large academic centers for help in the interpretation of difficult and problematic cases. Improved patient care can also be achieved by forwarding radiographic examinations on a teleradiology system to the primary referral center or a consultation institution prior to the patient's arrival. Within the hospital, teleradiology can provide images both to clinicians and to radiologists, in multiple locations, at the same time. Radiologists have access to images for diagnostic reading while clinicians have access for immediate care. When the radiology department is fragmented over distance, teleradiology makes images available to all radiologists, at any radiology location. The future of teleradiology is dependent upon the constantly improving technology of digital imaging modalities, film digitizers, interactive gray scale display workstations, image processing algorithms, hardcopy recorders, and digital data networks. Digital images are produced by CT, MRI, nuclear medicine, DSA, ultrasound, computed radiology and digitized radiographic film. This paper deals with some of the considerations of a digital data network, while considering the costs associated with acquisition and display of digital images.

Wide Area Networks

A teleradiology system serving sites separated by large distances must have at least one source of images, a display device at each location, and laser film printers where needed. An example installation with one referral center, two associate sites, and a radiologist's home is shown in figure 1. The main difference between sites is the quality of the display device, and the presence of image acquisition and printing equipment. In the case of the radiologist's home, a standard computer is capable of displaying nearly 512 x 512 resolution images, which may be enough for screening images on an "on-call" basis. In the referral center, the radiologist requires high resolution images on sophisticated workstations for diagnosis and subspecialty interpretation. The devices in figure 1, are connected by a wide area digital data network. The network requirements remain the same for both the "high resolution" and "low resolution" teleradiology systems. The following features of a teleradiology WAN are desired: (a) the WAN promptly accepts without excessive delay an intended transmission from a host computer; (b) the WAN is efficient, that is, no part of the network is idle for long periods of time; (c) image data arriving at receiving host computer is processed rapidly; (d) when a received transmission protocol demands a response, such as an acknowledgment, the delay in responding is minimal; (e) the topological connectivity of the WAN is modifiable in a reasonable amount of time; (f) the operation, connectivity, and maintenance of the WAN is transparent to the user; (g) the cost of establishing, maintaining, and operating the WAN is to be a minimum; and (h) the charges incurred by a user of the WAN should be a minimum. These requirements for a teleradiology WAN will determine the type of digital public service to be used.

Requirement (a), above, implies that each host computer provide buffer storage for queueing image data to be transmitted. This requirement also demands that the reliability of the WAN be high (uptime over 99 percent). Requirement (c) requires that each host computer provide adequate computer power. Requirement (d) implies that the WAN should adhere to standard communication protocols and be a high speed digital service. Requirement (e) demands that the connectivity of the WAN be programmable; that is, the digital service provide for a virtual WAN capable of being changed under software control. Requirements (b), (f), and (g) imply that the WAN should be a part of a public digital communication network so multiple users and applications are available to share the cost. Requirement (h) implies that the user on the teleradiology WAN only pay for access and use charges; that is, be able to employ a casual user tariff. The requirements for a teleradiology WAN network mandate the use of public digital service communication networks.

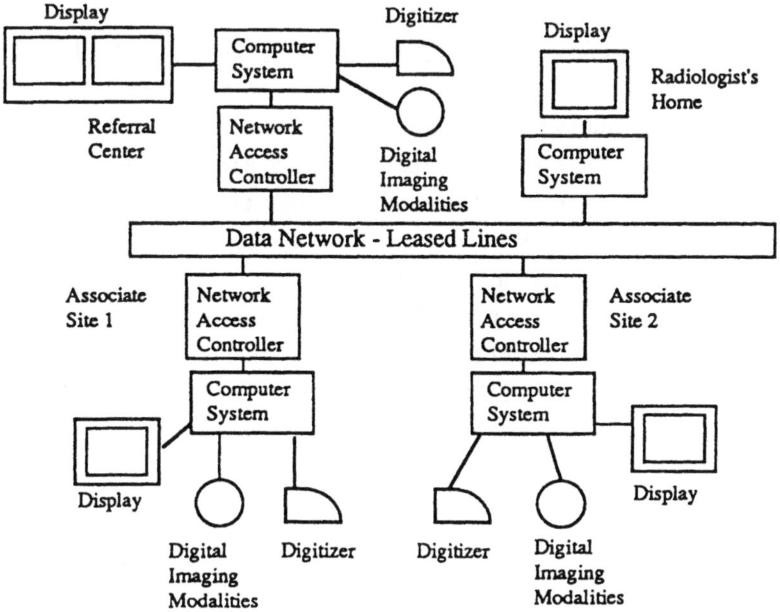

Figure 1: Four Node Teleradiology System

Types of Wide Area Networks for Teleradiology

There are three types of public digital service computer communication networks available. They are circuit switching, message switching, and packet switching. In circuit switching (1), a complete route of connected links from the source host computer to the receiving host computer is established prior to the initiation of transmission. These connected links remain "intact" and dedicated to this transmission until released by both the source and the receiver. In the same manner as the telephone network for voice communication, the route is established by a special signaling message that finds its way through the network from the source to receiver, seizing channel links in the route as it proceeds. Once the route is completed, a return signal is sent to the source to initial transmission of the image data. During transmission, the source and receiver have a circuit connecting them during the entire transmission. When transmission is completed, signaling on the network disconnects the communication links used.

The simplest circuit switch is one which has N-inputs and M-outputs. This type of circuit switch is called a space division switch (2). At the intersection of each input and output is an element called a crosspoint (called crossbars in the telephone network). Advances in Very Large Scale Integrated (VLSI) technology permit from 256 to 1024 crosspoints in a single chip. The individual crosspoints are capable of passing digital signals with data rates from 100 M bits/second to over 1 G bits/second. Switching times are on the order of a few nanoseconds.

A second type of circuit switch is a time division switch (3). Bytes of data from N input channels are time multiplexed together, written sequentially into a semiconductor memory, and then by computer control are read out and demultiplexed to the correct output channel. VLSI technology permits a single silicon chip to provide 1024 channel time division switches, where the individual channels are operating at 56 k bits/second or 64 k bits/seconds.

Most digital switching systems use a combination of space and time division switching. These systems provide digital services of 64 k bits/second (or 56 k bits/second), 1.5 M bits/second (DS-l), and 44.736 M bits/second (DS-3). Time and space switching will permit digital services in the G bits/second range needed by fiber optics links.

Message Switching

One characteristic of computer-to-computer communication traffic is its "impulse" characteristic. Communications will consist of bursts of high-data rates followed by almost no data transmission. This implies that the average use of such communication paths will be low. In message switching (also called "store and forward" switching) (4), there is no complete path established prior to starting transmission. Rather, the sending host computer will transmit a message to a switching center where it will be stored in a buffer. The stored message is kept in this buffer until the line to another switching center is available. When this line becomes available then the stored message is forwarded. This process is continued until the data arrives at the receiving host computer.

The advantage of this technique is that the link utilization is high, for one link may be involved in sending data to many host computers. The disadvantage is that long messages, such as digital images, require large storage buffers. Furthermore, even if a link becomes available, a switching computer will not begin to transmit a message until it has been completely received.

Packet Switching

A packet switching network (5 - 7) subdivides messages into small sizes, called packets, typically 1024 bits. The message is transmitted packet-by-packet. Packet switching is a store and forward system where each packet is stored in buffers at the nodes on the network and then forwarded. Packets may follow different routes on the network and may arrive at the receiving host computer out of the sequence in which they were transmitted. For this reason, all packets are numbered so that they can be reassembled back into the original message. Packet switching for messages of varying length provide the advantage of dynamic and fair allocation of the WAN's available bandwidth. The end-to-end transmission delay is smaller than that of message switching due to the division of the message into many smaller size packets and their transmission over multiple path lengths. Packet switching requires smaller storage at each node on the network than does message switching. Each packet includes information bits as well as additional bits called overhead information. The overhead bits are used to identify the destination of the packet, the source of the packet, user identification (for charging purposes), synchronization bits, and numbering bits (so as to properly reassemble the packets at the receiving host computer). Packet switching has the disadvantages that: (a) packet switching requires the transmission of both information bits and overhead bits; (b) the packet switching hardware must packetize the messages, add overhead, depacketize and reassemble the message. However, due to the pipelining effect of packet transmission (multiple packets could be transmitted over multiple links), it is possible that packet switching will transmit a message in less time than that of message switching.

Integrated Services Digital Network (ISDN)

The telephone network was developed as an analog transmission system. Replacement of the analog system by a digital system was broadened to include voice, data, and images. The advent of digital signal processing technology and computer systems, digital switches, and digital transmission is gradually replacing electromechanical switches and the analog facilities between them. Analog voice grade lines have been replaced with pulse code modulated (PCM) digital data voice grade channels, operating at data rates of 64 k bits/second. A channel bank is used to convert 24 analog voice channels using analog-to-digital conversion so as to provide a Tl service of 1.544 M bits/second. A channel bank is required at each end of the communication link to convert the analog signal to digital data and digital signals to analog signals. Digital switches are being incorporated into the toll central offices, the local central offices, and digital signaling between these exchanges. An ISDN standard was established by the Consultative Committee for International Telephone and Telegraph (CCITT) for the functional standards of ISDN. To improve signaling speeds for the set-up and take-down communication links, a separate channel was established, called the common channel signaling network (Figure 2). As the evolution towards ISDN (8) continues, local loops (connects local switching exchange to office/home) are being converted to digital links capable of transmitting two 64 k bit/second channels (a B channel signals at 64 k bit/second) and a D channel (a D channel signals at 16 k bits/second) for signaling rates. Thus the basic rate access on the local loops is 2 B + D, or 128 k bits/second plus 16 k bits/second for signaling. With ISDN, digital communications at 128 k bits/second (plus D channel control signaling of 16 k bits/second) is made available to homes and offices similar to the dial-up telephone capabilities. The primary rate access for digital links between exchanges is 1.544 M bits/second (23 B + D channels).

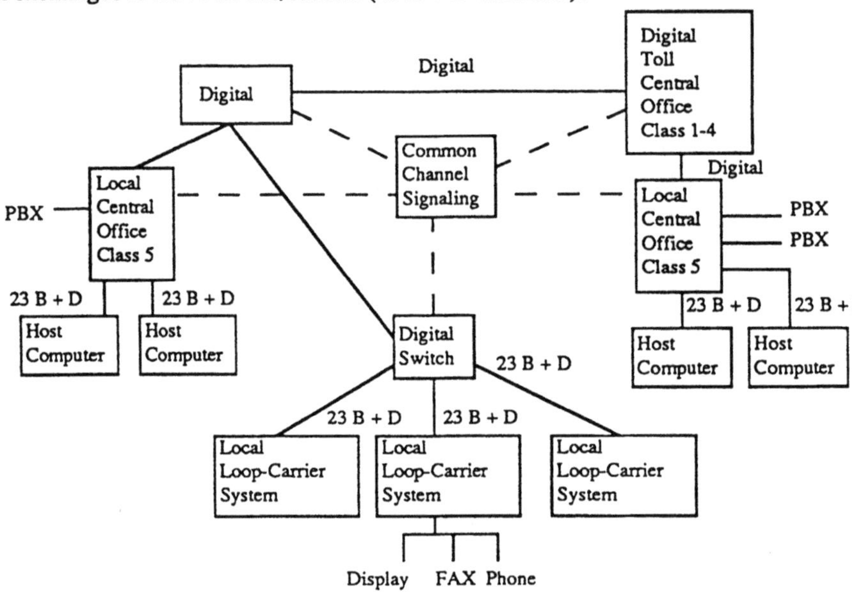

Figure 2: ISDN Network

The ISDN service will provide a wide range of applications in a public end-to-end digital communications network. An ISDN socket in the home or office would allow user connection of telephones, personal computers, display stations, and other devices. Existing circuit switched links, of which there are an abundance, provide the B channels of 64 k bits/second. Separate packet switches are added to provide the D channel service. The D channel is mostly used for signaling on the network. The true ISDN will evolve so that the B channels and D channels will be merged into the same switching equipment.

The amount of data to be transmitted in a WAN teleradiology network requires higher bandwidths than 128 k bits/second or 1.544 M bits/second. The ISDN standard supports clear channels which use packet switching for supporting fast facsimile, compressed video, and high-speed data. These channels are: HO (384 k bits/second), HII (1,536 k bits/second), and H12 (1,920 k bits/second).

Fiber Optics Networks

The primary application of optical transmission (9) has been to extend the available bandwidth. This has resulted in increasing the transmission rates. Current systems have signaling rates that operate between 2 G bits/second and 4 G bits/second. The result of this higher transmission rate is that: (a) the broad bandwidth services required for teleradiology applications are available; and (b) the cost of using a 56 k bits/second (or 64 k bit/second) switched circuit is significantly reduced. The widespread deployment of fiber optics links has served to increase expectations for broadband services. Agreement was reached in CCITT on the definition of a broadband integrated services digital network (BISDN) (2). This agreement for the Synchronous Digital Hierarchy (SDH) sets the transmission rates and signal formats for a digital broadband communications network. The transmission signal is called SDH-1 and has a rate of 155.52 M bits/second. Interfaces to the users will be a packet-like protocol called Asynchronous Transfer Mode (ATM).

For long haul communication networks, the Synchronous Optical Network (SONET) was recently adapted (10). The basic building block of the SONET signal hierarchy is called the Synchronous Transport Signal - Level 1. The STS-1 has a bit rate of 51.84 M bits/second and is synchronous with a network synchronization source. The levels of the SONET signal hierarchy are shown in Table 1. Products are under development for the use of SONET protocols. The OC-N refers to "N" increments of OC-1, where OC-1 is a 51.84 M bits/second optical carrier path. The SONET D/I is a drop and insert multiplexer that functions as a digital switching system. The DS-3 digital service (44.736 M bits/second) is the current digital primary interface for a fiber optic terminal (FOT).

Level Optical Carrier	Line Rate Mbits/sec
OC -1	51.4
OC -3	155.52
OC - 9	466.56
OC - 12	622.08
OC - 18	933.12
OC - 24	1,244.16
OC - 36	1,866.24
OC - 48	2,488.32

Table 1: Levels of SONET Signal Hierarchy

Implementation and Costs of a WAN Teleradiology Installation

The costs of implementing and operating a Wide Area Teleradiology Network (WATN) depend upon selected hardware, the desired level of maintenance, and communication charges. Hardware and Maintenance costs are fixed by the choice of a low resolution or high resolution type of radiology network. However, the communication costs of a WATN vary depending upon the type and number of images and the desired rate of transmission. We present the communication costs of a diagnostic type of WATN for a T1 carrier service (DS-1, point-to-point) and a switched multiple 56k bits/second service (N X 56k bits/second).

The University of Kansas Medical Center and three hospital sites have decided to work together and implement a diagnostic quality WATN. The characteristics of each of the three hospital sites is shown in Table 2. The radiology department at the 50 bed Army Hospital at Fort Leavenworth performs an average of 30 radiology procedures per working day, of which an average of 3 are GI/GU studies and the other 27 are conventional radiographs. Contract radiologists are available three afternoons a week costing an average of $275,000 per year and an addition $110,000 for on-call services. The radiologist will be on site to perform GI/GU studies, but the Commander of the hospital is investigating the costs of transmitting the rest of the studies to the University of Kansas for diagnosis. The radiology department at the 100 bed Army Hospital at Fort Riley performs an average of 100 radiology procedures per working day. Contract radiologists cost an average of $750,000 per year and are available five days per week and on an on-call basis. The Commander of the hospital is evaluating ways to reduce costs and provide quality services. The Pratt Medical Center is a 84 bed private practice serving five outpatient clinics within a 20 mile radius. The hospital has one radiologist who plans to retire. Radiology performs an average of 150 proceedings per working day. Teleradiology may be the only way diagnostic readings can be done in Pratt.

Remote Site	Distance in Miles to University of Kansas	Number of Beds	Estimated Mb/month Transmission (100%)
Fort Leavenworth	48	50	125,020
Fort Riley	100	100	348,082
Pratt Medical Center	168	84	407,422

Table 2: Characteristics of Three Remote Sites

Analysis of the types of images and number of images per procedure at each location produced estimates of the number of images to be transmitted per day and the number of Megabytes transmitted per month. T1 access charges/month were evaluated for both the sending and receiving site and the carrier charges compared between a T1 service and a switched multiple 56k bits/second service. The results are shown in table 3. In each case the costs of the service are estimated for 100%, 50% and 25% of total exams transmitted. In all cases, a switched multiple 56k bits/second service will cost less than a T1 carrier if 25% or less of the images are transmitted per month, and T1 carrier costs less if 100% of the images are transmitted per month.

Site	Mb/Month	Remote Site Access Charges	KU Access Charges	T1 Carrier Costs	NX56Kb/sec Costs
Fort Leavenworth	125,020 55,564 32,413	$1,700	$350	$4,182	$ 4,729 $ 3,241 $ 2,051
Fort Riley	348,082 174,087 89,355	$ 350	$350	$3,519	$ 8,163 $ 4,434 $ 2,619
Pratt Medical Center	407,422 203,711 102,178	$3,500	$350	$6,785	$12,580 $ 8,215 $ 6,039

T1 Carrier Costs and N X 56 Kb/sec Costs include both Site Access Charges

Table 3: Communication Costs for Referral Center and Remote Sites

The costs for hardware for the systems is shown in table 4. Each site will have a computer system supporting a laser film digitizer, grey scale workstation, optical disc archival, and network access. The University of Kansas will have a computer system supporting the same hardware with additional display capabilities and a laser film printer. The although the initial costs are high, each hospital administrator or commander teleradiology is a cost effective way to provide quality radiology service.

EQUIPMENT ITEM	ESTIMATED COSTS KU	ESTIMATED COSTS REMOTE SITE
Laser Film Digitizer	$72,000	$72,000
Laser Film Printer	$78,000	
Host Computer System	$55,000	$45,000
Optical Disc Archiving	$36.000	$18.000
Interactive Gray Scale Workstation	$70,000	$38,000
System Software	$26,000	$26,000
Network Access Controller	$18,000	$18,000
Maintenance Costs (10% of hardware)	$35,500	$21,700
TOTAL	$390,000	$238,700

Table 4: Hardware Costs for Referral Center and Remote Sites

Local Area Networks

A teleradiology system serving sites within the same building, campus, or city may relay on a local area network where the transmission media is typically owned and administered by the agency implementing the system. Teleradiology within an organization may be used to provide access to images in emergency rooms, operation rooms, intensive care units, and remote sites. Digital images may be viewed simultaneously at multiple sites, potentially improving patient care through immediate access to images by all who need them. The requirements are the same for both Local Area Teleradiology Networks (LATN) and WATN. If we assume the source of images, display capabilities, and storage space remains the same, the difference between the two is the network.

Types of Local Area Networks for Teleradiology

The three commonly used topologies for local area networks, a star, bus, and ring, are shown in figure 3 (11 - 12). Combinations of these three can be used to build almost any "shape" network configuration. The center computer in the star topology usually performs the function of circuit switching. When two computers on the legs of the star need to communicate, the center computer sets up a dedicated communication line until the connection is broken. This configuration is seldom used in LATN systems. However, the star topology is convenient when a controlling computer (in the center of the star) routes messages to slave terminals (not a true network). This configuration could be chosen if a single digitizer routes images to multiple display terminals. The star configuration without circuit switching is also used in combination topologies. The other two topologies are commonly used for LATN and are described in more detail.

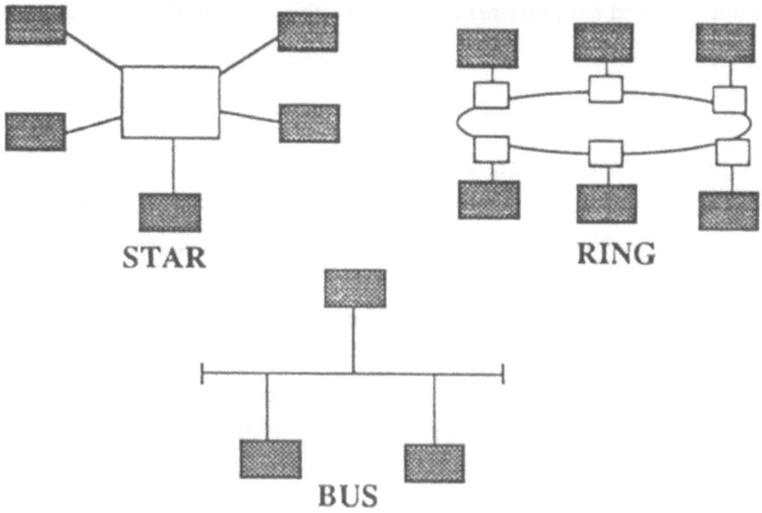

Figure 3: Local Area Network Topologies

Bus Topology

The bus topology is currently the most common type of LAN. It is a multipoint, packet switched network, with only one station at a time able to transmit a packet containing the address of a destination. Each packet is broadcast to all stations on the network and kept by the destination. Since multiple stations share a common data link, a method of obtaining access to the medium for transmission must be established. The original version of the most commonly used method Carrier Sense Multiple Access with Collision Detection (CSMA/CD) was patented by Xerox as part of its Ethernet local network. (13 - 14) When a station wishes to transmit a packet, it samples the carrier for usage. If the carrier is not busy, it transmits the packet, but continues to sample the carrier for possible collisions. If a collision is sensed, the station immediately stops transmission and issues a jamming packet. If a collision is sensed, another station has also attempted transmission and will also sense a collision. Both stations back off, wait a random amount of time, and try again. Propagation delay is defined as the time it takes for the beginning of a message to travel the greatest distance between two stations on the bus. Since it is important for a station to detect a collision before the end of the transmission of a packet, the time for transmission of a packet should be at least twice the propagation delay.

Although the bus topology is currently the most common local area network topology, it suffers from non-deterministic waits for access to the media and is limited by its implementation to a theoretical 10Mbps speed, although an actual speed of 2Mbps is more typical. Media for transmission includes twisted pair, thin-net coaxial , and thick-net coaxial cables.

Ring Topology

In the ring topology, as the name suggests, computers are connected in a circle with unidirectional communication. Each computer must have a repeater connected to two other repeaters forming the ring. The most common LAN ring control technique is the Token Ring, first proposed in 1969 (15). A bitstream representing a "token" is placed on the ring. A station wishing to send a packet to another station must wait until it captures the token before transmitting. After transmission, the station reissues the token giving the next station permission to transmit. Token rings have the potential of providing fair access to all stations, deterministic waits for transmission, and longer distance networks. Token rings can be either 4Mbps or 16Mbps, but suffer from high costs of implementation, complicated token administration techniques, and a loss of the network if one repeater fails. There has been no demonstrated improvement in the overall throughput of a token ring as compared to a bus (ethernet).

Fiber Distributed Data Interface (FDDI)

The FDDI protocol is still fairly new, but shows promise for faster transmission on fiber optic cable. This protocol still includes a ring topology, but can take advantage of the higher bandwidth of fiber (100Mbps). The main disadvantage at this time is the cost. Recent estimates put the cost of connecting computers with FDDI at $10,000 each.

Star-Ring Topology

One of the problems with the ring topology is the diagnosis of a non-functioning network. If there are many stations on the ring, locating a bad repeater may be a long and tedious job. The star-ring configuration uses a concentrator at a central location connecting multiple rings or subnets, each with fewer stations. The concentrator can include a build-in bridge/router between rings which can filter messages, sorting them into local and non-local traffic, and routing non-local packets to the correct subnet. The stations on a single ring have the advantage of deterministic waits to send, and the network administrator can isolate a non-functioning subnet in case of failure. The bridge/router limits the traffic on a subnet. The length of cable required for this topology is greater than with a single ring.

Speed vs. Storage

At this time, local area networks using commercially available hardware and relatively low prices can transfer images at about 2Mbps. If a radiologist wishes to view a PA and LAT chest radiograph digitized at 2K X 2K X 12 bits deep and the images are transmitted from a archival store to a display station over ethernet, it will take approximately one minute to load the two images on the screen, not considering any disk seek time. This is clearly too long to wait for images to appear, with probable low acceptance of a teleradiology system. If the images are preloaded onto local storage for viewing, the wait for disk access would be much less than the wait for network transfer. It may be possible to place more disk storage on each display station and optimize the time for transfer from disk to display monitor rather than increasing speed of the local area network. Since heavy traffic degrades network performance, analysis needs to be done to determine the most likely communication patterns and try to contain local traffic on subnets connected through bridges.

Implementation of a LAN Teleradiology Installation

The costs of implementing and operating a Local Area Teleradiology Network LATN depend on the costs of hardware, software, and maintenance. Since the owner of the network also owns the media, there are no carrier charges. We present the configurations for the implementation of a LATN between radiology and five ICUs, and within clinical radiology. We will also describe the preliminary plans for teleradiology in a remote cancer center and ambulatory care center.

ICU Teleradiology

At the University of Florida, approximately 43,000 portable radiographic exams are performed per year, many in the hospital's five ICUs. The clinicians responsible for patient care understandably want immediate access to images. The radiologists need the images for diagnostic reading. The conflict in needs could be solved by duplicating every portable exam, but film is expensive and storage of duplicates at already crowded nurses stations is difficult. The hospital purchased a teleradiology system from Dupont with one laser digitizer located in the radiology department and five display stations located in the ICUs. The network configuration is a modified ethernet on coaxial cable in the shape of a star with the digitizer at the hub. Instead of using CSMA/CD (the standard for ethernet), the computer controlling the digitizer inspects a queue of images waiting for transfer, polls a station to see if it is busy, then transmits the image if the station is idle. If the display station is busy, the computer will try another image in the queue. The display stations can display an image OR receive an image, but not at the same time. The digitizer can digitize images OR transmit an image, but not at the same time. Images can be digitized at a resolution of 1024 X 842 X 14 bits or 2048 X 1684 X 12 bits for a 14 X 17 inch image area. The digitizer takes about six seconds to digitize at low resolution, about 7 seconds to record the image to the local disk and about 13 seconds to transmit once the receiving station is ready. It takes about four times longer to digitize, store and transmit high resolution images. The digitizer and display stations each have 160 MBytes in storage for images and when the disk space is full, new images overwrite the oldest images. The cost of a digitizer is about $80,000 with each display station approximately $70,000.

Clinical Radiology

Two problems in clinical radiology were given highest priority for network solutions. The first was Nuclear Medicine with three SPECT imagers used both for clinical and research work, one Trionics TRIAD in the VA hospital, a Trionics TRIAD in Shands hospital, and a Trionics BIAD, also located in Shands. The VA hospital is located approximately 400 Meters from Shands Radiology, and the Radiologist needed images from all cameras for clinical review and research. The other high priority problem area was the Neuro-radiology clinical cluster, including two GE 9800 CT scanners and two MR imagers, one a Siemen's Magnatom, the other a GE-Signa. Images were being lost or borrowed and were unavailable when needed. It was decided to archive these images on an optical jukebox and use digital displays for the Neuro-radiology reading room. The image network also needed to support multiple research computers.

The installed network backbone is Ethernet (TCP/IP) running on a combination of fiber and thin-net coaxial cable. Ethernet was chosen for economic reasons, it is readily available and is relatively inexpensive. The fiber runs were installed to allow future conversion to higher bandwidth networks. Vortech's Medical Imaging Gateway (MIG) was installed to handle the acquisition, storage, and retrieval of MR and CT images. The interface to the system from General Electric's equipment is through ACR-NEMA, and additional cables were required to complete that section of the network. The current configuration is shown in figure 4. The network is divided into subnets with bridges to reduce overall traffic load. Although the average transfer rate is only 2Mbps, workstation disk space allows prefetching of images so retrieval of current images is rapid.

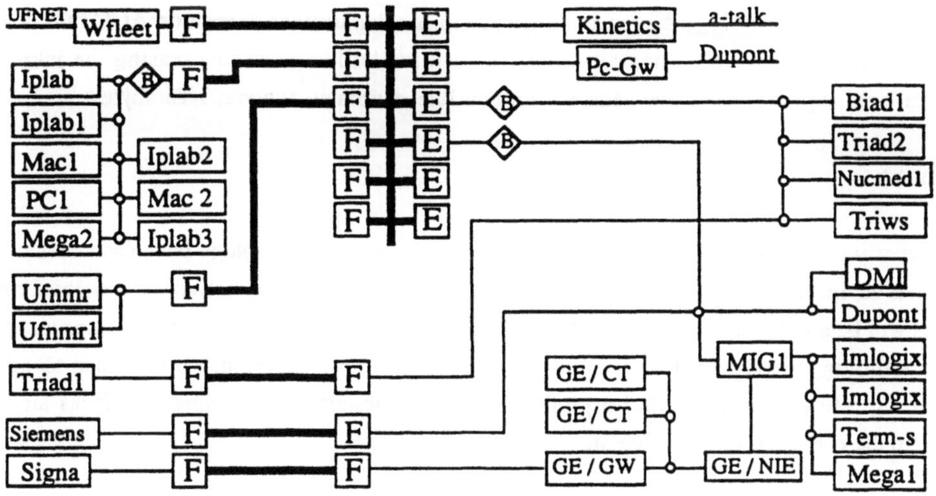

Figure 4: Clinical and Research Network Configuration

Cancer Center and Ambulatory Care Center

The new cancer center and ambulatory care center (ACC) buildings are planned for completion in 1991 and 1993 respectively. The buildings will be located next to each other, about 800 Meters from the current hospital. Cancer center patients frequently bring outside films for evaluation by radiology and have radiographic examinations during their treatment. Oncologists need to have access to all images during the course of treatment. Fiber-optic cables were installed between the sites, and twisted pair cables run to locations for display stations in the cancer center. A digitizer will be placed in the cancer center for capture of outside images, and examinations in the radiology department will be digitized for display in the cancer center. When a consultation is required, displays in the radiology department will be able to display the same images as the displays in the cancer center, allowing oncologists and radiologists to see the same image at the same time. Images will be stored locally on each display until treatment is complete.

The ambulatory care center presents a difficult challenge. We plan to have filmless radiology in the ACC except for mammography. All outpatient radiographic examinations (about 50% of all patients) will be performed in the ACC while inpatient examinations will be performed in the hospital. Images must be available to the radiologist reading the study in either location and to the clinician ordering the study. Each clinic area will be provided with multiple display stations. We plan to expand the technology currently employed in the department to include ACC radiology, installing higher bandwidth communications between the buildings and increasing the jukebox capacity to hold approximately two years of work on-line, estimated at about 3 Terabytes.

Summary

The University of Kansas and University of Florida Departments of Radiology have each developed approaches to image communications. In the future, both wide area networks and local area networks will be employed to implement teleradiology and PACS. As technology develops we may find the radiology department of the future linked by multiple communication lines to other departments, referring clinics and hospitals, and remote imaging sites, using interfaces largely invisible to the end-user, the radiologist.

References

1. Joel Jr., AE. "What is Telecommunications Circuit Switching?", Proceedings of the IEEE, 1977; 65(9): 1237 - 1253.

2. Bethold, JE. "High Speed Integrated Electronics for Communication Systems", Proceedings of the IEEE, 1990; 78(3): 486 - 511.

3. Tomita, S; Yokimatsu, KI; Mizuno, T; Miyaho, N. "Some Aspects of Time-Division Data Switch Design", Proceedings of the IEEE, 1977; 65(9): 1295 - 1304.

4. Maxemchuk, NF; Zarki, ME. "Routing and Flow Control in High-Speed Wide-Area Networks". Proceedings of the IEEE, 1990; 78(1): 204 - 221.

5. Roberts, LG. "The Evolution of Packet Switching", Proceedings of the IEEE, 1978; 66(11): 1307 - 1313.

6. Kleinrock, L. "Principles and Lessons in Packet Communications", Proceedings of the IEEE, 1978; 66(11): 1320 - 1329.

7. Tobagi, FA. "Fast Packet Switch Architectures for Broadband Integrated Services Digital Networks", Proceedings of the IEEE, 1990; 78(1): 133 - 167

8. Stallings, William, Integrated Services Digital Networks (ISDN), Tutorial, IEEE computer Society Press, Los Angeles, Calif, 1985.

9. Tomlinson, WJ; Brackett, CA. "Telecommunications Applications of Integrated Optics and Optoelectronics", Proceedings of the IEEE, 1987; 75(11): 1512 - 1523.

10. Hac, A; Mutlu,, HB. "Synchronous Optical Network and Broadband ISDN Protocols", Computer, 1989; 22(11): 26 - 34.

11. William Stallings, Data and Computer Communications, MacMillan Publishing Company, New York, 1988.

12. Tanenbaum, AS, Computer Networks, Prentiss-Hall Software Series, Englewood Cliffs, N.J., 1981.

13. Metcalfe, RM; Boggs, DR; Thacker, CP; and Lampson, BW Multipoint Data Communication System with Collision Detection. U.S. Patent 4,063,220, 1977.

14. Metcalfe, R.M., and Boggs, D.R. "Ethernet: Distributed Packet Switching for Local computer Networks." Communications of the ACM, July, 1976.

15. Farmer, W.D., and Newhall, E.E. "An Experimental Distributed Switching System to Handle Bursty Computer Traffic." Proceedings, ACM Symposium on Problems in the Optimization of Data Communications, 1969.

Three Tiered Network Architecture for PACS Clusters

Brent K. Stewart

Medical Imaging Division, Dept. of Radiological Sciences, UCLA School of Medicine, Los Angeles, CA 90024

Abstract: A three tiered network architecture is used at UCLA for our PACS project. Ethernet is used for connection to radiological imaging devices and as a global backup. FDDI is used to connect related computers in what is termed a PACS "cluster." UltraNet is used as a high-speed backbone to connect the PACS clusters and the various image display stations within them.

Keywords: Ethernet, FDDI, Fiber Optics, Network Architecture, Network Topology,.UltraNet

Introduction

After a digital image has been acquired, via a digital modality like computed radiography or through film-digitization, it must be transmitted to a storage device for archiving and later be available for retrieval, display, manipulation and review. This requires consideration of a digital communication network for images and image-related data. There are a plethora of networking solutions in the marketplace. Deriving a networking solution for a specific PACS system involves decisions regarding network architecture and topology, price/performance, and current and future bandwidth requirements.

PACS Clusters

PACS nodes which have similar access patterns are grouped into "clusters." We define a PACS cluster[1] as a coordinated group of computers which perform related tasks and which statistically access the same patient studies, e.g., Pediatric cluster, Neuro cluster, Chest cluster, etc. Present plans call for the radiology department to be divided into PACS clusters along sectional divisions. Each cluster includes radiological imaging devices, image acquisition-reformatting computers, an image archive computer (aka cluster controller) with associated optical disk library, and image display workstations. There are also two PACS database computers on the network which store duplicate information for fault tolerance.[3]

Image acquistion-reformatting computers are responsible for interfacing with specific radiological imaging devices, like CT and MR, and converting the image files into UCLA format. An image archive computer receives images from various acquistion-reformatting computers and archives these images to optical disk. It then forwards these images, and any related images retrieved from optical disk, to specified image display workstations. The image

NATO ASI Series, Vol. F 74
Picture Archiving and Communication Systems (PACS) in Medicine. Edited by H. K. Huang et al.

display workstations consist of host computer, parallel transfer disk and 2kx2k frame buffer-video monitors.

The architecture of a total PACS solution would contain many such clusters. The clusters connect to one another through a high-speed, wide bandwidth network. Our department is currently expanding our radiology service to a new Outpatient-Ambulatory Care (Medical Plaza - MP) center, which is across the street from the main medical center (Center for the Health Sciences - CHS). It is anticipated that a large amount of image traffic will occur not only within a cluster (intra-cluster), but between clusters (inter-cluster) both within and between the two buildings. A major effort is presently underway to connect these two buildings and the present and future PACS clusters within them with a high-speed digital network. In addition, we are seriously considering high-speed networking with five other medical centers within a 25 kilometer radius of UCLA.

Medical Imaging Requires a Wide Bandwidth

The timely communication of digital medical images requires a wide bandwidth due to the enormous amount of data in each image or set of images. For example, a Computed Radiography (CR) image (2kx2kx8-bit) comprises four megabytes of digital data. A Computed Tomography (CT) examination consisting of sixty-four 0.5 megabyte images (512x512x12-bit) comprises 32 megabytes. For PACS to be useful to the radiologist, we estimate that a CR image should be transferred within the network in less than one second. This places a minimum bandwidth requirement for the network of four megabytes/second (mBps). Naturally, for an extended network, many communication processes occur simultaneously. This requires a network with not only high-performance point-to-point image transmission, but more importantly, one that allows many simultaneous, wide bandwidth image transmission processes without significant degradation in the performance of any individual process.

Experiments with Various High-speed Networking Solutions

We have tested and compared three possible high-speed network solutions in our laboratory: Canstar, the Fiber Distributed Data Interface (FDDI) and UltraNet.[2,4] Canstar is a 100 megabit per second (mbps) rooted-tree topology network. FDDI is a 100 mbps ring topology network. UltraNet is a 1000 mbps meshed hub topology network. The UltraNet network consists of two major components: a hub with 1000 mbps theoretical bandwidth and a host adapter with 125 mbps theoretical bandwidth. The UltraNet host adapter inserts into the backplane of the host computer (Sun-VME) and connects to the hub with fiber optic transceivers at each end and a pair of fiber optic cables. The UltraNet network has a meshed star topology, i.e., multiple hubs can be interconnected in any configuration.

The point-to-point, memory-to-memory transfer performance of Ethernet, Canstar, FDDI and UltraNet were found to be 0.372, 0.518, 0.833 and 3.15 mBps, respectively. Recent testing has provided a 6.8 mBps rate for this type of transfer over UltraNet between two Sun Sparcserver 490s. In addition, a large-scale test of UltraNet using ten Sun workstations showed at most a two percent degradation in the data transfer performance of an initial client-server process as four additional client-server processes were added to the network. The sum total data rate routed through the UltraNet hub for the five client-server processes in this experiment was 13.9 mBps.

UltraNet was the only one of the proposed high-speed network solutions approaching our four mBps yardstick for point-to-point image transmission as well as not suffering significant degradation as independent communication processes are added to the network. In addition, UltraNet provides enough bandwidth headroom to allow our PACS implementation to expand beyond current requirements without network bandwidth limitations. Therefore, we have based our high-speed PACS network backbone on the UltraNet network.

UCLA PACS Fiber Plant

We have installed 64 multi-mode and 48 single-mode optical fibers over the one kilometer distance between the CHS and MP. Although most fiber on campus is regulated by UCLA Telecommunications, the fiber distribution within the CHS and Medical Plaza is solely owned by the radiology department. To-date, a total of 46 multi-mode and 14 single-mode fibers have been installed within the CHS and 60 multi-mode and 20 single-mode within MP. The intra- and inter-building fibers are connected through a patch panel at both sites. In addition, six multi-mode fibers run a distance of two kilometers to a remote MR imaging site. These six fibers are leased from the campus authority as they transit the campus fiber backbone.

Description of the UCLA PACS Network Topology and Function

Figure 1 shows the three networks used within the CHS and MP for our PACS project: a global Ethernet network, FDDI for the back-end computers and UltraNet for the front-end computers. For image transmission between our PACS computers, FDDI and UltraNet is used. We think it is not necessary to use UltraNet for connecting the acquistion-reformatting computers and cluster controller-archive computers because of the relatively slow transfer rates of radiological imaging devices. On the other hand, image transmission between the archive computers and the display stations requires a high transfer rate because workstation usage is demanding during clinical hours and in the event of workstation failure. In this case UltraNet is necessary.

Ethernet

We use Ethernet for transmitting images between radiological imaging devices and the image acquistion-reformatting computers because the rate of transfer from the radiological imaging devices is approximately 50 kilobytes/second (kBps) or less. For the CT imaging devices the data flow is through dedicated or semi-dedicated Ethernet lines to intermediate PCs and then over global Ethernet into Sun workstations for reformatting. Bridges serve to isolate the

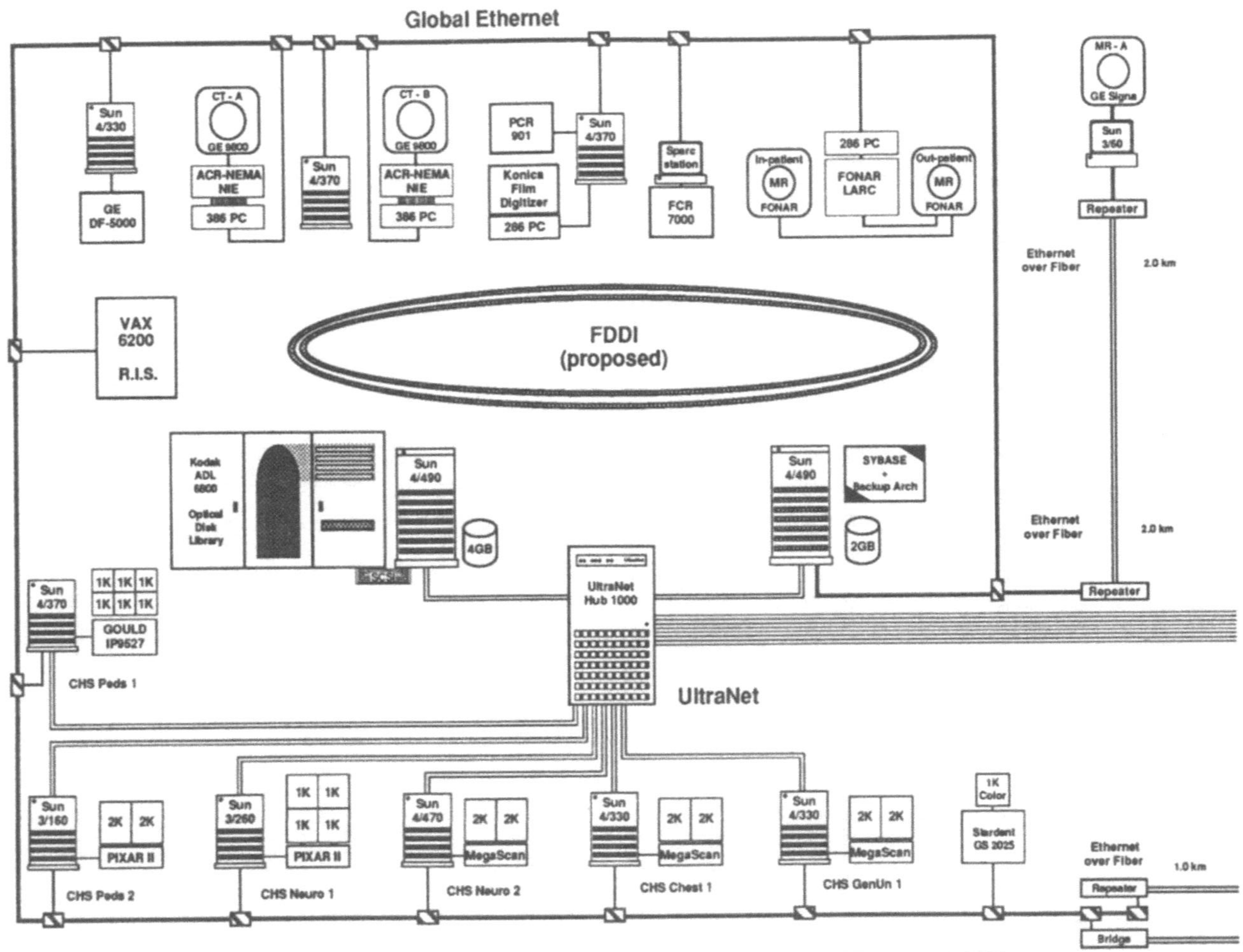

Figure 1(a). UCLA PACS Network: Center for Health Sciences and Remote MR Site

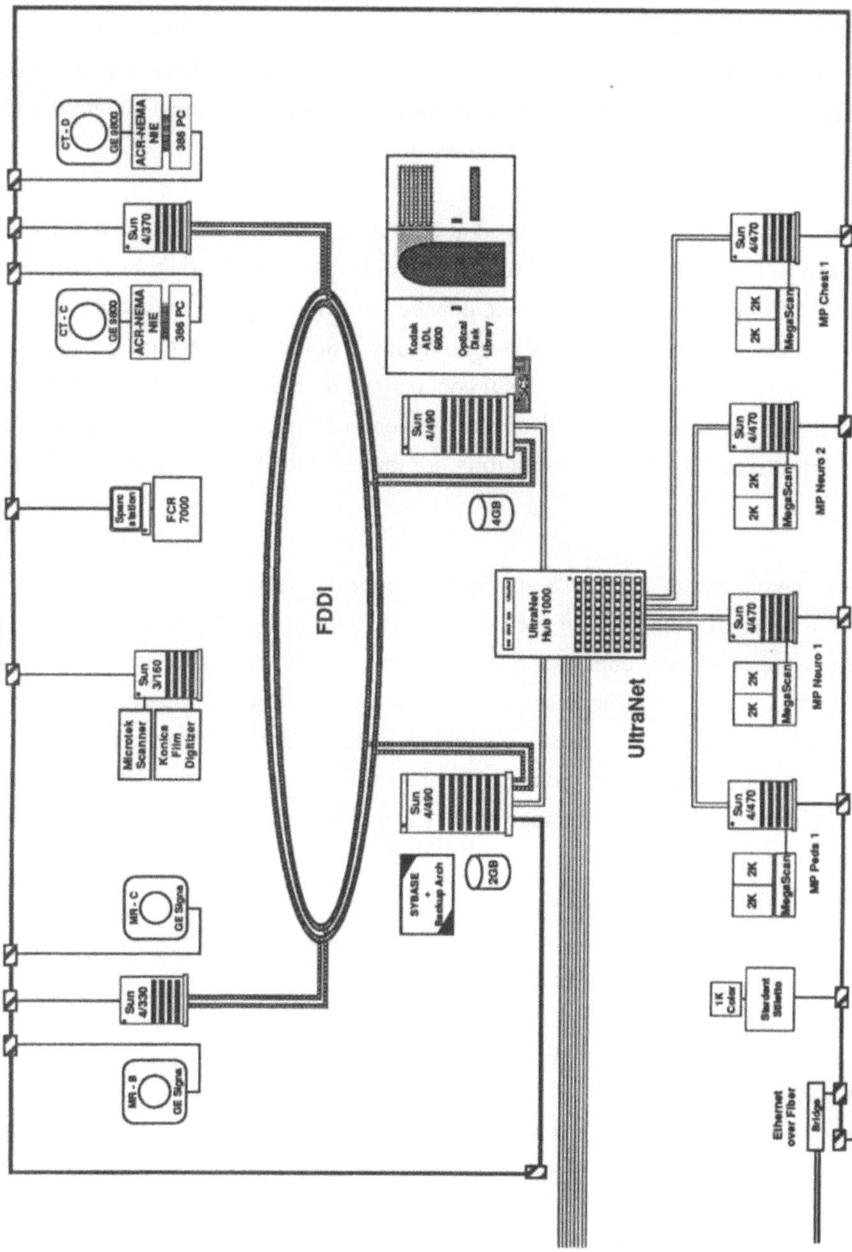

Figure 1(b). UCLA PACS Network: Medical Plaza

acquistion-reformatting nodes from the rest of the network. Ethernet over fiber is used between the CHS and MP. This link has redundant equipment which automatically switches over upon link failure. In the case of the remote MR imaging center, Ethernet over fiber is used to span the two kilometer distance to the Center for the Health Sciences. The global Ethernet network serves as a backup for the FDDI and UltraNet networks.

FDDI

Although FDDI has not been fully developed at our facility, we are intalling it in the Medical Plaza as shown in Figure 1. The CHS installation will be phased in during the first quarter of 1991. The FDDI rings serve as a high-speed connection between the acquisition-reformatting computers with the cluster controller-archive computer.. Some manufacturers of radiological imaging devices have indicated future FDDI connections with their equipment. In this event, the FDDI ring can directly connect all of the back-end computers: the radiological imaging equipment, the image acquistion-reformatting computers and the cluster controller-archive computer.

UltraNet

The front-end computers of the PACS system are connected throughout with UltraNet. One hub serves the needs of each building, providing intra- and inter-bulding connection between the cluster controller-archive computers, workstations and the two database computers. At present, only one cluster controller is resident within each building. However, as the system grows within each building, more cluster controllers and optical disk libraries will be added as funding permits and requirements dictate. The enormous bandwidth that UltraNet provides allows the flexibility for distributed image display and a centralized database archive that may function as quickly as a distributed one.

Fiber Optic Video Systems

In addition to the above digital networks, a video fiber network using single and multi-mode fibers connects the CHS, MP and the remote MR imaging site. A multi-channel fiber optic broadband CT/MR patient monitoring system has been established with 512 resolution. We are in the process of testing a 2k resolution fiber optic video system.

Acknowledgement

This work was supported in part by Public Health Service Grant No. R01 CA 39063, R01 CA 40456, and R01 CA 51198, awarded by the National Cancer Institute, Department of Health and Human Services, and by the UCLA Department of Radiological Sciences Research Fund.

References

1. Huang, H.K. Planning a Totally Digital Radiology Department. AJR 154, 635-639 (1990).
2. Huang, H.K. Radiologic Image Communication Methods. AJR 155, 183-186 (1990).
3. Stewart, B.K. PACS Database. In: NATO ASI: PACS in Medicine. Berlin: Springer-Verlag 1990.
4. Wong, A. Performance Comparisions of Image Communicaiton Methods. In: SPIE Medical Imaging III. Bellingham, Wash: SPIE 1990.

An Introduction to the ACR-NEMA Standards

Steven C. Horii

Georgetown University Hospital, Washington, D.C. 20007, U.S.A.

ABSTRACT The ACR-NEMA Standard was developed initially as an interface standard for the interconnection of two pieces of imaging equipment. The Standard defines a point-to-point, hardware connection with the necessary protocol and data structure so that two differing devices which meet the specification will be able to intercommunicate.

The Standard specifies neither a particular PACS architecture, nor a database structure. In part, these are the reasons why implementers have had difficulty in using the Standard for network communication in a full PACS.

The Working Groups formed by the Committee to work on the ACR-NEMA Standard have changed some of the "flavor" of the Standard. It was realized that connection of PACS with hospital and radiology information systems (HIS and RIS) is necessary if a PACS is ever to be successful. The idea of interconnecting these often differing computer systems has pushed Standards development beyond the scope of the original work.

This paper will present a brief overview and history of the "parent" standard, describe the structure of this standard, and discuss the "offspring" of the standard with regard to the activity of the ACR-NEMA Working Groups.

KEYWORDS: radiology, standards, interfaces, PACS

INTRODUCTION

Though the concept of PACS has been well embraced by the medical imaging community, this has been at a pace considerably slower than originally predicted despite the fact that satisfactory technology for implementation has been available for several years. Though perhaps the major reason for this slow adaptation is the cost of these systems, the inability to connect multiple-vendor imaging departments to networks (or network interface units, NIUs) is also a contributing factor.

NATO ASI Series, Vol. F 74
Picture Archiving and Communication Systems (PACS)
in Medicine. Edited by H. K. Huang et al.

This problem, the lack of hardware and software standards for image interchange, prompted the radiology (American College of Radiology, ACR) and manufacturer (National Electrical Manufacturers Association, NEMA) communities to form the ACR-NEMA Digital Imaging and Communications Standards Committee. The original goal of the Committee was to produce a standard which would allow imaging equipment from different manufacturers to exchange images and associated data.

HISTORY

At the first U.S. PACS meeting, held in Kansas City in 1982, acquiring image and image descriptive data from different manufacturers' equipment was already seen to be a problem [4]. Most groups doing PACS-related work were acquiring images for their systems either through digitized video or magnetic tape. The former method often results in a loss of dynamic range, and the latter method requires an understanding the tape format and pixel packing technique, which the manufacturers were not always willing to reveal. The potential users of PACS, the radiology community, asked the American College of Radiology to help find a solution to this problem. Engineers who worked for the manufacturers also began to realize that there was little to be gained by keeping image formats proprietary. The manufacturers were represented in issues of standardization by a number of groups, but the National Electrical Manufacturers Association (NEMA) had experience in standards related to medical imaging equipment. The two organizations first held exploratory talks in late 1982, and the result was the formation of the ACR-NEMA Digital Imaging and Communications Standards Committee in early 1983.

The original charter of the Committee was to develop an interface for the exchange of image and related information between a piece of medical imaging equipment and whatever was on the other side of the interface [1]. With this charter, the Committee restricted its range of activity. The resulting standard would not be a standard for a full PACS, a database, or a network. However, the ability to bring digital image information out of a piece of equipment over standard hardware using a standard format was felt to be a worthwhile goal.

The Committee formed three Working Groups, each addressing a particular aspect of standardization. The Working Groups were assigned to hardware and protocols (WG I), data format (WG II), and system and performance specifications

(WG III). Coordination of the Working Groups was assured by both Committee action and the participation of members in more than one group.

The Working Groups began their tasks by examining existing standards since it was the intent to use other standards if possible. Hardware standards examined included IEEE 488, Ethernet, Intelligent Peripheral Interface (IPI), SCSI, DEC® DR11W and Unibus®, and Multibus. Data formats of AAPM Magnetic Tape, LANDSAT, NATO magnetic tape, and the Japanese JAMIT were examined for possible useful data structures. While some ideas were drawn from these standards, none in particular was satisfactory for all the requirements defined. In addition, because of antitrust regulations in the United States, NEMA must operate under certain constraints. This prohibited the use of certain proprietary interfaces, or interfaces which would unfairly benefit a single manufacturer. The effort of examining standards did not end when the Working Groups concluded that a new specification would have to be written. The MAP/TOP standard, for example, was discussed even after the ACR-NEMA draft was written.

Because of rapidly evolving PACS research, it was thought necessary to get a standard written as quickly as possible. The Working Groups went through numbers of proposals by circulating them to each other and the engineering staffs at the parent companies. Within two years, a standard had evolved which was thought to meet all the requirements set forth by the Committee. Following Committee approval, the document was approved by vote of the member companies and ACR representatives. The approved draft was distributed at the Radiological Society of North America meeting in 1985 and published as a NEMA Standard (ACR-NEMA 300-1985) in the same year [2].

ACR-NEMA 300-1985: THE PARENT STANDARD

This Standard, titled "Digital Imaging and Communications" includes the data format, protocol, and hardware interface necessary for transmission of data between two pieces of medical imaging equipment. The location of the interface plane is defined to be between the two devices as shown in Figure 1.

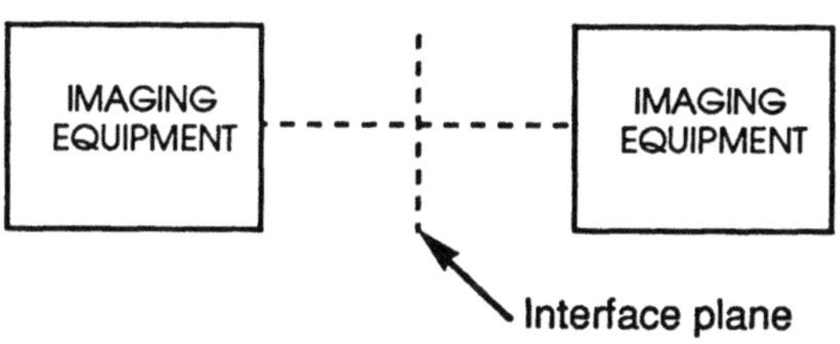

Figure 1: The location of the ACR-NEMA interface plane

Though not intended as a network standard, it was certainly anticipated that for use in a PACS environment, connection to a network of some form would be essential. In this case, the device on one side of the interface would be a network interface unit (NIU), the network in use then determining the nature of the NIU. It was recognized that such a unit would have to have significant local processing power, probably on the order of a microcomputer. However, given the magnitude of the interfacing problem, this was thought to be a reasonable approach. A diagram of how the ACR-NEMA interface might be used in a network environment is shown in Figure 2.

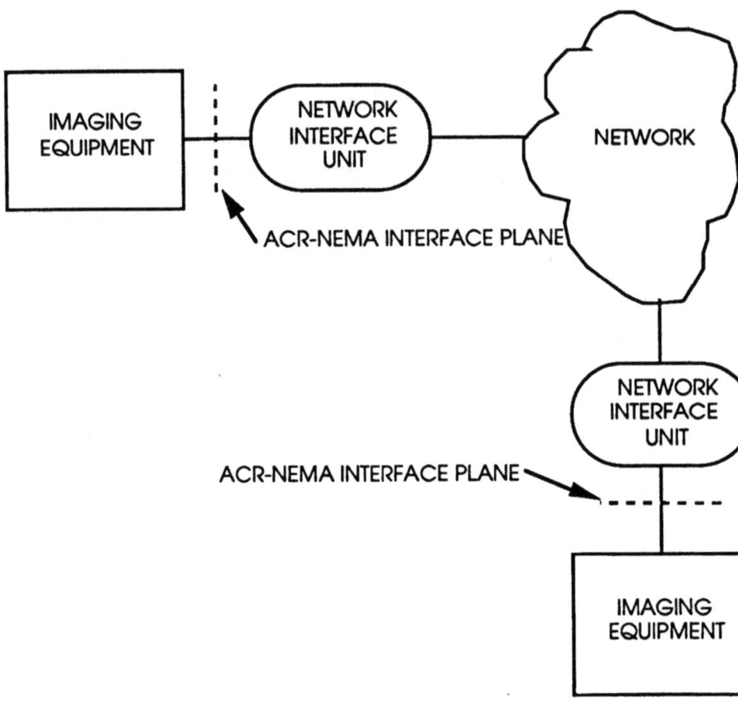

Figure 2: Location of ACR-NEMA interfaces with a network

In designing the interface, the Working Groups used the International Standards Organization Open Systems Interconnect (ISO-OSI) Reference Model [5]. In this model, the functions of a network communications interconnection are split into well-defined layers. Each layer performs a specific set of operations, providing services to the layer above and below. On each side of an interface, the same layers are performing the same functions, so to the programmer, the layers can be thought of as communicating directly with their equivalents. This "peer-to-peer" communication actually occurs in a vertical direction in the layers, and the true connection is only at the physical layer.

The ACR-NEMA Standard builds upon the ISO-OSI Reference Model in that specific layer functions were defined. The ACR-NEMA document refers to named layers and divides functions accordingly. An attempt was made to adhere to the layer functionality of the ISO-OSI model, but the ACR-NEMA Standard is not ISO compatible. In addition to differences in definitions of layer-to-layer connections (called interfaces in ISO terminology), there is one convention used in the ACR-NEMA Standard which violates an ISO-OSI concept (this is discussed later). The conceptual correspondence between the ISO-OSI layers and ACR-NEMA layers is shown in Figure 3.

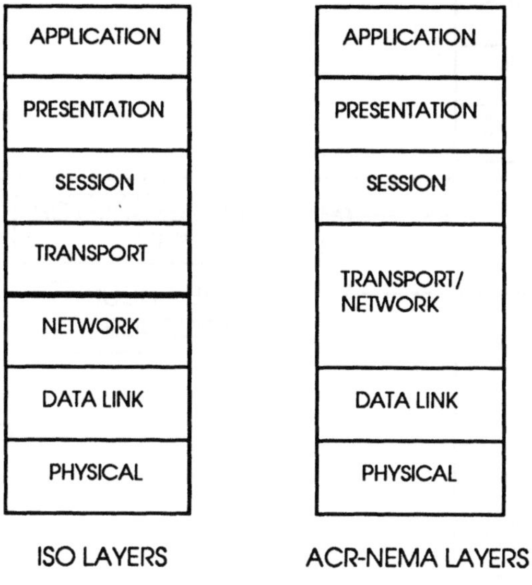

Figure 3: Similarity of ISO and ACR-NEMA layers

In Figure 4, the flow of data through one "stack" of layers is shown. The discussion which follows will trace data downward through this stack.When medical imaging studies are performed on a patient, the demographic information about that patient, the data on how the images were acquired, and the images themselves form a set of data amenable to display and interpretation (a data set in ACR-NEMA terminology). In the Standard, this data set is further associated with commands, and the whole is called a message.

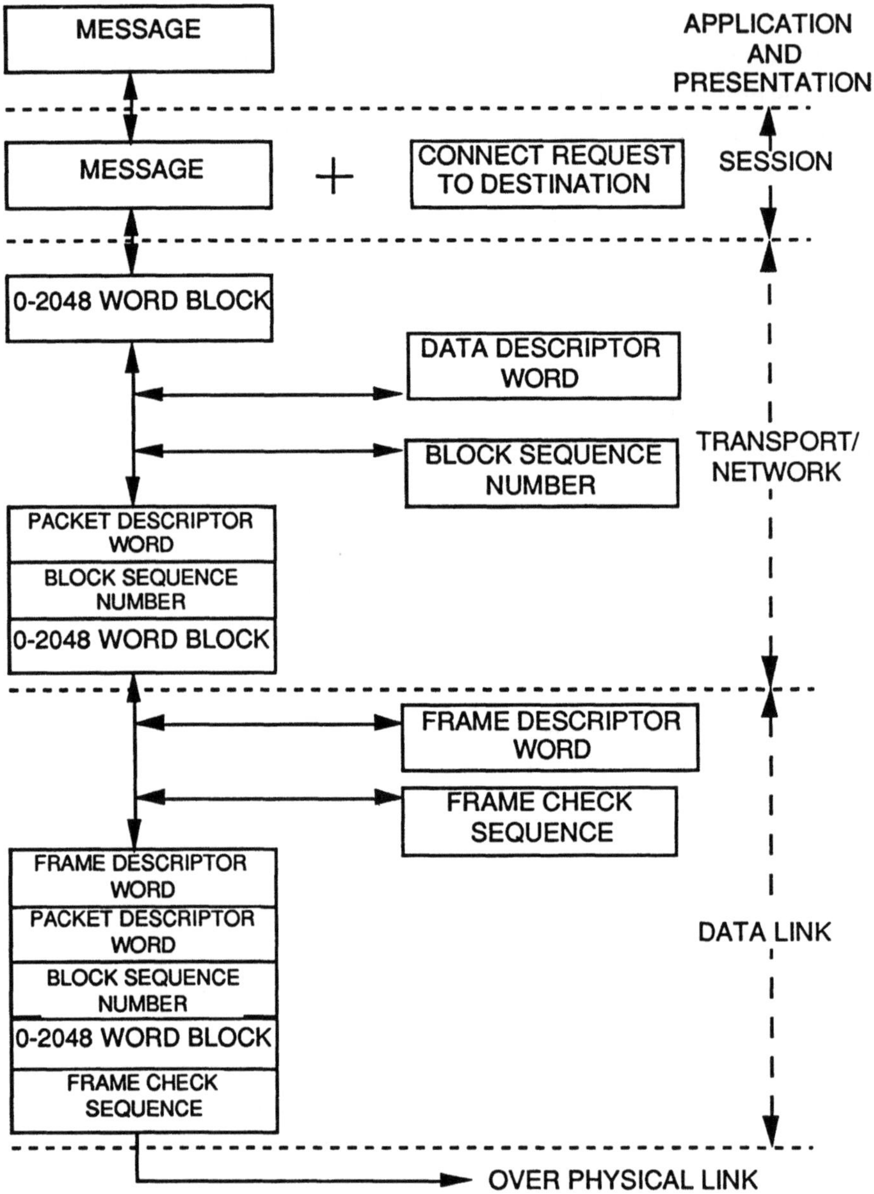

Figure 4: Data flow through the ACR-NEMA interface

The commands indicate certain fundamental operations on the data: SEND, GET, FIND, and MOVE. SEND is the basic command, allowing an imaging device to output its data. GET allows a requesting device to obtain a specific data set. FIND is a primitive search tool which will locate all data sets which match the information requested. MOVE allows one device to request a second to send data to a third. In the Standard, assembly of the message from individual data items and commands, and formatting of the result of this assembly takes place in the Application and Presentation layers. The aformentioned problem with OSI compatibility results from the upper layers, which build the message from commands and the data set, making a distinction between data types. Commands and data travel in different types of packet, but this is not encoded until the lower layers. This means that the upper layers have to look into the packet to determine the type upon receiving a packet, or have to tell the lower layers whether what is being passed down is command or data. One of the areas of investigation by the Working Groups is how to either eliminate this problem, or work around it, while remaining compatible with existing implementations.

The message format itself consists of data elements collected into groups. Each data element (for example, patient name) consists of two numeric fields forming a key, equivalent to the name of the data element. This is followed by a length field and the element value. The key (which is fixed length) consists of the group number and element number (in hexadecimal). The length field is a 32 bit integer which gives the length of the value field in bytes. The value field is therefore of variable length, the only restriction being that lengths must be an even number of bytes (padded if necessary). Note that this 32 bit length allows for some very large elements. Though questions were initially raised about this size, recent work is proving its value. Figure 5 shows an example of the structure of an ACR-NEMA element.

Group	Element	Length	Value	Meaning
0008	0010	000C 0000	4341 2D52 454E 414D 3120 302E	"ACR-NEMA 1.0"

Figure 5: Structure of ACR-NEMA data elements

Groups were defined in order to collect elements in a logical fashion. This was not entirely successful as it seemed some elements could be located in two places. Since elements are not duplicated (except with specific elements in group

0000 hex), this resulted in imperfect groupings. However, the group-element keys are unique. The structure also provides for expansion in that some "room" was reserved between assigned group and element numbers to allow for additions. To provide for user-defined or manufacturer specific data, a structure of shadow groups was defined. This also provided a mechanism by which new groups or elements could be used until considered for incorporation into revisions of the Standard. The defined groups of the Standard are given in Table 1.

Table 1
ACR-NEMA Standard Groups

Group Number	Type of Information	Description of Group Contents
0000	Command	Information which provides for end-to-end communication capability between two pieces of equipment
0008	Identifying	Information which provides unique image identification parameters to the user
0010	Patient	Information related to the patient
0018	Acquisition	Information related to the image acquisition device and imaging procedure
0020	Relationship	Information relating the image to the location within the patient and to other associated images
0028	Image presentation	Information related to the manner in which the image can be presented or displayed in a consistent and reproducible way
4000	Text	ASCII text
6000-601E	Overlay	Information related to overlay pixel data associated with an image
7FE0	Pixel data	Image pixel data

Though it appears that the Standard contains a great many elements which need to be sent with each image, in fact, some elements are specific for certain types of images. For many medical images, the collection of elements will be a small fraction of the total message size. Also, the elements were assigned types based on how important they are for interpretation. For example, pixel format information is essential whereas equipment manufacturer name is probably not. The former is considered a mandatory data item (a type 1) and the latter is optional (type 3). There are also type 2 elements which are considered mandatory, but which conceivably may not be known at the time of the examination. Unlike type 1 elements, which must be sent and must contain valid data, type 2 elements may be sent with a zero length, indicating that the value is not known. Defaults exist for many elements and fall into type 1D and 2D categories. The difference between these is that the default is always assumed if a type 1D element is not sent. It is not possible to send a zero length (indicating "unknown") for a type 1D element whereas this is possible for a 2D (in which case the default is not assumed). Type 3 data elements are those which are not essential for interpretation, but which may be important for certain medical images.

Once the message is assembled, it is sent down to the Session layer. The Standard has provided "hooks" in this layer for a multiple virtual channel protocol. In effect, virtual channels allow the single physical connection to support multiple logical connections. Connection control words and message source and destination fields are defined in this layer for incorporation in layers below.

The Transport/Network layer takes the message with the information from the Session layer, and fragments the message into packets. Packets consist of a block of the message (0 to 2048 bytes) plus a packet descriptor word and block sequence number. The packet descriptor word contains the information on channel number as well as packet type (channel control) and packet service class (i.e., whether to acknowledge each packet or not). Packet structure is shown in Figure 6.

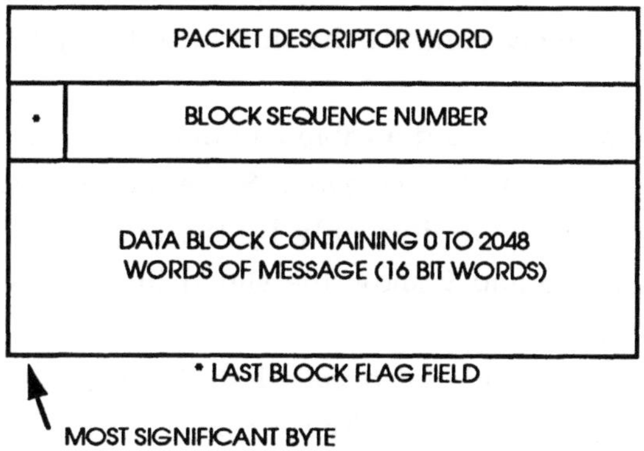

Figure 6: ACR-NEMA packet structure

The now packetized data proceeds down to the datalink layer. Each packet is enclosed by a frame descriptor word and frame check sequence to form a frame. The frame descriptor word consists of a frame type (e.g., data and status) and a parameter field (length or status depending on frame type). The frame check sequence is a checksum which allows for error detection. Figure 7 shows the structure of an ACR-NEMA frame.

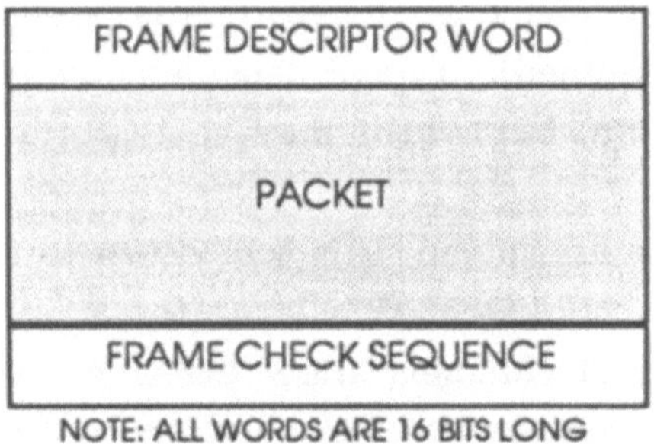

Figure 7: ACR-NEMA frame structure

Frames are sent to the Physical layer which handles transmission across the interface. The physical interface consists of 16 asynchronous parallel lines with control signals. The drivers for the signal lines are EIA RS485 standard, driving differential pairs for high noise immunity. The connector is a widely available 50-pin micro-type and must be female on the equipment. Since the interface is designed to operate in a master-to-master environment, the control lines cross over in the cable. The master-to-master operation is with respect to the electrical signals. Because of the protocols used, when one side of the interface opens a channel, that side effectively controls the interface, though the other side may close the channel.

With the specified drivers and a cable length of 15 meters, hardware speeds of 8 megabytes per second can be achieved. Since the interface is asynchronous, however, much slower rates (limited only by the one second idle timeout) are accommodated. Figures 8 and 9 show the signal lines of the interface and the interface electrical connector.

For the transport/network (also called the"tnet"), datalink, and physical layers, very specific protocols exist. These consist of state, cable activity, and timing diagrams. The protocols for the presentation and application layers are not as rigorously defined, though the structure of the message and the action of the commands are thoroughly specified.

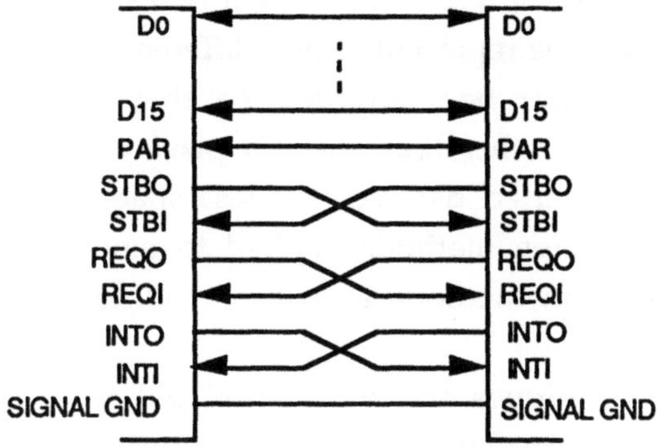

Figure 8: Signal lines of ACR-NEMA interface

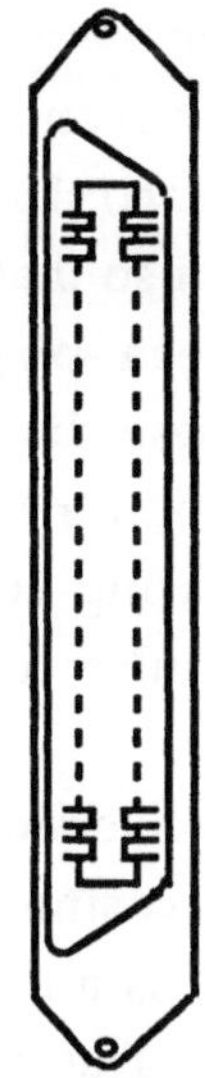

Figure 9:
ACR-NEMA
50-pin connector

On the other side of the interface, the frames would proceed back up through the layers, eventually resulting in a reassembled message.

While the Standard at first seems complex, in fact, a minimum for compatibility has been defined. This states that of the commands, only SEND need be implemented. Also, only a single channel need be supported. The full information on minimum requirements has been published by NEMA [3].

THE OTHER STANDARDS: DATA COMPRESSION

Following publication of ACR-NEMA 300-1985, it was recognized that additional work was necessary as new requirements were made known to the Committee. New Working Groups were formed to examine these issues. Initially, data compression was assigned to Working Group IV, exchange media to Working Group V, and validation issues to Working Group VI.

The vast volumes of data generated in medical imaging departments put a strain on archival media, even optical disk. The communications systems which must handle the volume are also burdened. To help alleviate this problem, data compression has been proposed as a solution. This paper will not address the benefits of lossless *versus* lossy data compression as this was also not the task of

WG IV. Rather, the Working Group sought to allow for data compression and decompression in a manner consistent with the existing Standard.

It was recognized that data compression consists of two steps. In the first the data is operated on so that it can be represented in a different way. For example, differences between pixels might be computed so that the data may be represented by these differences rather than the full pixel values. The second step consists of coding the converted data so that less space is needed for its storage. In reviewing compression methods applied to monochrome medical images, three popular conversion methods and three coding techniques were identified.

Supported conversion techniques are: differential pulse code modulation (DPCM), discrete cosine transform, and the pyramid (and its relative, the S) transform. DPCM is a differencing method which represents original data by a set of differences. The discrete cosine transform is a real valued transform for which fast algorithms exist. The pyramid and S transform are also differencing methods but compute sets of differences (in horizontal, vertical, and diagonal directions) from pixel data. The pyramid and S transforms have the useful property of producing viewable images, reduced in resolution, at each step.

Once converted, the coding methods take advantage of redundancy in the data to do the compression. The compression standard supports run length, Huffman, and Lempel-Ziv methods. Depending on how the data are quantized, the compression may be bit-preserving (lossless) or non bit-preserving (lossy). An example is in quantizing the cosine transformed data. If the quantization extends out until transform values vanish, there will be no high-frequency information lost. Truncating the transform data before this point will cut off such high-frequency information. Loss of the high frequencies corresponds generally to a decrease in edge sharpness and diminishing of fine detail.

Another popular compression method which is lossless but does not fit well into the above list of methods is perimeter encoding. In this technique, which is used for nonrectangular images, the zero values outside the actual image area are not saved. A map describing the number of pixels per line is sent instead. This method is also supported by the compression standard.

The design of this standard is such that methods may be combined if such combination will yield more compression. The compression standard consists of new groups and elements which allow a device to describe how compression was done. For proprietary algorithms, the standard provides a structure for a manufacturer to send these data about the compression used. The structure of the compression standard was designed to be flexible and allow for future growth.

The compression standard was voted on by members of the ACR and NEMA and was approved. The document is currently available from NEMA as ACR-NEMA Standard PS-2 at nominal cost.

EXCHANGE MEDIA

Working Group V of the ACR-NEMA Committee was charged with evaluating exchange media for possible standardization. When the Working Group was formed, the most popular exchange medium was one-half inch magnetic tape. Though there was considerable interest in optical disk, there was no industry standard for it. The interest in archival as well as exchange media came from the manufacturers who preferred to write one set of software for generating exchange or archive volumes.

Since there was no consensus on optical media standards at the time WG V began working, the group decided to consider magnetic tape first. Since the charter is to consider exchange media in general, as industry moves towards standards for optical media (for example, the 5.25-inch medium) the Working Group may continue to evolve standards.

As the full ACR-NEMA Committee had done, existing standards were first reviewed. The AAPM Magnetic Tape Format [6] was a very good possibility, but did not address all the issues required for exchange. The manufacturers thought that too much modification of existing operating system drivers would have to be done. LANDSAT, NATO, and Japanese JAMIT were among the other standards reviewed. None was thought to be as good as the AAPM standard for medical use. The Japanese standard, in particular, had fixed-length fields. If possible, the desire was simply to write out an ACR-NEMA data set onto the exchange medium. While manufacturers had widely differing formats for data, a common element proved to be use of the ANSI standards for magnetic tape. These covered both the physical media standardization and common labels and file structures. Using the ANSI standards, it was possible to fit the ACR-NEMA data set structure into the ANSI standard file structure with a "wrapper" consisting of ANSI labels.

As of this writing, the draft of the magnetic tape standard has been balloted, but one negative vote required ACR-NEMA to address the associated comments. This has been done, and the standard is being re-balloted. The Japanese PACS standards organization has approached ACR-NEMA WG V with a proposal for an exchange standard based on 5.25-inch erasable optical disk. This proposal is currently undergoing evaluation, and it does seem to have a fairly wide industry backing.

VALIDATION

Working Group VI was formed to examine issues related to validation of the "parent" standard. This includes both validation of the Standard itself, and assistance of those implementing or testing hardware and software for the Standard. The Working Group serves as the clearinghouse for questions about the Standard and for suggestions regarding new data groups or elements. This group also examines issues related to revision of the Standard, and defines how revisions will be compatible with prior versions. The basic guideline has been that no element would be changed or replaced unless found to contain errors.

The first revision of the Standard (version 2.0) has passed the vote of the ACR and NEMA members, and is now available from NEMA as ACR-NEMA 300-1988. This revision includes new elements for definition of an image identification hierarchy, elements added for ultrasound, and commands and elements added to support laser film writers and other hardcopy devices. All changes are compatible with version 1.0 of the Standard.

In addition to the Version 2.0 of the Standard, WG VI has published a number of engineering bulletins, errata lists, and responses to questions pertaining to the Standards. One of the very successful activities of WG VI in 1989 was to hold a workshop for implementers. This meeting resulted in a number of clarifications of the Standard as well as some proposed solutions for problems. The results of this meeting were published by NEMA, and distributed to those who ordered copies of the Standard. A recent (May 21-23, 1990) hardware workshop was conducted in which six manufacturers (De Jarnette Research Systems, General Electric Medical Systems, Merge Technologies, 3M, Siemens Medical Systems, and Vortech) brought test hardware implementations of the Interface Standard to Georgetown University, and tested all possible pairwise connections. Of 15 possible connections, all were successful in that bidirectional data exchange was demonstrated. In some instances, testing was successful up to the application layer of the Standard. The current activity of WG VI has been to form an *ad hoc* working group to iron out remaining differences in implementation interpretations so that future testing is likely to result in communication through the application layer for all who participate. It is anticipated by the members of this *ad hoc* group that such work will not take a long time, given the degree of success achieved in the first round of testing. For any manufacturer or institution who wishes to be included on the mailing list of this WG, or who would be willing to participate in the next round of tests, the NEMA Staff Executive (David Snavely, (202) 457-1965) should be contacted.

Two additional Working Groups have been formed to examine other issues which have been brought to the attention of the Committee. Multidimensional data was assigned to WG VII, and interfaces between hospital information systems (HIS), radiology information systems (RIS), and PACS are being developed by WG VIII. The original three Working Groups have been deactivated, though some of the members are now active in the newer WGs.

MULTIDIMENSIONAL DATA

With the increasing interest in images which are multidimensional, either in spatial dimensions or with parametric features, Working Group VII was formed to examine how such data could be accommodated in ACR-NEMA 300-1985. The Standard currently specifies that images are rectangular with the number of dimensions allowed to be "2".

This Working Group has some working proposals for member use, but has suffered from difficulty assembling all members for meetings. Some of the newer work of WG VI may supersede the activity of this group. The Working Group is on inactive status, but can be reactivated when needed, and is nonetheless seeking suggestions from interested users or manufacturers.

HIS/RIS/PACS INTERFACES

The newest of the Working Groups is defining a standard for interface of HIS, RIS, and PACS. Unlike the parent Standard, the product of WG VIII will be an upper level standard. It is thought unlikely that, in the heterogeneous computing environment formed by these systems, the hardware parts of the Standard would find much acceptance. It will most likely build around the parent Standard so that ACR-NEMA messages (or parts) may cross from HIS to RIS to PACS and back.

The necessity of this effort is based on the requirement of having certain RIS- or HIS-based data in the PACS (patient demographic information, for example), and getting PACS data (such as time of examination) back to the RIS or HIS. Allowing direct exchange reduces the problem of operator errors in transferring information between systems. Currently, much patient information has to be re-entered in the PACS. Errors lead to database inconsistency which can result in an inability to access data and in duplication of files. A similar situation pertains to PACS data which must be copied into the HIS or RIS.

The WG began by defining an entity-relationship model of radiology in order to understand what kinds of data would exist at different places in a radiology department and what the relationships between entities (such as "patient" and "report") are. This effort resulted in developing the necessary new data set groups and elements, the groups being thought of as the entities and the elements as the attributes of the entity. The current effort is to develop a functional or operational model of radiology so that the necessary commands can be similarly defined. Work in these areas is being done with careful attention to the standards work of the IEEE Medix group as well as the HL7 and HISCC. These groups are working on standards which will affect the HIS and RIS sides of the interface.

FUTURE DIRECTIONS

The original ACR-NEMA Standard has undergone one revision, and has spawned two new standards with a third on the way. All of these have retained the structure of the parent Standard, and are compatible with it in various ways. However, as user requirements mature in PACS environments, even more changes to the Standard are anticipated.

The most ambitious effort at present is directed by Working Group VI in conjunction with Working Group VIII, and consists of examining proposals to define a version of the Standard which would operate more easily over networks. A joint development by Philips and Siemens, the Standard Product Interconnect (SPI) is one model being used. Another is based strongly on the work of WG VIII, and addresses a combination of the application level work being done by that group along with the changes necessary in the lower layers so as to support network interconnection. A definite possibility for the future is to "split" the Standard. The most recent proposal is do retain the existing Standard in its entirety for point-to-point connections and to define additional protocol so that the upper layers of the Standard (application and presentation) could be fitted onto existing standard layer stacks. These layer stacks then interface to a variety of media (Ethernet, FDDI, et cetera) providing an ACR-NEMA interface which can operate over networks. Figure 10 illustrates this idea. Please note that this is preliminary, and should in no way be taken to represent ACR-NEMA standard structure

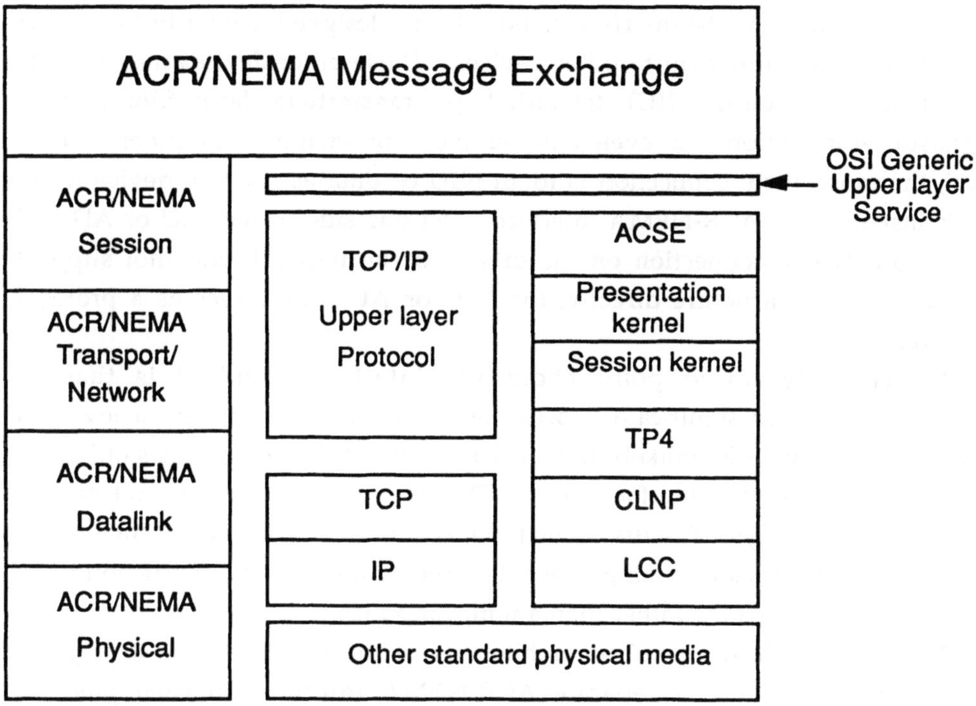

Redrawn from ACR-NEMA Working Group VI Minutes of 5 September 1990
Proposal contributed by Charles Parisot of GE Medical Systems

Figure 10: A proposal to extend the ACR-NEMA interface to other lower layers

Sanctioning the use of the ACR-NEMA data structure over non-ACR-NEMA interfaces creates a problem in that it has the potential to result in a multitude of hardware-incompatible devices. However, many of the ACR-NEMA Committee members see this change as a necessity as it will allow the ACR-NEMA data structure to be more widely used. The hardware interface will still be available, and for applications which require its speed, is still highly desirable.

A frequently asked question concerning the Standard is what ACR-NEMA implementations are available. As of this writing, implementations for VME bus to ACR-NEMA, IBM PC/AT bus to ACR-NEMA, and Digital Equipment Corporation Q bus to ACR-NEMA exist, though some are admittedly prototypes. The Japanese manufacturers have indicated that they are working on both full

ACR-NEMA implementations as well as a minimum requirement version they call MIPS.

THE USES OF THE ACR-NEMA STANDARDS

The 300-1985 and 300-1988 interface standards are designed primarily to connect image generating equipment to other devices. Those other devices may include network interface units (NIU), teleradiology transmitters, laser film writers, diagnostic workstations, or even another piece of imaging equipment. In a PACS, the most likely connection is to an NIU or data acquisition device (AD). In this instance, the ACR-NEMA interface is on one side of the NIU or AD and the PACS network connection on the other. If the network does not support ACR-NEMA data structure directly, the NIU or AD then serves as a protocol converter.

A frequently raised point about the interface Standard is that its performance dictates significant processing power, and that unless a low cost version is available, it is unlikely to be used on equipment which costs only four to five times what the interface does. This is a valid criticism, and can be addressed in two ways. The first is that low-cost interface hardware, such as the PC/AT card version, is becoming available. This would allow construction of an NIU or AD based on a PC rather than a minicomputer.

The second solution involves architecture of the PACS. While it may not be cost-effective to use a separate ACR-NEMA interface on each piece of equipment, it may be economical to use a single interface with multiple devices connected to it. Examples are in ultrasound and nuclear medicine. In the former, multiple video outputs (or digital, if available) could be connected to a central computer which would manage the data from the machines so as to keep the images sorted by case. This computer would have a single ACR-NEMA interface for PACS connection. Nuclear medicine would be similar, and is more likely to evolve this way since many nuclear medicine departments are already networked to central computer systems. These approaches are not very different from systems which currently use a shared laser printer with a computer front end.

The data compression standard will be of value in reducing PACS storage and network bandwidth requirements. Even a lossless compression of 2:1 doubles archive capacity. The use of the lossy compression techniques requires that those who will be using the expanded images accept this limitation. Since the techniques are often computationally intensive, to be of value for reducing network bandwidth, some form of hardware for fast compression and expansion will be required. Research on lossy compression and hardware for performing the algorithms is presently underway. It may be possible to use compression selectively, compressing images only after they have been interpreted or only if they are being sent to non-diagnostic workstations.

The exchange medium standard, which currently supports magnetic tape, will find use in inter-institutional exchange. Though engineers have been predicting the demise of magnetic tape for some time, it manages to remain heavily used, even if not very popular. The emergence of a standard for optical disks, particularly the 5.25-inch format, will eventually result in phasing out magnetic tape for exchange. Industry has had more success in developing support for this smaller medium than for the 12-inch size. Another potential use of the exchange medium standard is for use in a per-patient or personal health data system. In these proposed systems, each patient's medical record information is written on a small, removable medium (probably optical). The idea is to allow the patient to carry his or her data around to different healthcare facilities as needed. The necessity of a standard for this application is self-evident.

Communication between the different information systems in a medical facility could benefit from use of the standard which is being developed by WG VIII. The HIS/RIS/PACS standard will of necessity be an upper-layer one. It will, along with standards efforts in the HIS and RIS communities, help promote the concept of an electronic medical record. This idea continues to grow in popularity as the advantages of a record which is amenable to database operations become more clear. As healthcare becomes more regulated, the production of reports demonstrating efficacy or documenting the necessity of treatment will be an increasing burden if paper records have to be searched manually to retrieve the necessary information.

CONCLUSION

In the seven years since its formation, the ACR-NEMA Committee has succeeded in producing and interface standard which is beginning to be used in a number of PACS. This standard has undergone one revision, and changes for the next version are being studied now. In addition, they have written a standards for image data compression and exchange media, and are working on a standard for HIS/RIS/PACS interfaces. These efforts are guided by the radiologic community as well as by representatives of industry. The collaboration between those who establish requirements and those who must build equipment to meet them has been very productive.

The ACR-NEMA Committee welcomes comments on all of its efforts. In particular, WG VI is interested in hearing from those who are building ACR-NEMA interfaces and WG VIII in those who are working on other standards for healthcare information management. The various Working Groups can be contacted through:

National Electrical Manufacturers Association
2101 L Street, NW
Suite 300
Washington, DC 20037 U.S.A.
Attention: Staff Executive, MedPACS Section
(202) 457-1965

ACKNOWLEDGEMENT

The author wishes to acknowledge the members of the ACR, the manufacturer members of NEMA, nonmember contributors, and NEMA staff for the work on the Standards. The ACR members included: James Lehr, M.D., Gwilym Lodwick, M.D., John Loop, M.D., John Moore, Ph.D., Alan Rowberg, M.D., Roger Shannon, M.D., Bob Thompson, Yen Wang, M.D., and Jason Zielonka, M.D. Manufacturer members and nonmember participants included: Edgar Alzner, Ph.D., Bill Bennett, David Best, Ph.D., Hartwig Blume, Ph.D., Jack Bowie, Todd Brackett, Samuel J. Dwyer, III, Ph.D., Allan Edwin, Paul Fromme, David Glassman, Ph.D., Shankar Hegde, Ph.D., David Hill, Ph.D., Bob Hindel, Ph.D., John Hoffman, Will Hunt, Bob Lawrence, Daniel Lessard, Robert McCarthy, Royal Morse, Owen Nelson, Ph.D., Herman Oosterwijk, Marcia Paisner, John Perry, Fred Prior, Karen Randall, Ph.D., Alan Rothlauf, Andrew Sherman, Ph.D., Chris Spilker, Kelly Stronger, and Bill Wessel. NEMA staff members Laura Lee Murphy, Cindy Fuscoe, and David Snavely made much of the work possible. The author wishes also to thank all those not named above who contributed their time and energy to the development of the ACR-NEMA Standards.

REFERENCES

1. ACR-NEMA Dig. Imaging and Comm.Stds. Ctee., Minutes of Feb. 7 meeting, p 6, February, 1984.
2. ACR-NEMA Dig. Imaging and Comm. Stds. Ctee., ACR-NEMA 300-1985: Digital Imaging and Communications, November, 1985.
3. ACR-NEMA Dig. Imaging and Comm. Stds. Ctee., Application Guide: Minimum Requirements for Compatibility with the ACR-NEMA Digital Imaging and Communications Standard, 1987.
4. Duerinckx, A.J. (ed.): SPIE Proceedings, Vol 318, parts I and II, Picture Archiving and Communication Systems (PACS) for Medical Applications, January, 1982.
5. International Organization for Standardization, Document ISO 7498-1983, Information Processing Systems - Open Systems Interconnection - Basic Reference Model, 1983.
6. Maguire Jr., G.Q., Baxter, B.S., Hitchner, L.E., An AAPM standard magnetic tape format for digital image exchange, SPIE Proceedings, Vol. 318, PACS for Medical Applications, pp 284-293, 1982.

In addition to these specific references, different versions of this paper were published previously.

Horii, S.C.: The ACR-NEMA Standards. Administrative Radiology, Feb. 1990, pp 68-75.

Horii, S.C.: The ACR-NEMA Standards: A tutorial on their structure and use. In: Arenson, R.L. and Friedenberg, R.M. (eds.): *Computer Applications to Assist Radiology*. Symposia Foundation, pp 405-422, 1990.

Horii, S.C.: The ACR-NEMA Standards and their implications for teleradiology. SPIE Proceedings, v. 1355: Telecommunication for Health Care: Telemetry, Teleradiology, Telemedicine, pp 152-163, 1990.

Integration Issues in PACS

K.K. Chan, Ricky K. Taira, and H.K. Huang

Department of Radiological Sciences, UCLA School of Medicine, Los Angeles, CA 90024-1721, USA

Abstract: A successful PACS (Picture Archiving and Communications System) implementation requires an eclectic integration of a number of key technologies. Among these are equipment interfaces, communications, storage, and display. Coincident with this, the software architecture must support a distributed system of heterogeneous structures, provide for protocol and format conversions to a unified system standard, be scalable to accommodate expansion, and provide a measure of fault tolerance. In this paper we survey the current state of the UCLA PACS components and architecture.

Keywords: archiving, communications, image management.

Introduction

Contemporary PACS (Picture Archiving and Communications Systems) implementations are invariably composed of networked systems of heterogeneous computers. Even within one computing system, the software architecture will usually consist of multiple cooperative processes using various forms of interprocess communications to coordinate their activities. Given the state of current computing technologies, this distributed processing approach is both necessary and advantageous.

Radiologic imaging devices, from which data first enters the system, are usually standalone systems with a host computer and multiple embedded micro-controllers. These are used by the manufacturer to run the system software and to respond to real time events. At the other end of the system, are display workstations situated at widespread locations. Assuming digital image transmission, these workstations must be configured as local computer systems integrated with the appropriate display and networking capabilities. Interactive response to user interfaces is handled much better by dedicated local processors, rather than one powerful central processor switching between many different tasks. (We do not discuss here the possibility of multiprocessing systems coupled with long range analog image transmission.) Between these two ends, various other subsystems must handle archival, communications, database, and other

NATO ASI Series, Vol. F 74
Picture Archiving and Communication Systems (PACS) in Medicine. Edited by H. K. Huang et al.

PACS functions. It is the task of the system integrator to tie together these distinct and disparate subsystems into a single functional PACS.

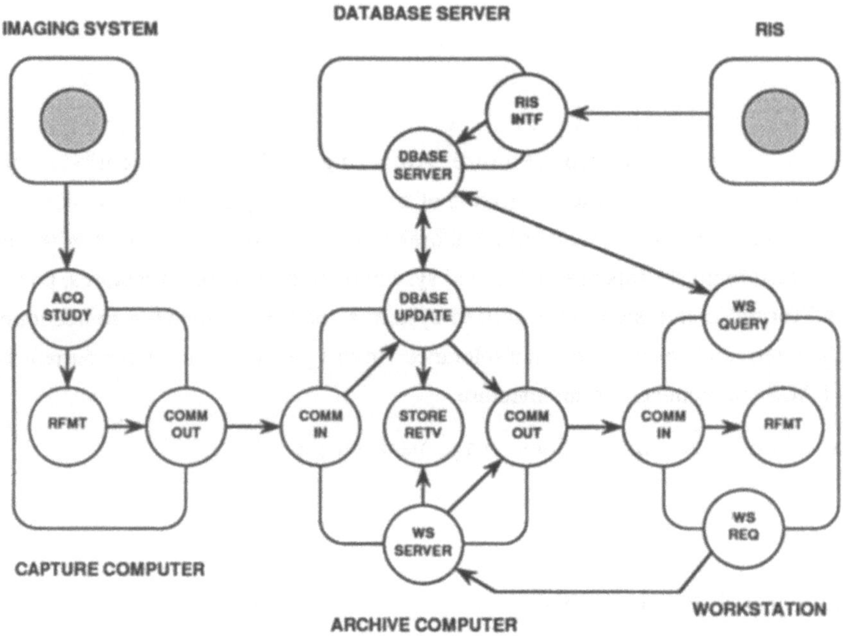

Figure 1. Dataflow in the UCLA PACS. Boxes represent computer systems. Circles represent major software subsystems. Arrows indicate the flow and exchange of information among subsystems.

Figure 1 shows the major functional components and processing flow in the UCLA PACS. In the remainder of this paper we shall briefly survey each of the components with particular emphasis on system integration issues. Detailed descriptions of each subsystem are given in other papers contained within these proceedings.

Acquisition

Image acquisition is perhaps the most tedious and currently, least reliable, part of a PACS. In the typical environment, there is usually a large installed base of existing equipment which must be interfaced to the PACS. These imaging devices vary widely in system design, and are made by a number of different manufacturers, each with proprietary interfaces, communications protocols, and data formats. ACR/NEMA standard [1] interfaces and data formats are not yet a widespread reality in the industry. Indeed, it is likely and hopeful that the 50 pin ACR/NEMA plug standard will be abandoned in favor of more modern, versatile, and cost effective network style interfaces. There have been proposals [2] to restrict and migrate ACR/NEMA protocols to the upper three layers of the ISO model (application, presentation, and session), while using more widespread industry standards such as TCP/IP and ethernet (or FDDI) for the lower four layers (transport, network, data link, and physical).

Currently the UCLA PACS encompasses 4 CT (computed tomography) scanners, 5 MR (magnetic resonance) scanners, 2 CR (computed radiography) units and 2 laser film scanners. Interfaces may be divided into two classes. These are peer to peer network level interfaces, and master slave device level interfaces. In either case a capture computer is interposed between the imaging system and the rest of the PACS. An example of a network interface would be an MR scanner which communicates [3] with the capture computer using TCP/IP and ethernet. A device level interface would be used for a laser scanner. In this case a DR11W interface board is plugged into the capture computer which then controls the operation of the scanner by sending appropriate signals over the DR11W link.

A capture computer has three primary tasks. These are acquisition of data from the imaging systems, conversion of data formats from proprietary to PACS standards, and transmission of data to the archival system. The UCLA PACS data format is intended to be fully ACR/NEMA compliant. Data acquisition includes the capture of patient information and procedural parameters used for the study. These may be acquired automatically from the imaging system or interactively via technologist input. Incomplete study information can be a major source of errors in the system. Capture computers are multitasking UNIX based machines which may service more than one imaging system. Thus there may be more than one instance of acquisition, reformat, or communications processes (see Fig. 1) at a time.

Acquisition may be initiated by either the imaging system (push operation) or the capture computer (pull operation). In the push mode the capture computer provides an immediate response to the imaging system so that procedures are not delayed. However if the capture computer is not ready, it may be possible to lose data. Therefore, we consider the pull mode preferable when possible. In this case, data can be obtained from the imaging system when the PACS is ready. This also reduces the role of the technologist, who does not have to explicitly initiate a transfer to the PACS.

Data buffering is also an important factor. At times when the archival system is unable to accept data, the capture computer must store the data temporarily, allowing the imaging system to continue normal operation. The buffering capacity can be sized according to needs and economics. We find that one day's storage is the minimum acceptable capacity. In the case of imaging systems with network interfaces, software reconfiguration can provide a substitute capture computer in the event of failure.

Networking and IPC

Networking in the UCLA PACS uses a three level hierarchical structure [4]. The lowest in cost and performance is a global ethernet which links all PACS nodes together. Local ethernets are used to connect to various imaging systems which have network level interfaces. The protocol on these local nets may be either TCP/IP or vendor proprietary. The global ethernet is used for exchange of command and control messages, error logging, performance monitoring, and database transactions. Image data are transferred from capture computers to archival computers using either ethernet or FDDI (fiber distributed data interface). This hierarchy is necessary because each capture computer may service several imaging systems and each archive computer may service a number of capture computers. FDDI increases the bandwidth into the archive computer and decreases the processing required for network operations. Optical bypass switches are highly desirable in the FDDI ring topology. Highest in the hierarchy is Ultranet (Ultra Technologies, San Jose, CA), a proprietary Gbit/sec hub-based fiber network. This is used for data transfers between different archive nodes and between archive nodes and workstations.

Interprocess communications (IPC) within the UCLA PACS are layered upon socket level interfaces to TCP/IP and UDP/IP protocols. This provides a uniform mechanism for coordination of processes running on either the same or different machines. PACS commands and requests are exchanged using ascii text messages to facilitate communications between machines with different binary standards. TCP/IP is used for all transactions which must be reliable (ie. file transfers and exchange of commands). UDP/IP is used for network wide PACS performance monitoring. These protocols have been implemented on systems ranging from PC's to supercomputers, using media ranging from slow serial links to high speed fiber, and over distances from across a desktop to across the country. The widespread industry support for these networking mechanisms make them valuable tools for incorporation of heterogeneous computing components and expansion to wide area networks.

As a precaution against system crashes, jobs within PACS are scheduled through disk resident priority queues. In the event of system failure, incomplete jobs are restarted from the beginning. Jobs are considered complete and acknowledgements sent, only after data have been flushed from system buffers and physically written onto magnetic disk. In the event of failure due to lack of resources, component failure, or incomplete peer response, jobs can be scheduled for retry at a later time, or retried with different resources (such as an alternate archive node).

Archive and Database

Archive computers within the UCLA PACS have five primary functions (see Fig. 1). These are: (1) accepting data from capture computers; (2) updating the PACS global database with information regarding these data; (3) archival storage and retrieval of data; (4) routing data to appropriate workstations; and (5) servicing workstation requests for data. Currently the UCLA PACS has two archive nodes, each supporting 680 GByte capacity optical disk libraries, and 4 GBytes of magnetic disk cache. Caching decisions are performed by PACS based on time of recent access and need for comparison studies for a patient.

Input and output communications processes are implemented as UNIX concurrent servers. Thus there may exist multiple instances of each process at one time. Archiving per-se is handled by a storage/retrieval process which services a priority queue. These archival operations may be queued (via IPC) by either the workstation server process, or the database update process. The update process parses and interprets the file header fields of newly arriving images, then performs the following operations: (1) updates a variety of global PACS database tables; (2) determines the display workstations where the study is most likely to be reviewed; (3) determines necessary comparison studies to retrieve from long term storage; and (4) queues a study for archive. Routing to workstations is based on anatomical region, modality, patient age, patient status, radiologist, and referring clinical service. Prior studies for a patient are selected for comparison retrieval based on correlation of anatomical region, modality, time of study, and study class (admission, discharge, pre-operative, or post operative).

A central integration issue is the logical design and physical implementation of the database management system (DBMS). The design should include an external data model (local user view) for each stage of the radiology process (request, registration, acquisition, case management, interpretation, report generation, and etc.). These models must incorporate the varied viewpoints of the entire radiology staff. Then, a conceptual data model can be developed which integrates the data and processing needs of each local view to produce a global data structure capable of supporting all PACS operations. Logical tables in the UCLA PACS are defined and organized in *third normal form* [5]. These tables include patient demographics, study description, image description, study diagnosis, and etc. [6]

The UCLA PACS uses a commercial DBMS from Sybase Inc. (Emeryville, CA). This product is based on a client-server architecture [7]. In this design, a server process tuned for database operations, stores and maintains all PACS text information. The server machine is responsible for all database transactions including inserts (with protection and integrity checking), retrievals (including complex table joins), table updates, and record deletes. All other PACS nodes (archive, workstation, etc.), operate as database clients. Clients interact with the server process via remote procedure calls. These DBMS network messages are formatted as SQL (structured query language) statements. For fault tolerance, the DBMS server is mirrored onto a second machine which operates as a *hot-backup* in the event of failure of the primary server [8]. Only text and study descriptions are stored within the database. Image data are stored separately and are located using pointers stored within the database tables.

Integration problems can occur when other DBMS's such as RIS (Radiology Information System) and HIS (Hospital Information System) exist in the department. Integrating these heterogeneous systems is a difficult problem requiring translators, formatters, message passing, and duplication of information. Our existing RIS system (Maxifile IV by Dimensional Medicine, Inc.) is currently a closed system. Our initial integration of RIS and PACS uses one-way communication packets to copy selected RIS fields into the PACS DBMS. The RIS computer will forward diagnostic reports, study requisition information, and patient scheduling information to PACS as soon as they are generated.

Workstations

The hardware architecture of workstations are discussed by Cho [9]. Our current strategy is to implement two types: high resolution (2.5K x 2 K) high performance displays for primary diagnosis, and lower performance low resolution ones for clinicians, consultation, and other purposes. Currently 2K display systems are available from only a few vendors at considerable cost. Numerous low cost 1K solutions are available in the market.

Workstations in the UCLA PACS have three primary interfaces to the rest of the system (see Fig. 1). First, the workstation must be able to receive images that have been routed by the archive computer. These enter through a server process (COMM IN in diagram) which then queues them to a reformatting process. Reformatting involves converting the data into a form suitable for efficient image display (such as storage on parallel transfer disk), and update of a local database which tracks images present on the workstation. Two other functions are required. The workstation must be able to query the global database for information about prior studies; and it must be able to request particular studies from an archive. These are implemented with client procedures which talk to the database server and archive computers respectively.

Summary

We have presented a PACS architecture based on distributed heterogeneous processing that is scalable and incorporates elements of fault tolerance. A structured exchange of data and commands is used to isolate vendor proprietary protocols at their point of entry into the PACS. For efficiency, communications, and performance monitoring, multitasking capabilities are a requirement for PACS computers. Capture computers are used to buffer the flow of data from imaging systems to PACS. Image data are stored on distributed archive nodes and text data are stored in a centralized DBMS which is mirrored for reliability. A hierarchical network heavily based on industry standards provides a base for incorporation of heterogeneous systems and future expansion. As demand increases, nodes can be added up to the limit of network capacity. This capacity can be increased with segmentation and clustering of image traffic. Workstations can receive or request data from any archive. Future database enhancements will serve the needs of researchers, which include mass screenings of studies based on correlation of diagnosis, patient history, demographics, clinical findings, radiographic techniques, and other factors.

Acknowledgements

This work was supported in part by Public Health Service Grant No. R01 CA 39063, R01 CA 40456, and P01 CA51198, awarded by the National Cancer Institute, Department of Health and Human Services, and by the UCLA Department of Radiological Sciences Research Fund.

References

1. ACR-NEMA Digital Imaging and Communication Standards Committee, *Digital Imaging and Communications ACR-NEMA 300-1988*, NEMA, Washington D.C., 1985.
2. S.C. Horii, D.G. Hill, H.R. Blume, D.E. Best, B. Thompson, C. Fruscoe, and D. Snavely, *ACR-NEMA Standards Activity: an Update*, Proc. SPIE **1232**, 293-298(1990).
3. W.S. Weinberg, M. Loloyan, R.K. Taira, K.K. Chan, and H.K. Huang, *Automatic Acquisition of CT, MR, and US Images for PACS*, Proc. NATO ASI: PACS in Medicine 1990, Evian, France, (this volume).
4. B.K. Stewart, *Three Tiered Network Architecture for PACS Cluster*, Proc. NATO ASI: PACS in Medicine 1990, Evian, France, (this volume).
5. T.J. Teorey and J.P. Fry, *Design of Database Stuctures*, Prentice-Hall Inc., Englewood Cliffs, N.J., 1982.
6. B.K. Stewart and R.K. Taira, *Database Architecture and Design for PACS*, Proc. NATO ASI: PACS in Medicine 1990, Evian, France, (this volume).
7. C. Malamud, *INGRES: Tools for Building an Information Architecture*, Van Nostrand Reinhold, N.Y., 1989.
8. R.K. Taira, K.K. Chan, B.K. Stewart, and W.S. Weinberg, *PACS Reliability Issues*, Proc. NATO ASI: PACS in Medicine 1990, Evian, France, (this volume).
9. P.S. Cho, *Display Workstation: Hardware Architecture*, Proc. NATO ASI: PACS in Medicine 1990, Evian, France, (this volume).

PACS Reliability Issues

Ricky K. Taira, Kelby K. Chan, Brent K. Stewart, and Wolfram S. Weinberg

Department of Radiological Sciences, UCLA School of Medicine, Los Angeles, CA 90024-1721 U.S.A.

Abstract: Reliability is perhaps the most important performance parameter of a PACS system. Any system downtime may seriously affect patient care. The following reliability issues are discussed: (1) hardware redundancy (networks, archives), (2) database backups, (3) monitor routines for local host processes and network status, (4) uninterruptable power supplies, and (5) in-service training of all radiology technologists.

Keywords: Fault tolerance, hardware redundancy, reconfiguration, database integrity.

1. Introduction

Reliability is perhaps the most important performance parameter of a PACS. Any system downtime may seriously affect patient care. A reliable PACS will build user confidence and accelerate system acceptance. Development efforts should prioritize reliability over excessive performance and/or user features which are used infrequently. The development of a PACS which sometimes works is relatively easy. The challenge is to develop a highly reliable system at an acceptable cost to the hospital.

Figure 1 illustrates the difficulties of developing a fault tolerant system. A PACS will contain several different classes of computers differentiated by their designated function (image generation, PACS acquisition, archive, database server, workstation, monitoring, etc.). Each will have processes which run continuously. A radiographic study must traverse several processing stages before it becomes available at a PACS workstation. If any of these processes should fail, the acquisition chain faults. The probability of successfully acquiring a study is approximately equal to the product of the succcess probabilities of each processing stage. (Approximately since some processes share the same computer hardware, networks, and software modules). The remainder of this paper discusses four key reliability issues: 1)

NATO ASI Series, Vol. F 74
Picture Archiving and Communication Systems (PACS) in Medicine. Edited by H. K. Huang et al.

software reliability, 2) database integrity, 3) hardware fault tolerance, and 4) prevention and recovery from operator errors.

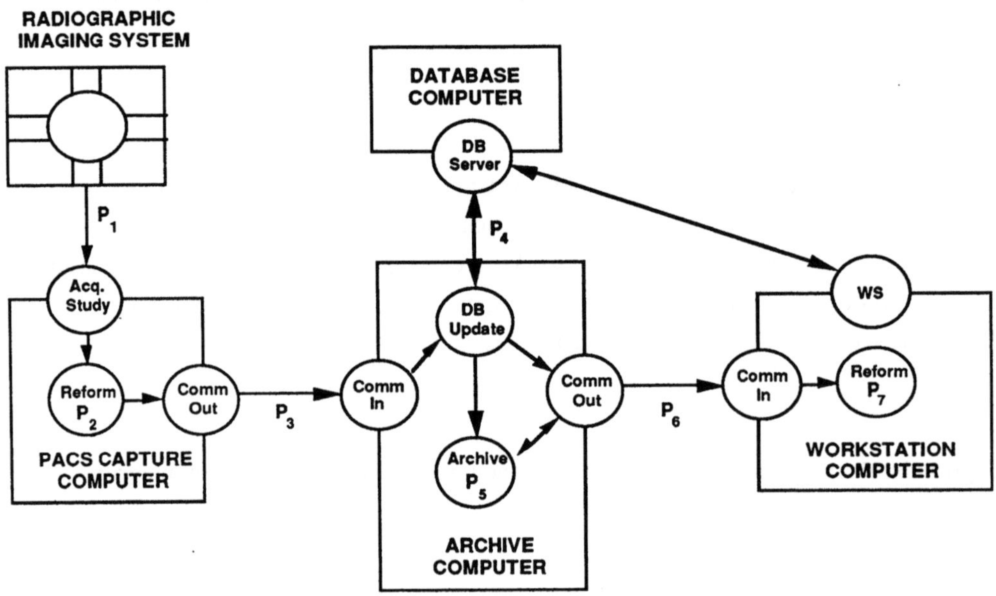

Fig. 1 - *Acquisition of new radiographic studies requires the successful completion of several processing stages.*

2. Software Reliability

Refinements, quality control, and maturity of the computer industry has made hardware / hardware configurations quite reliable (one should never assume however 100% uptime). Software reliability becomes increasingly difficult as programs and systems become more complex[1]. It becomes nearly impossible to check and debug every execution path that the software can traverse[3]. Common software errors result from faulty programming logic, inconsistent views of the PACS environment by various processes, faulty interprocessor communication, non-standard data formatting and/or packaging, data access/protection violations, inability to allocate necessary resources (disk space, communication channels, etc.), synchronization deadlocks, queue overflows, operator errors, and unannounced manufacturer updates. Resolving software errors involves three steps: (1) detecting errors as soon as possible, (2) confining errors to a particular software module / machine, and (3) analyzing the error so that appropriate corrective action(s) may proceed.

2.1 Error Detection and Logging

Error logs are an important record for understanding system reliability and quality of system software. All software modules should check the return status of important system and user function calls. This will localize software errors to a particular routine as well as prevent errors from propagating and contaminating other software modules[4]. Appropriate error recovery measures should then be taken. A primitive PACS routine which logs errors to a master PACS table will facilitate the analysis of system errors (frequency of each error type, time between errors, etc). Recommended fields to this table include: 1) source node ID, 2) destination node ID, 3) program ID, 4) routine ID, 5) check point in routine that failed, 6) generic PACS error number and message, 7) generic operating system error number and message (eg. UNIX errno), 8) specific error message describing parameter values and/or program state when error occurred, 9) recommended action for recovery, and 10) time stamp when the error occurred.

2.2 System Monitoring Routines

Each PACS node should have a monitoring (watchdog) routine which periodically checks for alarm conditions including: 1) whether all system processes are alive (i.e. there have been no software crashes), 2) if the maximum queue length of a process is exceeded, 3) whether sufficient resources are available (eg. magnetic disk space), and 4) time since last process transaction occurred. The list of processes to be check should be stored in a PACS system table. An interprocessor message facility should send alarm messages to a central system monitoring computer. The monitoring computer also periodically checks whether all system nodes are reachable via the network.

2.3 Software Error Recovery

Once an error has been detected and confined, appropriate corrective action should proceed. Five types of error recovery blocks are described.

1. **retry operation with same parameters and resources**: In many cases, reattempting an operation may result in a successful completion. This is particularly true of a transient condition where there are no actual hardware failures. The fault may be just a temporary condition (eg. someone unplugged a cable for a short while) or a data file was locked out by another process. Recovery occurs at different

programming levels according to the nature of the error. Errors in operations which try to allocate a resource (eg. file, communication socket, disk storage, etc) are often due to a *lock out* condition (resource allocated to another process or too many of the requested resource assigned by operating system). In this case, a retry of only the allocation operation is performed at a subroutine level. The period between retries is determined by the urgency of the operation. Retries should continue until one of the following ending conditions is encountered: a) operation successful, b) maximum number of retries reached, c) maximum queue time exceeded, or d) when a "high water" alarm condition is present (e.g. local storage space low).

2. **retry operation with alternative resources:** If retrying an operation with one set of resources fails, then it is necessary to identify the failed component, reconfigure the job with a new set of resources, roll back the job to a known state (usually the start of the job), then try executing the job again. This should be automatic. The determination of alternative resources should be either table driven via a system configuration file and/or defined using system-wide environment variables. Example resources which frequently fail include: 1) remote host computers, 2) communication networks, 3) magnetic and optical disk systems, 4) interface and processor boards, and 5) software server processes. Figure 2 summarizes the error recovery procedure.

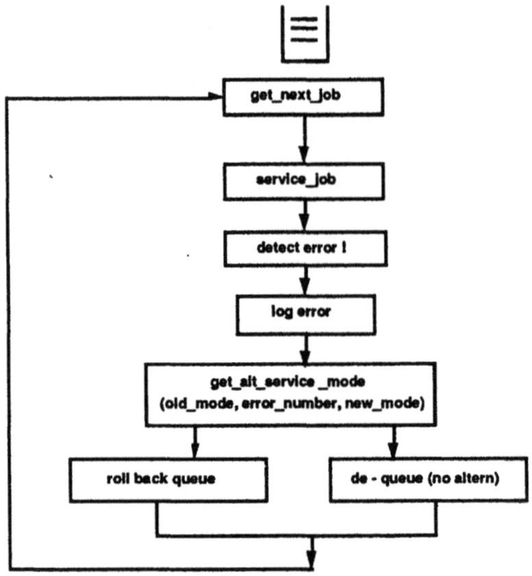

Fig. 2 - *Flowchart for reconfiguring job resources*

3. **send job to a resolution queue:** in many instances, alternative solutions are indeterminant. Two examples include inability to identify patient and inability to

determine destination workstation. These jobs are entered in a "to-be-resolved" queue which is serviced by a technologist from a PACS workstation. The PACS technologist must periodically service this queue.

4. **die and restart**: The error was fatal (logical programming error, synchronization deadlock condition reached, infinite loop). Some errors will persist despite all error checking pre-cautions. They typically have no symptoms except in very rare circumstances. The program should exit, perform a *sanity check* (eg roll back queues, release lock tokens, etc) and be restarted (perhaps by a monitoring process).

5. **PACS bootstrap**: All machines should be able to resume clinical PACS operation after a power down - power up sequence. This implies all necessary queue roll-backs, database synchronization, initial recovery procedures, and "daemon" process startups. The sequence in which these operations are performed is usually critical.

2.4 Operator Errors

Operator errors can be the source of substantial program downtime and/or system malfunction. Most operator errors occur during PACS image acquisition. This is because the operator typically must satisfy the data requirements of two heterogeneous data management systems (the global PACS DBMS and the local acquisition machine's DBMS). The database schema and format rules are different in general. The technologist's user interface is normally menu screens provided by the equipment manufacturer. PACS exclusive information then must be *snuck* into fields not preallocated by the manufacturer's software. Format checking and variable prompting thus cannot be performed. Clear instructions concerning proper data formats, description of information required, and description of location where information is expected should be provided for each acquisition station. Information formatting and inclusion errors can be minimized by periodically requiring all technologists to attend in-house educational seminars.

3. Database Integrity

A PACS database system is susceptible to the loss of a great deal of information with a relatively small scale event such as a disk crash. In planning of any database system, it is necessary to take precautions to ensure the integrity of the stored data. Three precautionary methods include: 1) backing up critical data files (i.e. duplicating files onto other storage devices), 2) periodically analyzing important files to ensure that they have not been corrupted, and 3) logging system transactions to output scratch files.

The frequency with which files are backed up depends on 1) file update activity, 2) whether the files can be reconstructed from other known data, and 3) relative importance of the data. In addition to backing up PACS application programs and data, operating systems should also be backup up since PACS hardware components are often operating system version dependent. The backup device may be an off-line media (magnetic tape), on-line device (mirrored disk), or a mirrored database server computer [6].

4. Hardware Reliability

Hardware reliability can be increased by: 1) implementing necessary environmental safeguards, 2) performing periodic preventative maintenance, 3) configuring the system with redundant hardware, 4) having available spare parts, and 5) using alternative protocols during times of hardware failure.

4.1 Environmental Factors

Computer systems which manage patient information require "clean" and uninterruptable power. Electrical power failures and surges can be a major source of system downtime if proper precautions are not taken. Installation of power conditioners, uninterruptable power supplies (backup battery systems), and emergency hospital power should be considered. Other environmental factors to consider include: 1) air conditioning systems with room thermostat sensors which automatically set off a central alarm and trigger a graceful power shutdown, 2) earthquake-proof stablizing structures, and 3) de-oxygenation (halon) systems in case of fire (instead of water sprinkler systems).

4.2 Scheduled Preventative Maintenance

Preventative maintenance of acquisition machines (CT, MR, CR, US, DSA, etc.) and PACS computers, communication equipment, etc. should be periodically performed during off-peak clinical hours. All critical PACS machines should be under service contracts with local companies. Mean time to repair should be less than one day.

4.3 Redundancy and Reconfiguration

Fault tolerant systems using hardware redundancy are safer than non-redundant systems[5]. Hardware fault tolerance also substantially increases the total cost of a system since it adds greatly to a system's complexity. Hardware fault tolerance is warranted when the consequences of system failure are high enough to justify the cost. Some common PACS hardware faults and their recovery solutions are described.

PACS capture computer down: Radiological imaging systems (eg. GE 9800 CT system) [5] that are connected to the PACS via a network, can re-routed studies to a designated backup capture computer. If the radiological imaging system interfaces to the PACS capture computer via a direct connection (eg. PCR-901 Computed Radiography system)[2] then it may be possible to do the following: 1) If the capture computer can initiate a study transfer from the imaging system to the PACS acquisition computer, then images can be acquired (although delayed) after the capture computer is brought back up; 2) If the radiological imaging system can initiate a study transfer, then a technologist can again send the data to the capture computer once it is up; 3) If the data transfer cannot be re-initiated by either machine (i.e. acquisition is a real time process / one time event only), then an alternative image acquisition method must be considered (eg. tape transfer, laser scan digital hardcopy film image, etc.).

Archive Down: If an optical archive is down, images may be sent to an alternative archive node if available. The storage media for archive machines should be interchangeable. If a second archive is not available, images may be sent to a bank of magnetic disks for temporary storage until the archive is repaired. This however may result in loss of images if a disk crash should occur.

Database Server Down: Database server machines are critical for PACS operations (case/image identification, routing information, etc.). A "hot backup" machine may be employed. This backup mirrors all system database transactions onto a separate host. A simple change of an environment variable or configuration table allows all database client machines to switch to the backup database server if necessary.

Communication Network Down: Multiple network connections to each PACS node is desirable. For example the UCLA PACS system employs both high speed networks (FDDI and Ultranet) as well as a global Ethernet network for serveral machines[7]. Failure of a particular network triggers an attempt to send data over a different network circuit.

Disk Crash: One should always be prepared to recover from a disk crash. Versatile software should allow the system to continue operating from another disk. Spare scratch image storage disks should be available and made easy to integrate into software routines. Important database disks should be mirrored.

Processing and Interface Boards: Spare boards should be available in house or available within a few hours from a service maintenance company.

Workstation Monitors: Spare CRT should always be available. Viewing monitors will frequently burn out especially if screen saver software is unavailable.

Acknowledgements

This work was supported in part by Public Health Service Grant No. R01 CA 39063, R01 CA 40456, and P01 CA 51198, awarded by the National Cancer Institute, Department of Health and Human Services, and by the UCLA Department of Radiological Sciences Research Fund.

References

1. R.L. Enfield, "The Limits of Software Reliability," *Technology Review*, : 36-43, April 1987.
2. B. Ho, C. Morioka, L. Kaabi, R.K. Taira, A. Hayrapetian, and H.K. Huang, "Acquisition Interface for Computed Radiography," *Proceedings NATO ASI - Picture Archiving and Communication Systems in Medicine*, Evian, 1990.
3. J. Martin, *System Design from Provably Correct Constructs*, Prentice Hall, Inc., Englewood Cliffs, New Jersey, 1985.
4. O. Serlin, "Fault Tolerant Systems in Commercial Applications," *Computer*, **17**(8): 19-30, 1984.
5. D.P. Siewiorek, "Architecture of Fault-Tolerant Computers," *Computer*, **17**(8): 9-19, 1984.
6. B.K. Stewart, and R.K. Taira, "Database Architecture and Design for PACS", *Proceedings NATO ASI - Picture Archiving and Communication Systems in Medicine*, Evian, 1990.
7. B.K. Stewart, "Three Tiered Network Architecture for PACS Clusters", *Proceedings NATO ASI - Picture Archiving and Communication Systems in Medicine*, Evian, 1990.

HIS and RIS and PACS

Albert R. Bakker

BAZIS, Schipholweg 97, 2316XA Leiden, The Netherlands

Abstract: In this paper first the historical background is presented of the concepts HIS, RIS and PACS. Next the present situation of the realization of the concepts is considered; the conceptual relation between the three is discussed. Accepting the existence of the three concepts first the relation between HIS and RIS is analysed, next the relation between PACS and HIS/RIS is considered. It is concluded that such a relation is vital for the realization of a full-scale PACS. Moreover it is concluded that not only data has to be exchanged but also information. The need for a generic description of an interface between the two is emphasized.

Keywords: integration, HIS, RIS, PACS/HIS relation, prefetching

INTRODUCTION

Each of the three concepts HIS,RIS and PACS deals with medical data be it partly of different nature, for the further fruitful introduction of information technology in the hospital it is worthwhile to consider the mutual relation of these concepts. Each of the concepts has its own history. Because of the evolution of their realizations the risk for overlap and miscommunication increases. This paper will discuss the relations both on the conceptual and on the implementation level, not to give a final solution but to stimulate efforts to come to a widely accepted view on this relation that can be used as a guideline in further developments.

THE CONCEPTS HIS, RIS AND PACS; THEIR HISTORY

The concept HIS and its history

The concept integrated Hospital Information System emerged at the end of the sixties. It was a reaction to the isolated computer applications developed for the hospital, that did not yield the expected benefits. By considering both data and functions in an integrated way it was expected that the benefits would be easier to achieve.

According to Collen "the goal of a hospital information system is to use computers and communications equipment to collect, store, process, retrieve and communicate patient care and administrative information for all hospital affiliated activities, and satisfy the functional requirements of all authorized users" [1].

NATO ASI Series, Vol. F 74
Picture Archiving and Communication Systems (PACS) in Medicine. Edited by H. K. Huang et al.

As characteristics are mentioner at another place [2]:
- coherent storage of data in a central database
- access to data, for authorized users, from the work location, whenever needed, presentation of data as wanted by the user
- the system takes care of some coordinating tasks.

Note that in these definitions of an HIS no restriction is made to alpha-numeric data.

The realization and introduction of HIS-ses has been confronted with many problems. Afterwards we can state that in the early days of HIS the technology was not yet sufficiently developed to allow for a rather complete HIS. The users of first attempts to realize an HIS were confronted with serious limitations these were related to hardware, software and organization. Hardware problems lead to long response times, insufficient storage capacity, too many break-downs of the system. Software problems were related to poor user interface, rigid program structures (difficult to adjust and to maintain), limitations in DBMS. It also came to light that the organizational consequences of such a system were insufficiently considered like: access rights, responsibilities, back-up procedures.

The workload of a full-scale HIS was seriously underestimated.

In initial implementations of the concept, at least in the USA, the emphasis on finance and administration; the medical and nursing functions being underdeveloped. In Europe integration got more attention from the start on and more attention was given to HIS-functions for the medical support departments (clinical chemistry, pharmacy, radiology etc.).

Nowadays in the western world the majority of the large and medium-size hospitals make use of some sort of HIS. The functionality of these HIS-ses is still far from the ideal. Even after more than 20 years of HIS developments much is still unclear, the confusion of tongues is a major hurdle for fruitful cooperation in HIS development. Within the scope of the International Medical Informatics Association (IMIA) its working group 10 is dealing with HIS; in May 1988 a working conference was held with as topic: "Towards New Hospital Information Systems" [3]. In this conference recommendations were made on further HIS development; some of these are relevant for PACS and RIS developments as well.

The concept RIS and its history

With the appearance of computers in hospitals it was recognized that they might be a support for information handling in the radiology department as well. Perhaps because of the limited support offered by HIS for these departments the concept of a Radiology Information System emerged. There is no principal difference with the concept of HIS, only the scope is limited to a specific department. Especially in the USA the concept RIS became popular, the RISC consortium made a major contribution to translating the concept in global specifications. Where most of the early HIS projects were set-up for mainframe computers, the RIS projects were from the start on using mini computer equipment. RIS in practise is restricted to alpha-numeric data. Typical functions cover: patient registration, appointment scheduling, reporting, patient tracking, film tracking, diagnosis coding, teaching and management info.

At present most of the implementations of an RIS are systems on their own with some link to the HIS of the hospital. Besides that some HIS-ses offer embedded RIS subsystems with a high degree of integration with other subsystems.

The concept PACS and its history

The concept PACS (Picture Archiving and Communication System) emerged around 1980 and could be described as [4]:
- storage of digitized images being produced by different imaging modalities in a central image database (archive)
- images generated by different modalities accessible for the users (of different disciplines) by means of workstations that also offer image processing facilities
- fast transmission of images both from the acquisition devices to the image database and from the image database to the workstation.

The idea to realize a PACS was triggered by two developments: on one hand the introduction of digital modalities that could directly supply images that could be handled by information technology, on the other hand developments of information technology (storage facilities, networks, high resolution displays and software).

Like in the early days of HIS, several groups started to work on realization of PACS without a thorough analysis of the capacities and facilities needed. This lead to launching of immature solutions that caused dissatisfaction amongst users. The development of PACS is clearly in an earlier stage than that of His or RIS.

AN OVERVIEW OF THE FIELD

A quick inventory of the field yields the following: It is well accepted now that a modern hospital should have an HIS. There is still discussion on scope and structure of an HIS leading to a situation that a hospital has to choose for either dependency on one vendor in case it wants a reasonably integrated solution or choose for a set of rather loosely coupled subsystems. Standardization is barely needed and is getting considerable attention now.

An open HIS architecture is still far away; perhaps the RICHE project within the scope of the EC ESPRIT program might lead to such a structure.

The situation as to HIS vendors is different in the different countries: in the USA the main vendors are dedicated systems houses (with an exception for IBM with its PCS system); in Europe we see besides such dedicated systems houses also cooperative structures of hospitals; in Japan the main computer industries are the main suppliers of HIS. Furthermore there are many small companies offering departmental systems.

To give a feeling for the volume of a reasonably developed HIS here some figures on Leiden University Hospital, The Netherlands. It is a 875 bed hospital with over 300,000 outpatient visits a year. In the databank data on over 800,000 patients are stored, 1040 terminals are operational, the number of messages (strings of characters delimited by a CR) amounts to over 800,000 a day. The volume of HIS use has been growing at a rate of about 20% per year, it is expected that this growth will continue for at least the next 5 years [5].

At the IMIA working conference 'Towards New Hospital Information Systems'[3] it was concluded that HIS is not restricted to alpha-numeric data but is by nature multi-modality, images are 'just' a type of data, so PACS is a subset of HIS.

There are several vendors at present offering RIS departmental systems with a wide variety of functionality. The integration with the related HIS being often restricted to retrieval of patient identification data from the HIS and reporting of examinations carried out. These departmental systems are offered by both vendors of imaging modalities and independent systems/software houses.

Besides these departmental RIS solutions several HIS vendors offer RIS functionality as embedded system within their HIS. PACS developments are clearly in an earlier state of evolution than HIS and RIS; systems are being developed by both university groups and vendors of imaging modalities. Till now PACS implementations have a limited scope (part of the hospital, or special types of images, or radiology department only).

The management of the huge amount of image data that will have to be stored in a full-scale PACS is still a bottleneck as are the response times to be expected. Most studies on PACS cost conclude that PACS is still significantly more expensive than the film based system. Dataprotection issues (carefully considered for HIS) till now hardly got attention; especially the availability issue and the software quality are urgent.

Gradually the need for close cooperation between PACS on one side and RIS/HIS on the other is recognized; this became quite clear during the 8th Europacs Conference May 1990 in Trieste.

For the time being it seems the best policy to accept that PACS on one side and RIS/HIS on the other side are different worlds; coexistence seems advisable, no iron curtain !

On the relation between HIS and RIS

At first sight the RIS functions seem to a large extent separated from the other HIS functions. However the need for integration will become clear however from the following considerations:

- users in the clinic will need access to all available medical data on their patients; for them it is irrelevant whether the data are collected within different systems, so there is a clear need to make also data on radiology exams available (e.g. reports) through the HIS terminals. This may be realized by storing the results in the HIS database (together with the other patient data) or by giving the HIS access to the RIS database.
- the RIS will need patient demographic data, since these data can be expected to be available within the HIS they should be retrieved there. An own patient demographic file in the RIS leads to the risk of inconsistencies
- the radiologist may be interested in additional medical data on the patient; these data should be retrievable from the HIS at his RIS terminal
- the RIS should be able to accept requests for radiologic examinations entered at His terminals in the wards
- data on film tracking are not only of interest for the radiology department; also the clinical user (at his HIS terminal) may be interested, so another need for integration
- patients often have to make an appointment for more than one examinations; e.g. an ECG and a radiologic exam. For him it is important to have these well coordinated. Separate appointment scheduling programs will make this very difficult
- for clinical research it should be possible to interrogate the 'medical database' for questions combining different types of data including radiology data.
- patients want to get an integrated invoice if any
- for management of the institution data on production in relation to diagnosis or DRG should be available across the border of RIS/HIS.

The need for integration between RIS and HIS can be expected to increase with the further evolution ofRIS/HIS use.

On the relation of PACS and RIS/HIS

There are two reasons to consider the relation. First the user, be it the clinician, the radiologist or the hospital administrator, should not be hindered by the separation of the systems. Second to realize an acceptable performance using the presently available technology the PACS image management software will need RIS/HIS information.

The clinical user will need access to both alpha-numeric data from RIS/HIS and image data; it is unacceptable for him to be forced to use different workstations; so it should be possible to view RIS/HIS data on clinical PACS workstations. The radiologist may need additional medical data of the patient in the diagnostic process; from his workstation he should have access to these data together with the images. In the PACS database huge volumes of data will have to be stored (at least several Terabytes). With present technology this can only be realized with a layered set-up of the databank where the lowest layer is the archive (a huge storage capacity but relatively slow, e.g. a jukebox of optical disks), the next layer is an intermediate buffer layer may holding images with a high probability to be retrieved (medium capacity; medium speed; e.g. a set of large Winchester disks) and the highest level are fast buffers in the workstations, containing the images to be almost directly needed.

For such a storage structure to function satisfactorily it is necessary to be able to judge the probability that images will be retrieved and bring those to the intermediate buffer or the buffer in the workstation before they are requested.

In designing an algorithm for prefetching of images the following characteristics will be important: the hit rate, indicating the percentage of images found in the fast memory on one hand and on the other hand the utilization rate indicating the percentage of prefetched images that are really used. The algorithm will have to asses the probability an image will be requested. This probability is influenced by the following patient related parameters: patient status (admitted, waiting for admission), current appointments (both for outpatient clinics and for radiology), diagnoses and other medical data, treating specialism. Also image related parameters will have an effect: type of exam, report, previous use of the image. The algorithm will also have to take into account the current status of the image stores (free capacity, active volume, traffic load).

It is clear that data wil be needed from HIS (the medical data), the RIS (exam data) and PACS (the status of the image database), so communication between the three systems should be possible. When different systems have to communicate the problem of differences in data definition has to be dealt with. When the RIS is an embedded subsystem of the HIS communication between the two will be rather simple. In the communication with the PACS the problem of data definition will come up anyhow.

Giving the PACS user access to patient data from the HIS may be realized by emulating an HIS terminal and supplying the data in a transparant way; so supply data. When the prefetching algorithm uses RIS/HIS data these should be interpretable, so information will have to be supplied. It might be advisable to separate the algorithm in two parts: the clinical propability that the image will be requested (to be calculated from RIS/HIS data) and a technical part to be calculated from PACS data. When this could be achieved the interface between PACS and RIS/HIS might become rather simple, the clinical probability to be calculated within RIS/HIS and transmitted to the PACS.

References

[1] Collen M.F., HIS Concepts, Goals and Objectives. In Bakker A.R. et al (eds), Towards New Hospital Information Systems, North Holland, Amsterdam New York Oxford Tokyo (1988), pp.3-9.

[2] Bakker A.R., Kouwenberg J.M.L., Ottes, F.P., HIS and PACS Integration Aspects. In: MEDINFO 89 Part I, North Holland, Amsterdam 1989, pp.377-381.

[3] Bakker A.R., The Future of HIS; conclusions and recommendations of an IMIA Working Conference. PROCS eight annual Joint National Congress, The Practice of Medical Informatics. Hammond Ed.W., Washington Ph.D., May 1989, American Association for Medical Systems and Informatics (AAMSI), San Fransisco 1989, Vol.7 pp.230-234.

[4] Lodwick, G.S. (et al), Pictorial Information Systems and Radiology Improving the Quality of Communications. In Hoehne K.K. (ed.), Pictorial Information Systems in Medicine, Springer Verlag, Berlin Heidelberg New York Tokyo (1984), pp.1-29.

[5] Leguit, F.A., The Case: Leiden University Information System. In: PROCS of the IFIP IMIA Working Conference, Towards New Hospital Information Systems, Nijmegen The Netherlands, Bakker, A.R., Ball, M.J., Scherrer, J.R., Willems, J.L. (eds.), Elsevier Science Publishers B.V. (North Holland) Amsterdam, pp. 321-326, ISBN: 0-444-70502-3.

Clinical Experience with a Multimedia Workstation at the Ottawa Civic Hospital, Canada

M. Goldberg*, G. Bélanger**, J. Tombaugh****, R. Dillon****, J. Robertson*, M. Coristine**** , J. Mastronardi***, C. Beeton***

*University of Ottawa Medical Communications Research Centre
Department of Electrical Engineering
Ottawa,Ontario, Canada K1N 6N5

**Ottawa Civic Hospital
Department of Radiological Sciences
Ottawa, Ontario, Canada K1Y 4E9

***Bell-Northern Research
P.O. Box 3511, Ottawa, Ontario, Canada K1Y 4H7

****Carleton University
Department of Psychology
Ottawa, Ontario, Canada K1S 5B6

Abstract: The design and implementation of a multimedia workstation capable of handling images, voice, graphics and other computer data for radiological applications is described. This workstation is part of a multimedia communications system that was implemented by a multi-disciplinary, multi-institutional team of researchers. The system was installed in a large teaching hospital affiliated with the University of Ottawa, Canada and underwent extensive clinical evaluations. In the paper we describe the design and evaluation process for the user interface to the workstation.

Keywords: Medical image processing, user interface, human-machine interaction, PACS.

Introduction

At the University of Ottawa Medical Communications Research Centre we are looking at the PACS opportunity from the perspective of improving the *communication* of Radiological information (images and diagnostic reports) between the radiologist and the client physicians. To that end, we have developed an experimental multimedia communication system (called *IRIS*) which supports the capture and distribution of digitized X-Ray images and voice reports as integrated "electronic" patient folders [5]. It also supports on-line consultation between the radiologist and attending physician by providing a multimedia "shared space" through synchronized workstation operation. In order to assess the clinical value and acceptance of these service concepts, the system was deployed in the Ottawa Civic Hospital to provide a real-time Emergency/Radiology link.

The multimedia radiological communication system is the result of a large multi-disciplinary, multi-institution research and development effort. The Industrial partner, Bell-Northern Research was responsible for the technological issues and contributed hardware as well as systems engineering expertise to the joint project. The University of Ottawa Department of Electrical Engineering undertook much of the early research and was responsible for its implementation in the system. The Carleton University Department of Psychology managed the clinical trial evaluation and helped

NATO ASI Series, Vol. F 74
Picture Archiving and Communication Systems (PACS) in Medicine. Edited by H. K. Huang et al.

design the user interface. Finally, the Departments of Radiology and Emergency of the Ottawa Civic Hospital provided the initial system specifications and participated in the preliminary trial procedures and the extensive clinical field trial of the multimedia radiology communication system. This work was funded by the Telecommunications Research Institute of Ontario (TRIO), Bell Canada and Bell-Northern Research (BNR).

We will very briefly describe the system and its functionality in the next section. In section three we will describe the methodology used for designing the user interface for the multimedia workstation. Finally , we present some results on user evaluations.

Description of IRIS

IRIS is a multimedia system capable of handling patient files consisting of radiological images, documents, and voice messages. The system can be used in one of three modes:

1. The radiologist can read patient files and dictate a verbal report which then becomes part of the patient file.
2. The clinician can review a patient's radiological images and listen to the radiologist's report.
3. The clinician may leave a verbal query for the radiologist.
4. The clinician can have a conference (teleconsultation) with the radiologist through a shared visual and auditory workspace created on two workstations.

Operations are initiated by the use of a mouse, which controls a cursor on the two monitors in the workstation: a control screen and a high resolution display. To begin a session a physician uses the mouse to point to and click on a button on the control screen. The system then displays the names of all current patients. By pointing and clicking on a patient name, the corresponding file is opened and a list of documents, images and reports contained in the file is displayed on the control monitor. The mouse can then be used to select the contents of the file: documents and images are displayed on a high resolution monitor and voice reports are played back. The physician can move the cursor to the high resolution monitor and perform various manipulations - contrast changes, zoom, reverse video, measurements, and reorientation - by pointing to the corresponding icon (a symbol that represents an action). In addition the physician can create verbal messages and request a consultation. Patient data, images, demographic information and documents are entered on a separate data entry workstation.

The *IRIS* system is an integrated voice/data network system consisting of four nodes - three clinical workstations (*Image Entry, Radiologist,* and *Attending Physician*) and one server station. Each clinical workstation is built around a Compaq 386/20 microcomputer containing an EGA compatible display controller and monitor which serves as the workstation "Control Screen" . The "Image Screen" is a portrait-oriented, high resolution monochrome monitor (Philips) driven by an Imagraph

IP 1210/10 display controller. It provides an 8-bit pixel image of resolution 1280 (vertical) by 1024 (horizontal) with a 2-bit graphics overlay plane. Due to limited network bandwidth, the size of image files, and the requirement for rapid display, a copy of all images is normally available in the local image store at each workstation . Each workstation is equipped with a handsfree telephone (Northern Telecom Meridian 2317 digital set). The voice terminal provides a link to the Voice Mail system for dictation and playback of voice reports. It also supports voice calls between workstations for consultation sessions.

In addition to the above equipment, the Image Entry workstation is also equipped with two devices for digitizing films and documents:

- a Konica laser film scanner is capable of digitizing all standard film sizes up to 14" by 17". The maximum image resolution is approximately 2000 by 2400 for a 14" by 17" film. All films are digitized to 1024 grey levels (10-bit pixel resolution).

- a Hewlett Packard Scan Jet document scanner (not shown in figure 1) provides the capability of digitizing document pages up to 8.5" by 11" at a resolution of 100 dots per inch. This is used for capturing Examination Requisition forms and other pertinent patient documents.

The data network is the Hospital's SYTEK 6000 broadband data LAN. The IRIS system uses a dedicated frequency channel which provides a nominal channel bandwidth of 2 megabits per second. The voice network consists of a dedicated mini PBX, the Northern Telecom SL-1ST, which links the voice terminals at the workstations to each other or to the Voice Mail system (part of the server station). Connections between the terminals and the PBX are made over the Hospital's existing twisted pair wiring.

The server station consists of a PC/AT-386 compatible microcomputer, which is a network file server for all system control and status files as well as patient database files and a Northern Telecom Meridian Mail /SP Voice Mail system, which is capable of storing up to 24 hours of digitized voice messages.

User Interface Design

A major goal of the IRIS project is to provide a system which is suited to the medical professionals who will be using it. We used a five-stage iterative design process which involved radiologists and attending physicians in all the phases [1,6].

1. Goal Setting. What are the goals for user performance? To set these goals we had to quantify criteria, such as system flexibility and user satisfaction, that are often considered by designers to be

soft or nonmeasurable. Three major goals were identified. First, the system must be easy to learn; physicians should be able to use all features with less that 1 hour of training, and on-screen information must be presented and controlled in a manner that is consistent with the ways in which physicians think about and organize their work. Second, the system must be powerful and quick; physicians must not be frustrated by cumbersome techniques of command selection, and they must be able to produce as many reports with the same diagnostic accuracy by using the system as they can with the present film and paper methods. Third, the system must be satisfactory to users, or they will not use it. We measured satisfaction through observation, rating scales and comments.

2. Identifying issues. What aspects of the design require special consideration?. How will these aspects be tested in practice? What tasks would the system have to perform? How can supporting materials be made as realistic as possible? During development we discussed potential problems in implementing the system and identifying areas of concern. As the clinical trial progressed we spent several weeks observing radiologists and clinicians working. We then discussed our findings with them to verify our conclusions.

3. Interface prototypes. How should features of the design be presented to the physicians who will be using them? We used simplified representations of design features (sometimes by using other systems and prototyping software and sometimes simply with pen and paper) to show clinicians how these aspects of the design would work.

4. Usability testing. How well does each prototypical feature meet the needs of physicians? How easily can they use the system, and how satisfactory do they find it to be? Assessing the prototype required creating structured testing procedures and detailed question lists. We recorded performance measures, comments, and suggestions on prototype systems.

5. Evaluation. Should a particular feature be changed or accepted? If it needs improvement, how should it be altered? If a particular improvement is technically impossible, what compromises can be made? Each change must be incorporated into a new prototype and the testing procedure repeated until the users are satisfied and the goals are met.

Emergency/Radiology Trial

A clinical trial of IRIS was conducted at the Ottawa Civic Hospital with the Emergency Department and the Department of Radiological Sciences between April 4 and May 12, 1989. During the trial, 319 active Emergency Department cases (905 films) were processed using IRIS. Radiologists read each case from the digitized images and formed a diagnostic impression. However, before reporting the case, the x-ray films were also examined. A report was then dictated using IRIS, so that the images and report were immediately available at the emergency physician's workstation. Observers stationed at the clinician and radiology workstations recorded detailed information on how the images were read and provided assistance to the doctors and attending physicians as requested.

After the clinical trial, 9 of the 12 radiologists participated in an extensive debriefing interview in which they were asked about the applicability of the IRIS system as a clinical tool, and about their satisfaction and assessment of the image enhancement features of the system. In the interview, a number of rating scales designed to address specific areas of concern to the design team were presented to the nine radiologists. Each item was rated on a scale from one to five. A score of one indicates a very high degree of dissatisfaction and a score of five represents a very high degree of satisfaction. To aid in decisions about modifications to the system, specific usability goals for satisfactory ratings were established. It was decided that a feature would be regarded as acceptable when the mean rating was 4.0 or greater. At the other extreme, a feature with a mean rating of 3.0 or lower would be considered for redesign. Average ratings falling between 3.0 and 4.0 would be considered marginal and other considerations such as comments made in the interviews would be used to decide whether design changes were required [2,3].

The main findings are as follows:

1. Radiologists would be willing to make diagnosis using IRIS for most cases involving bones but were hesitant, in general, for cases involving soft tissues.
2. Resolution does not seem to be a problem, but improvements to contrast enhancements are needed.
3. There is a definite need for more that one image monitor and the ability to quickly switch between a number of images.
4. The availability of the voice report with the image is an important asset, although there are problems with the user interface of the report generation mechanism.
5. The consultation facility was rated as being very important, although some improvements to its implementation are needed.

Radiologists examined the digital images on the image screen to formulate a diagnosis, then before dictating a report, they examined the analog films. In 30 of 319 cases reported during the clinical trial of the IRIS system some discrepancy was noted by the observer [4]. The discrepancies were one of three forms: (i) a diagnosis or finding evident on the film but not on the image, (ii) a query on the diagnosis reported by the radiologist, and (iii) increased difficulty in reaching a diagnosis or increased doubt about the presence/absence/or significance of a finding using the digital modality as compared to using film media. These 30 cases were selected for a follow-up study and further analysis.

Three physicians with experience in using IRIS were asked to reread the cases. Ten of these discrepant cases presented no difficulty to any of the readers. We can safely assume that the earlier reported difficulties probably arose because of lack of training and usage of IRIS. The other twenty cases persisted and points to the need for further improvements to the image enhancement algorithms and further studies to measure the limitations of the system. The most troublesome cases involved the diagnosis of osteoporosis and the identification of calcifications.

Summary

Overall IRIS provided a successful radiological communication service during the clinical trial. Although further development is needed, physicians were satisfied with using images in the majority of cases. The user interface was easy to learn and use. Physicians enjoyed the image enhancement facilities. They considered IRIS a viable tool for hospitals and especially hospitals with associated clinics or where there are networks of small hospitals.

References

1. Bélanger, G. *et al.* : Evaluation of a Workstation by Clinicians. SPIE Medical Imaging III Conf., Vol. 1093, pp. 214-219, California, Feb. 1989.
2. Coristine, M. *et al.* : The Use of IRIS Image Enhancement Facilities on Digital Images by Physicians During a Clinical Trial at the Ottawa Civic Hospital. Presented at SPIE Medical Imaging '90 Conference, Newport Beach, February 1990.
3. Coristine, M. *et al.* : Physician acceptance of the IRIS user interface during a clinical trial at the Ottawa Civic Hospital. Presented at SPIE Medical Imaging '90 Conference, Newport Beach, February 1990.
4. Goldberg, M. *et al.* : Analysis of Discrepancies Observed Between Digital Images and Film During a Clinical Trial of IRIS. Presented at SPIE Medical Imaging '90 Conference, Newport Beach, February 1990.
5. Mastronardi, J. *et al.* : Design of a Multimedia Communication System for Clinical Trials at the Ottawa Civic Hospital. SPIE Medical Imaging III Conf., Vol. 1093, pp. 230-243, California, Feb. 1989.
6. Tombaugh, J. *et al.* : Goal setting and user testing to ensure a PACS interface satisfactory to radiologists. SPIE Medical Imaging III Conf., Vol. 1093, pp. 345-351, California, Feb. 1989.

PACS Clinical Experience at Georgetown University

Steven C. Horii, Seong Ki Mun, Betty Levine, Ben Lo, Brian S. Garra, Robert K. Zeman, Matthew Freedman, Clifton Leftridge, Dieter Schellinger, John Keyes, Larry P. Elliott, Robert Fielding, Harold Benson

Department of Radiology, Georgetown University Hospital,Washington, D.C. 20007, U.S.A.

ABSTRACT Georgetown University Hospital has been operating an image management and communications system (IMACS or PACS) for three-and-a-half years. This work was initially funded under the Army Medical Research and Development Command Digital Imaging Network Systems (DINS) project. The system was taken from a research system supporting only radiology tasks to one extended to clinical use, and has been used in clinical work for two-and-a-half years.

This paper will summarize our PACS clinical experience and will describe the operational features implemented and those still necessary.

KEYWORDS: PACS, radiology, clinical experience

INTRODUCTION

During the past three and one-half years, we have taken the Georgetown PACS (A T & T /Philips CommView) from a large scale prototype to an operational, clinically useful system. There is still a large amount of work remaining before the system can be said to have completely replaced conventional methods of operating the imaging side of a radiology department. We have learned a great deal from these years, and will be applying what we know to the further growth of the system.

Most of the work which needs to be accomplished in order to take the system further is not in adding hardware. The development of more tailored software is far more important. The technical aspects of this project have been presented in three major sources [1,3,8] and so will not be reviewed here.

PACS in Abdominal Imaging

The Abdominal Imaging Division of the Georgetown Department of Radiology is responsible for performing and interpreting gastrointestinal (GI), genitourinary (GU), biliary tract, and certain interventional procedures. In addition, the Division is responsible for most thoracic CT

NATO ASI Series, Vol. F 74
Picture Archiving and Communication Systems (PACS) in Medicine. Edited by H. K. Huang et al.

studies. The imaging techniques used by the Division include plain radiography (without and with contrast agents), analog and digital fluoroscopy, CT, MRI, and ultrasound. An ultrasonic biliary lithotriptor was added to the interventional armamentarium in 1989. Of the equipment in the Department used by Abdominal Imaging, both CT machines, the MRI system, four ultrasound units, and one fluoroscopic system (in the biliary lithotripsy suite) are interfaced to the IMACS. The film digitizers are also available for input of plain films.

The Division was one of the original two (Neuroradiology being the other) to have workstations installed. The early unit was a two-screen display workstation which was replaced by a four-screen enhanced graphics display workstation (EGDW) when it became available. Most recently, the Division workstation was the first to be upgraded to the high-performance EGDW (informally known as the "turbo" workstation).

The workstation in the Division reading room is one of the most heavily used. The major change from occasional use to routine use was the result of the changes provided in release 3 of the IMACS software. Most important was the ability, at the workstation, to restore cases to the database from the optical archive. This automatic restore proved to be a very attractive feature, and resulted in very heavy use of the system by the radiology residents and fellows. They discovered very quickly that retrieving cases on the workstation was far faster than trying to get the films from the file room. A typical reduction in time was fifteen minutes from the file room (if the films could be found) to five minutes at the workstation. The addition of the high-speed display and large local storage afforded by the "turbo" upgrade increased use even further since these features reduced the delay in displaying images and eliminated the problem of a full local disk.

The increased capacity and speed, coupled with connection of the fourth ultrasound machine to the IMACS, led the Ultrasound Section to begin primary reading at the workstation. The local disk capacity is now large enough to store more than a full day's ultrasound cases. The worklist feature also simplifies the retrieval of cases, and to some extent, automatically routes ultrasound examinations to the workstation. This eliminates the task of having to load the workstation manually, much as one has to load a film alternator. The Ultrasound Section is still producing film, but these are being used as a check on display performance. Thus far, the pathology seen was displayed equally well on the workstation and on film. Most recently, routine use of the workstation for primary interpretation has been suspended while problems with system reliability are solved.

The CT section is using the workstation primarily to retrieve prior cases for comparison, and occasionally for review with referring physicians. Primary reading is inhibited at this time by the video frame grabbed nature of the images. Though the resolution is seen to be adequate, the lack of the full digital data prevents window and level adjustments through the entire range of tissue attenuation values.

The abdominal MRI section uses the workstation in a similar fashion to CT, but has been operating a digital interface for approximately nine months. This has reduced technologist workload by changing the acquisition process to entry of the patient and exam identification followed by a file copy command. It is no longer necessary to acquire images individually.

Evaluation Processes

One of the first evaluation studies to be performed on the system was done prior to installation of the EGDW 4 in ultrasound. Seventy-eight cases from routine ultrasound studies which had been acquired on the IMACS as well as on film were read in a double-blinded fashion by Drs. Garra and Horii of the Ultrasound Section. They took turns reading from the workstation and from the EGDW 4 in Neuroradiology (the first such workstation installed). They then compared their readings, and found no differences in diagnoses or findings[2]. Timings done of their reading sessions from film and at the workstation also showed no significant differences in total times. Image presentation time was slightly longer for the workstation reading, but this was offset to some degree by a somewhat longer time to load the film alternator (when the case was not already loaded). A current study is repeating more refined timings for the "Turbo" EGDW *versus* the film-based reading times. The preliminary results show that reading from the workstation takes approximately 30 seconds longer per case. This is thought primarily to be the result of the time it takes the workstation to display the entire ultrasound examination, whereas the entire study is usually available at the alternator without having to move to another alternator panel.

A study of the suitability of digitized film and digital fluoroscopy for endoscopic retrograde cholangiopancreatography (ERCP) was performed by Dr. Zeman (Director of the Abdominal Radiology Division), and the results reported at the 1989 Radiological Society of North America (RSNA) Annual Meeting[6]. Fifty proven cases were randomly taken from the files, and the spot films and photospots (those images photographed from the image intensifier output) were digitized on the DuPont LFD. The images were stored in routine fashion on optical and magnetic media. Dr. Zeman then interpreted these studies at the workstation without knowledge of the original reading. He judged both diagnostic quality as well as image preference (film or CRT image). The results showed no significant differences between the original films and their CRT-displayed and interpreted counterparts in either diagnostic quality or user preference. As a result of this study, an ongoing prospective study has been undertaken to examine digital fluoroscopic images of the gastrointestinal (GI) tract viewed on the workstation compared to photospot films of the same areas.

PACS in Neuroradiology

The Neuroradiology Division was the first of the sections on IMACS to upgrade to the EGD workstation. Early evaluation for A T & T was performed on this workstation, and the ultrasound reading study (see above) was done using the EGDW. Most of the use of the workstation initially was to review occasional MRI or CT studies. The EGDW upgrade made it possible to view up to 80 images simultaneously, and coupled with the teleradiology link to Montgomery Imaging Center (see the section on teleradiology) provided the first use for primary interpretation. The neuroradiologists have been the longest continuous users of the system, and are still using the workstation despite termination of our MIC link.

The EGDW has been upgraded further to the EGDW "turbo" configuration. This allows at least a full day's MRI studies and CT head studies to be kept on the system. The workstation in the Neuroradiology reading room serves primarily for MRI review and comparison. The digital interface to MRI, though beneficial to all MRI users, will be of greatest initial value to Neuroradiology because of the large fraction of MRI case volume which is either brain or spine evaluation. Training of new Neuroradiology Fellows and residents has become routine, and use of the workstation is encouraged while they are on the service. This training serves to increase overall usage of the system as the residents rotate to different services and carry their knowledge of workstation use with them.

A summary of the use and acceptance of the system by Neuroradiology was presented at the SPIE Medical Imaging III meeting in 1989 [5]. The neuroradiologists like the system, and do use it. The only drawback they currently see is that many of the newer MRI studies yield more than 80 images (100 to 120), so that it is no longer possible to view whole examinations at once. They have been the only group so far to request that future workstation designs consider six or eight display monitors.

Evaluation Process

Much of the evaluation consisted of monitoring performance of primary reading from the workstation compared to film. The MIC studies, though read on the EGDW, were still filmed, and the films were sent to Georgetown. The neuroradiologists could then review the films to be sure that they had not missed any findings. After some months of this, with the discovery that no findings had been missed on the workstation reading, most of the film reviews were skipped.

In addition, discussions and interviews with the neuroradiologists were held to assess their reactions to the system. They were generally positive, though all thought that speed could be improved. Preloading their workstation with current images reduced much of the complaint about slow image presentation. There were also some complaints about reliability of the system, particularly regarding the MIC teleradiology link. However, much of this reliability deficiency was the result of problems (for example, power outages and non-IMACS vendor errors) outside of the IMACS. The perception, though, was that of a system which did not work as reliably as might be wished. After the problems with MIC were solved, and the hardware there upgraded, reliability of the link became very good, and was no longer complained about.

For the Neuroradiology Division, the IMACS proved an acceptable clinical tool. In both the teleradiology application and the local reading use, the system proved to be time saving. In the former, this was the result of viewing the images on the same day they were obtained, in some instances within an hour of their being acquired if decisions about the completeness of a study were needed. In the latter, the value has been in retrieving comparison cases. This is also primarily a time saving feature since the neuroradiology MRI reading room is two floors above radiology.

The support of MIC by neuroradiology was also important. This is further described in a later section, but was so well accepted by the MIC radiologist that we were requested to keep the equipment at MIC until the very last possible day of our affiliation.t

PACS in General Radiology

The General Radiology Division (consisting of Musculoskeletal and Chest radiology) was the last of the radiology divisions to have an EGDW installed. The bulk of the work done in General utilizes screen/film exposures for image generation. On the basis of department-wide workload, the largest number of plain films are interpreted by the General radiologists. Some musculoskeletal MRI and CT is also read as is most of the emergency room work.

Some informal training of the residents was done immediately when the workstation was installed. Formal training for physicians was done a short while later with follow-up individual instruction as requested. Without computed radiography on-line, the use of the IMACS by the General staff has been low. In anticipation of interfacing the CR system, though, the EGDW has been upgraded to the "turbo" version.

Most of the use of the workstation in the general reading room has been for review of musculoskeletal MRI or CT. It is also used by the pediatric radiologic technologists as a preview station for the digitized PICU or ICN films. This latter function is done at the workstation in the general reading room because of its proximity to the film digitizing area. This use has suggested to us that some form of fast, simple display for confirmation of film digitization would be a useful tool.

For the last six months, we have been evaluating phosphor plate radiographic systems (also called computed radiography or CR systems) as a replacement for digitizing films. We have placed a Fuji AC-1 (to be replaced by an AC-2 when available) and an Agfa ADC into operation, and are currently working on interface methodology for these units. Early results with portable radiography confirms what other CR users have found, that such use is clinically acceptable.

Based on our experience with digitizing films for the intensive care units, we developed an implementation plan for CR. We will begin CR use by performing portable and emergency room studies on CR plates.

Evaluation Process

Despite the fact that no CR is yet available on-line, we moved very early to have the general radiologists give us opinions on image quality. We had a number of review sessions with Dr. Matthew Freedman, the Director of Thoracic Radiology. We displayed digitized film and some CR images (loaded in from tape) and allowed him to manipulate them on the workstation. His position is that the workstation should be useable for primary interpretation if he has access to the full digital data set. If the images are generated at 2 x 2.5 k pixels, a display of 1 x 1.2 k pixels would be satisfactory if the 1 x 1.2 k screen can be used as a movable window into the full data set.

We also supported a study for NIOSH. We digitized 250 films of patients with occupational lung disease. Along with these, we digitized the International Lung Organization (ILO) reference film set, since standard NIOSH procedure is to have these references available for comparison whenever a patient's films are read. Dr. Freedman recently qualified as a NIOSH "B" reader, and will be doing some viewing of the studies on the workstation. He helped us with quality assurance during the digitization process by examining some of the digitized images.

Prior to a demonstration done for the Madigan Army Hospital Radiology and Command Staff, we loaded an EGDW with CR images from Philips CR sites. These consisted of a variety of clinical images. We also digitized some film images of skeletal trauma. Dr. David Curtis from George Washington University and Walter Reed Army Medical Center evaluated these images on the workstation. He was quite pleased with the results, especially the dynamic range of CR for skeletal imaging which can render soft tissue quite well. He emphasized that being able to adjust the window width and level to see soft tissues, and then bone, could direct the radiologist to sites of injury which otherwise might be undetected.

We have been interested in the gray scale resolution necessary for reading from the CRT displays. As a result, we undertook a study of the number of bits of gray scale required for locating catheters and evaluating pulmonary vasculature on adult portable chest films. The results were that six to seven bits of gray scale were needed at a minimum to see catheters well, and five to six bits to evaluate pulmonary vessels. Overall, seven bits were needed as a minimum in order to see all findings. There were no significant differences between the observers in the number of bits required[4]. We note that the study should be considered preliminary since it was done with a small number of images, a small number of readers, and with portable chest films. These films may not have the same display requirements as conventional chest images because they are frequently of a quality lower than that of conventional chest films. In general, this is because the patients are very ill, and cannot stand erect or fully expand their lungs.

In General Radiology, much of what happens over the next six months is tied to interfacing of the CR systems. Though CR is a proven technology for much of radiography, many of the users are reading the hardcopy the system generates. We will be interested in whether we can move to primary reading from the workstation, and have planned our studies accordingly. We will proceed in a fashion much as we have for ultrasound with comparisons of workstation and film interpretation.

PACS in Nuclear Medicine

Nuclear Medicine was an original workstation installation site, though the hardware placed initially was a single-screen display workstation. This was used quite regularly, primarily for review of CT, ultrasound, or MR images of patients who were undergoing nuclear medicine studies. When we went through our major hardware upgrade, the workstation was replaced by a single-screen EGDW.

Since Nuclear Medicine is located in a different building and three floors up from Radiology, it is very time consuming for the nuclear medicine physicians to come to radiology to retrieve or review cases. Since many nuclear medicine studies are functional, correlation with the anatomic imaging provided by other techniques is necessary for proper diagnostic interpretation. The Nuclear Medicine Division is the heaviest intradepartmental user of correlation studies, and use of the workstation reflects these needs.

Evaluation Process

Because Nuclear Medicine is not connected to the IMACS as an image source, we have not carried out studies aimed at interpreting nuclear medical images on the workstation, though the nuclear medicine physicians are conducting such studies with their internal system. We have conducted an ongoing process of interviews with the nuclear medicine physicians, and have involved them in our planning and feedback to the IMACS vendor.

In general, they have been pleased with the workstation, and have noted that it is a time savings compared to retrieving film. There are still some complaints about the user interface for the workstation as they have become accustomed to a graphical user interface on their local system. Prior to the remote optical archive case restoration software, however, they objected to the short time images were kept on-line. Most patients do not have all of their imaging studies done in rapid temporal sequence, rather, the studies are done as part of the process of evaluating disease. This means that several days may elapse between the performance of a CT study, for example, and the decision to do a nuclear medicine SPECT scan. Prior to expansion of the database magnetic disk, this meant that the CT study was unlikely to be on the database when the nuclear study was performed. If someone was not available to load the necessary optical disk, the nuclear medicine physicians had to fall back to the old method of retrieving the film, or use the unpopular route of performing the study without correlation. This latter alternative is so detrimental to good diagnostic procedure that it is thought better to delay reading or performing an examination rather than do so without proper correlation information.

The combination of an expanded database magnetic storage facility and software for automated operation of the optical disk jukebox has eliminated the problems which the nuclear medicine physicians had retrieving studies from the IMACS. There remains the occasional problem with a study not being on the system because it was never acquired. Our quality assurance procedures have reduced this to a very small number of cases, and the use of direct digital interfaces to CT and MRI will reduce this to a nearly negligible number.

We have begun to look at the problem of connecting Nuclear Medicine to the IMACS as an image source. In Abdominal Imaging and General Radiology, there is considerable interest in being able to use the nuclear studies for correlation. We have already successfully connected Sun Microsystems Sun 3 and SPARC series machines to CommView via the Ethernet gateway. The computer system used in Nuclear Medicine has a MicroVAX central processor connected to Ethernet. It is this path that is likely to be tried for the first interfaces.

PACS in the Pediatric Intensive Care Unit

The pediatric intensive care unit (PICU) was among the early non-radiology sites considered for a workstation location. The choice was based on the combination of a relatively small number of beds but with an urgent demand for film availability when portable films are done. The nature of operation of the PICU is one of a very busy service for a small number of patients. For this reason, the physicians who work in the unit are reluctant to leave for any length of time. As for any intensive care unit population, the patients are very ill, and constant attention is a must. This is particularly true of pediatric patients in whom the changes of clinical status are often much more rapid than in adults. The department had considered installing an alternator for films on the floor, but that would require radiology personnel to load and unload it, and a radiologist to review cases there.

We projected that a review workstation (a PC-based results viewing station or RVS) would suffice for their applications, and Ethernet cable was pulled to the PICU. When the workstation was installed, the physicians began using it immediately. However, they found that the single screen made comparison of images difficult, and the RVS performance only just met their needs. As a result, a two-screen EGDW was put in place of the RVS, and this solved these problems. Since we had pulled fiber optic cable at the same time as Ethernet, it was a simple matter to switch communications media for the higher-performance workstation.

PACS in the Intensive Care Nursery

The intensive care nursery (ICN) was the second of our Pediatric Division units to have a workstation installed. Its environment and demands are very similar to those of the PICU.

From the outset, we had decided to install a two-screen EGDW in the ICN area, as the time demands were thought to be slightly more urgent than for the PICU. Training of the ICN personnel was conducted in several sessions. The Director was trained by Dr. Horii, and a separate session was held to train the house officers and medical students. A follow-up formal training session was conducted for the ICN staff by the IMACS vendor personnel when the new software release was installed.

Once the release 3 software was installed, the pediatrician usage increased since they could then compare two images side-by-side, and could also restore any case to the system from the optical disk archive. One remaining problem is that neonatal head ultrasound studies are not routinely placed on the system. The reason for this is that these studies are done at the bedside, and without a video link back to the Radiology Department (four floors down and in the adjacent building) there is no simple way to get the images into the IMACS database. We have tried digitizing films on occasion, but find the small size makes this a suboptimal method. We have installed a video disk recorder/player which is small enough to sit on top of the portable ultrasound machine and records the video images on 2" floppy disks. These can then be played back through the recorder/player and into the video input port of the acquisition module. Our early experience is that this will be a satisfactory method of capturing the portable images.

Evaluation Processes

As for the PICU, evaluation of the system utility is being monitored by encouraging feedback from the pediatrician users. We have carried out one set of interviews with the personnel concerned, and will repeat this process as features are added or changed. Our results so far indicate some difficulty in learning to use the system, but that when used results in a time savings.

The addition of the video disk recorder and further training of the ICN personnel should result in better satisfaction with the system. Approximately 30-40 per cent of the images obtained in the ICN are neonatal cranial ultrasound examinations, so that adding these to the database should substantially improve utility of the system for the ICN physicians. We continue to monitor use and acceptance through interviews with key staff members from the ICN.

PACS for Teleradiology - Montgomery Imaging Center

The GUH Radiology Department operated an outpatient imaging center in Rockville, Maryland until November, 1989. Montgomery Imaging Center (MIC) is some 13 miles (about 20 km) north of GUH, and has a complete array of imaging equipment. Of primary interest to GUH was the MRI and CT equipment. Though this was used for all applications, it was the responsibility of the GUH neuroradiologists to read the MRI and some of the CT on neuroradiology referral patients. This volume was not sufficient to warrant a full-time neuroradiologist at MIC, so the initial method of reading was to have the images brought to GUH by courier, or in extreme cases, have a neuroradiologist from GUH drive out to MIC. The former method meant a minimum delay of one day in interpreting the studies, and the latter method, while reducing the delay, was unpopular with the neuroradiologists as it took them away from their responsibilities at the Hospital.

As part of the IMACS, an acquisition node (AM) and display workstation were installed at MIC. Communication to GUH was via leased T-1 service. The AM was connected via video frame grabber to the MRI and CT equipment. This operated well, and all neuroradiology MRI studies and some CT examinations were transferred to GUH. The equipment was quite reliable, though problems outside of the IMACS scope interrupted service. These were eventually solved, and the hardware was upgraded to a combination workstation which incorporated both display and acquisition hardware. Use of the system by Neuroradiology, including primary interpretation of the studies on the EGDW at Georgetown, became routine.

Because of a change in the management at MIC, Georgetown did not renew its affiliation contract with the Center. The responsibility to read MIC cases ended in November, 1989, and the IMACS equipment was removed at that time. The utility of the system, however, was such that the neuroradiologists and the GUH radiologist assigned to MIC insisted that the IMACS equipment be kept in place and operational until the very end of our contractual arrangement with MIC.

Evaluation Process and Results

The clinical evaluation process centered around speed of image transmission and image quality. The use of the system, for interpreting neuroradiology cases, meant that the daily case load was not nearly high enough to cause us to worry about saturating the T-1 capability. In fact, the transmission was timely enough that the system could be used to make decisions about studies before the patient left MIC.

The neuroradiologists thought the image quality high enough that they did primary reading of the studies from the EGDW. These were backed up by checking against the films when they were delivered, but experience showed that this step was not necessary, unless for some reason the case was not completely acquired. The only complaint was that for some MRI studies, the workstation could not display all of the images at once (if there were more than 80 images per study). There was also initially a problem in that the frame grabbed image from MRI did not include the slice position number. This is necessary for reading spinal MRI since the slice level may not be apparent from the appearance of the anatomy. The problem was solved by adjusting the frame grabber so that it took more of the image. The result, though, meant reducing the number of images on a screen to 12 (from 20).

This experience was one which came to true closure during the IMACS evaluation period. Our final conclusion, based on the use by the Neuroradiology Division, is that this teleradiology application was entirely successful.

PACS for Teleradiology - The On-Call Workstation

The Abdominal Imaging Division radiologists are required to provide 24-hour coverage for emergencies which require their imaging services (primarily ultrasound and CT). This is accomplished by having each member (including Fellows) share the emergency call on a week-long, rotating basis.

A feature of the IMACS which impacted this coverage was the availability of a dial-up modem through which one could access the IMACS database. During the latter half of 1989, we tested this feature by placing a RVS in Dr. Horii's home. After some initial testing to confirm operation and determine transmission speed and image quality, he started using the workstation whenever he was on-call for ultrasound.

The coverage for ultrasound includes a technologist who is also on-call. The technologist will come in to the Hospital to perform the examination, and the attending radiologist is then called when the study is complete. Often, the resident who is on-call for general coverage of radiology, and who is present all night will look at the study, and provide a preliminary report. In some 40-50% of the cases, this proves sufficient, and at most, the resident or technologist will consult with the on-call attending radiologist by phone. The remainder of the time, the attending radiologist must look at the study him- or herself. Since the attending radiologist does not stay in the Hospital when on-call, this means a trip in.

The RVS has been used by Dr. Horii when on-call since July, 1989, and has saved him a trip in to the Hospital approximately 80 percent of the time. On the remaining occasions, the study to be done is complex enough that he must be present to perform it himself. An example is a study in which an interventional procedure, such as an abscess drainage, must be done. Another use of ultrasound which would prevent teleradiology (perhaps more properly telesonography) from being used is if the study depends on seeing the real-time part of the examination. Typically, ultrasound employs equipment which can scan and display images at near video rates. Transmitting this would require tremendous bandwidth, and is not practical at present for simple communications.

Evaluation Procedure and Results

The first step in evaluation was to determine transmission speed and image quality. For the teleradiology RVS, data compression (including lossy compression) is available. Dr. Horii had the same set of images transmitted with no compression, and with medium and high compression. He found the uncompressed and medium compressed images adequate for interpretation. The high compression images showed a distinct loss of edge sharpness, and were thought to be unacceptable for interpretation. Dr. Horii decided to use medium compression.

Timings with the modem used showed that a 512 x 512 by 8 bit medium compressed image could be transmitted in approximately 45 seconds. The modems operate at 19,200 baud. A typical ultrasound case requires about 30-45 minutes for transmission of all images. Dr. Horii thinks this to be satisfactory, though he would like to try the newer modems which can run at up to 38,400 baud. His typical, late-night round trip time to the Hospital (not including the time to do or read the study) is about an hour.

Clinical evaluation has been to review the film images of all cases read on the RVS. At the time of this report preparation, Dr. Horii has found no instances in which findings were missed on the RVS and seen on film[7].

As a result of this success, a second RVS was placed at the home of the Director of Ultrasound, Dr. Garra. He has been using the system since July, 1990, and has found results similar to Dr. Horii's. In an effort to reduce transmission time, Dr. Horii has begun experimenting with having all images transmitted with high compression, and after viewing these, requesting any important images re-transmitted at medium or low compression. This reduces transmission time by about 50 percent, and a review of the images so far has not shown any degradation significant enough to mask findings. However, the cases so far have involved abnormalities which are easy to see, and further work is necessary to determine if this mode of use is more generally satisfactory.

CONCLUSION

The DINS work at Georgetown was successful in demonstrating that a PACS can be clinically useful to radiologists and referring physicians. The main benefit to referring physicians is in time savings. For the radiologists, availability of images and the use of teleradiology for daily and on-call work are beneficial aspects. Much work remains to be done before PACS are easily cost-justified, but rapid progress is being made.

ACKNOWLEDGEMENT

This work was supported in part by the U.S. Army Medical Research Acquisition Activity of the U.S. Army Medical Research and Development Command under contract number DAMA 17-86-C-6145. The views and opinions expressed in this paper are those of the authors and should not be construed to to represent official Department of the Army positions, policies, or decisions unless so designated by other documents.

REFERENCES

1. Arenson, R.L. and Fridenberg, R.M. (eds.): S/CAR '90: *Computer Applications to Assist Radiology*, pp 350-361. Symposia Foundation, Carlsbad, CA, 1990.

2. Davros, W.J., Garra, B.S., Ingeholm, M.L., Mun, S.K., Zeman, R.K.: The impact of PACS on a clinical ultrasound service. Scientific Exhibit number 1245, 1988 AIUM/WFUMB Meeting, Washington, D.C., October, 1988.

3. Elliott, L.P., Mun, S.K., Horii, S.C.: Digital Imaging Network System (DINS) Evaluation Report, MITRE Contract number N-55201, pp 208-212, March, 1990.

4. Freedman, M., Horii, S., Garra, B., Braudes, R., Lo, B.,Mun, S.K.: Evaluation of Digitization Dynamic Range in Laser-Digitized Adult Portable Chest Radiographs. Annual meeting of the RSNA, November, 1989.

5. Horii, S.C., Muraki, A., Mallon-Ingeholm, M.L., Mun, S.K., Clark, L., Schellinger, D.: A complete image management and communications network for the Neuroradiology Service of Georgetown University Hospital. Proceedings of the SPIE, v. 1093, Medical Imaging III, pp 154-158, 1989.

6. Horii, S.C., Garra, B.S., Zeman, R.K., Lo, S.-C. B., Davros, W., Silverman, P.M., Cattau, E.L., Fleischer, D.E., Benjamin, S.B.: Endoscopic Retrograde Cholangiopancreatography: Prospective Evaluation of Traditional Film-based Images, Digitized Images, and Digital Fluoroscopy. Annual meeting of the RSNA, November, 1989.

7. Horii, S.C., Mun, S.K.: Clinical Utility of Teleradiology in a PACS Environment. SPIE Proceedings v. 1234: Medical Imaging IV, 1990.

8. Mun, S.K., Benson, H., Horii, S., Elliott, L.P., Lo, S.-C. B., Levine, B., Braudes, R., Plumlee, G., Garra, B., Schellinger, D., Majors, B.:Completion of a hospital-wide comprehensive image management and communication system. Proceedings of the SPIE, v. 1093, Medical Imaging III, pp 204-213, 1989.

Clinical Experience.
16 Months of HU-PACS

Goro Irie

Radiology Department, Hokkaido University Hospital,
Sapporo 060, Japan

Abstract: Sixteen months' experience with HU-PACS (Hokkaido University PACS) is reported. This report describes the hardware and software aspects of HU-PACS in detail. Evaluation of the system and its working status are also reported. Usefulness of PACS is concluded after all.
keyword: PACS, clinical PACS, HU-PACS

1. Outline of HU-PACS

The concept of HU-MIS (Hokkaido University Medical Information System) is shown graphically in Figure 1. As it is shown there, PACS is one of the three main components of HU-MIS.

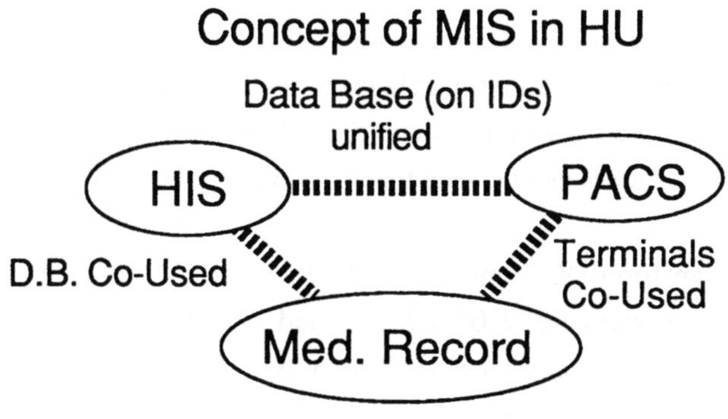

Fig 1

In HU-MIS, PACS handles all radiological images and the Medical Record System includes not only written medical records, but also all medical images other than radiological

NATO ASI Series, Vol. F 74
Picture Archiving and Communication Systems (PACS)
in Medicine. Edited by H. K. Huang et al.

images, such as echography, endoscopy, microscopy, EEG, ECG and EMG. A part of the HU-PACS has been implemented and been used as a clinical tool since June 1989.

The network covers only the radiology department and the orthopedics and general medicine clinics with a total of 20 image terminals.

The goal of the HU-PACS will be reached after implementation of the Medical Record System, which will be in 1993 at the earliest. About 200 image terminals and corresponding upgraded networks, computers and optical discfiles are then expected to be used by both PACS and the medical record system together. The goal of HU-PACS was set in this way because every one of the hospital should feel its merits and give favorable judgement on its cost performance balance.

2. Configuration of HU-PACS

The hardware aspects of the system are shown in Figure 2. Where imaging machines are interfaced to the PACS by each specific way, ACR-NEMA Standard is not used yet. In the Figure 2. the inner most optical loop LAN (O.L.L.) and the optical discfile (O.D.F.) attatched to MIS host computer are not installed yet, waiting for budget allocation for the Medical Record System. All X-ray images are to be taken through FCR system and sent to PACS via 13 read out machines of it.

Soft ware aspect of the HU-PACS is characterized by followings:

i) Coupling with HIS

This coupling keeps off the load of data base management from the PACS computers and informations on the reservation of patient-visit helps to realize prefetching or down loading of images beforehand. This function with magnetic discs distributed to the image terminals and the data compression makes users feel the data transmission very fast.

ii) Data compression

The discrete Cosine Transform(DCT) method with blocks of 256 pixels is adopted for data compression. Compression rates are 1/10 for X-rays and 1/3 for CT and MRI. Degradations of image quality due to these compressions have been proved to be clinically acceptable by our research works and in our clinical practice.

iii) Application programs on the terminals

As shown in table 1, many programs to process the images are implemented on each terminal. Those with asterisk are used frequently. These programs make users feel that CRT diagnosis is better than old film-view-box system.

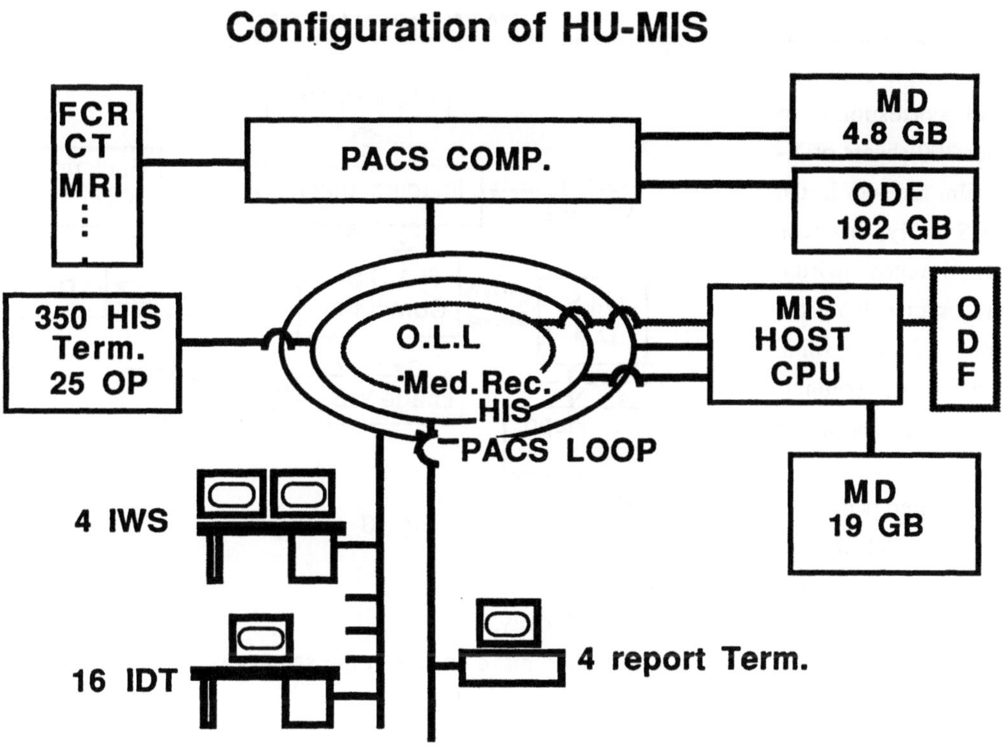

Fig. 2

Table 1. Image processing programs on each image terminal.

*1. Settings of window width and level of CRT image
*2. Measurements of distance and angle on CRT image
3. Rotation of image (90°, 180°, 270°) and conversion to mirror image
*4. Black and white conversion
5. Setting of gamma characteristics
6. Edge enhancement
*7. Magnification with interpolation and 2D scrolling
8. Weighted addition and subtraction of images
*9. Minification of images and multi image display

3. Data flow

Fig.3 shows the flow of image data roughly, where dotted lines are for the new images and solid lines for old ones.

About 30% of all X-ray images produced which corresponds to about 300 sheets of X-ray film are sent in the PACS daily, because only images which were ordered from the section of the hospital where image terminals are installed should be sent in.

DATA Flow of HU-PACS

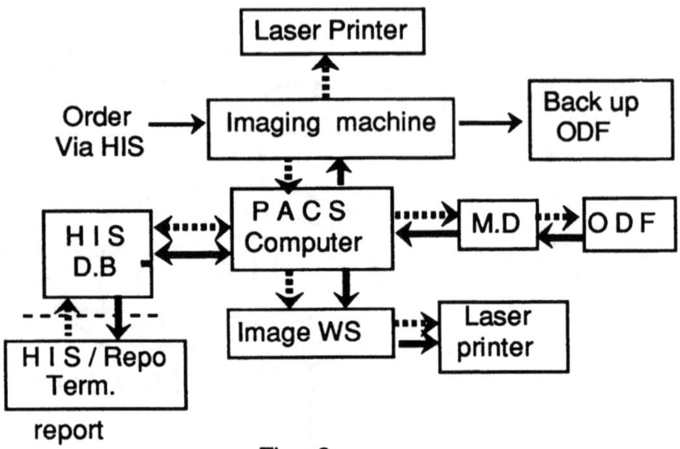

Fig. 3

Hundred percent of CT and MRI images corresponding to 500 sheets of X-ray film are sent in to the PACS daily.

Angiographic images are to be handled more or less in the same way as CT and MRI. Angiography using FCR will be practiced from march '91.

Other newly made images which are eligible to be sent to PACS goes to MD of PACS computer and after up-dation of data base in HIS sent to the MD of an image terminal which was specified to the system when the imaging was ordered.

Images on the MD of PACS computer stay there, about a week, though it is written on the optical disk file later in the day.

Old images on the optical disc file can be accessed from any image terminal. It takes 40~60 sec to see the image on CRT of the terminal.

4. Evaluation

i) There were some troubles in the beginning, in the first 2 months, mainly caused by bugs in the interfaces between imaging machines and PACS. Soft ware of the terminals are always improving according to the monthly meeting of the user and the vender (NEC).

ii) Doctors other than radiologist have said no claims on the access time. Images of CT, MRI and angiography are sent in on real time basis to MD of IWS which is a kind of image terminal used by radiologist. Radiologist, therefore, has to wait until the end of sending one image or in special case one series of images for his own access.

iii) A survey showed that about 75% of doctors (orthopedician in this case) feel quality of the CRT image as good as or better than the image printed on film. If a doctor does not mind to take time and to maneuver the terminal to make a good quality of CRT image, there would not be much clinical problem in quality of CRT image. No problem on image qulity of CT and MRI is claimed even by radiologist.

iv) Function of image terminal

Here again not much claims are heard from doctors other than radiologist. Function of the terminal which are needed by radiologist are somewhat different from the ones needed by other doctors because it is radiologist who has to use the terminals every day for their conference and report making where reference to old images are often necessary. The terminals with two CRTs namely IWSs are not good enough for these radiologist's purpose in the number of CRT and access time and are still under improvement.

5.Working status of HU-PACS

Fig. 4 and 5 shows monthly number of image taken in the PACS since June '89 for X-ray and since Sept. '89 for CT and MRI image after their interface problems were solved. In Fig. 4 images are classified by modality and in Fig. 5 by the department who ordered them.

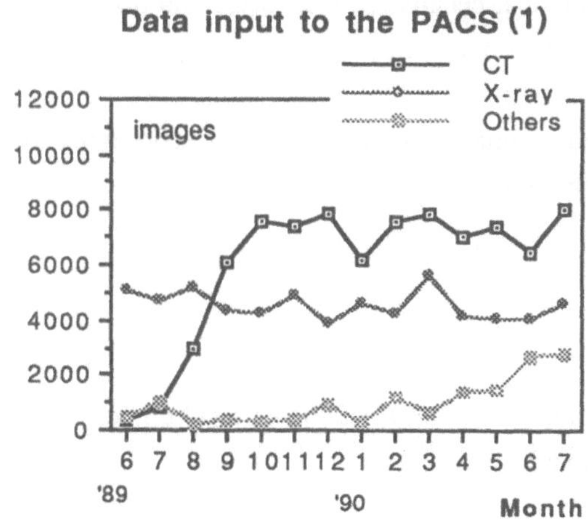

Fig. 4

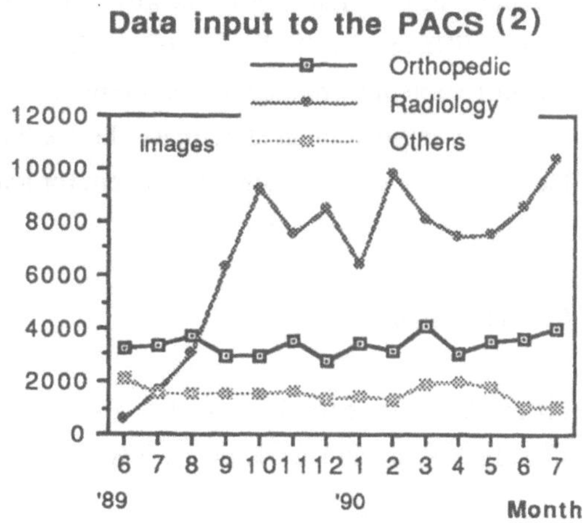

Fig. 5

Fig.6 and Table 2 show accessed rate of image via PACS for all images. All images are printed on film and distributed to the clinics at the same time that may be the reason of low access rate. This situation is expected to be changed and the users will be more dependent upon filmless PACS when the supply of film images were stopped after reformation of our medical insurance system.

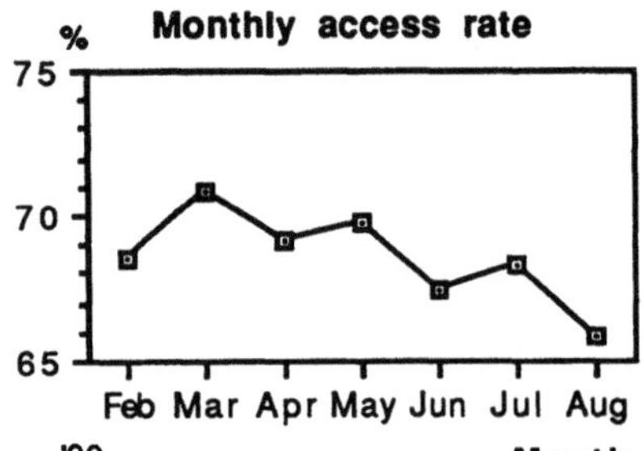

Fig. 6

Table 2 Total access

	total	1access	2	3	3<
Image	99041	63114	2085	1730	1155
Access %	-	63	2	1.7	1.2

6. Problems

There is no essential problems to persue the PACS project in both technical and clinical aspects if budgetary support and staffs cooperation are available. There are, however, some minor problems among which ones related to maneuvability of image terminal are the most important. Problems of image terminal became appearant when they are used for image conference. Adoption of too many CRTs, however, is not practical because of their price and occupying space. Efficient multi display of minified images on two or three CRTs and quick, direct magnification of them may be the solution.

Many improvement should also be continued on image quality and access efficiencies as had been done on film-view-box system for long period of time.

7. Conclusion

After 16 months experience of HU-PACS it is concluded that PACS is a useful clinical tool, already though there are some aspects to be improved.

PACS - CLINICAL EXPERIENCE AT UCLA

Hooshang Kangarloo

Department of Radiological Sciences, UCLA School of Medicine, Los Angeles, CA 90024, U.S.A.

Abstract: Implementation of picture archiving communication systems (PACS) at UCLA began with the evaluation of the systems in Pediatric Radiology, one of the 11 sections of the Department and the Coronary Care Unit, one of the 14 ICU's in the Hospital. We have now completed PACS development for all CT's and MR's which allows communication between a newly developed outpatient facility, the Hospital and the remote research facility in addition to Pediatric Radiology and the Coronary Care Unit. The following are some of the advantages of PACS from a clinical standpoint. [1] Conferences with clinicians have been more effective by spending more time on each case, but less total time for each conference; [2] Clinicians are satisfied with PACS at the remote site, but it is clear that the radiologists' interpretation must accompany the images; [3] PACS allows for the development of interactive teaching of the students; [4] PACS allows for a new method of radiology practice by analyzing the image and providing a more comprehensive, quantitative consultation, otherwise not possible with analog systems.

Key Words: Picture archiving and communication system, pediatric radiology, intensive care unit computer instruction, image procession.

Introduction

The traditional practice of Radiology has been inefficient regarding a number of issues related to patient care, teaching and research. For example, it is impossible to keep the Radiology file room organized and at the same time make the images available to clinicians taking care of patients. A common problem addressed to a Service Chief is a complaint from clinical colleagues when they cannot find their patient's films. Patient care can also suffer when one group of clinical consultants have possession of the films, thus others cannot see the images

NATO ASI Series, Vol. F 74
Picture Archiving and Communication Systems (PACS) in Medicine. Edited by H. K. Huang et al.

even when they are necessary in patient care. These management problems are more intense in a tertiary care environment and a teaching hospital.

Residency and student training in Radiology could be more effective if an interactive teaching environment was available, allowing students to think in solving problems and test their answers to questions related to diagnostic issues.

Research may also enhance if the data could be readily available and possible to organize in a way to test various hypotheses.

For this reason, it was clear to us, like many others, that a long term solution was necessary not only for the appropriate management of the Department, but also a better method for teaching, research and patient care [1].

1. *Pediatric PACS* [6]

1.1 From the planning stage it was clear that it is very difficult to implement a new system such as PACS for the entire Department without extensive testing. Since our department is organized in 11 sections (pulmonary, musculoskeletal, mammography, neuro, cardiovascular, genitourinary, gastrointestinal, pediatric, emergency room, nuclear medicine and ultrasound), it seemed logical to begin implementation in one section and proceed to the next and so on. When electronic imaging in all sections were developed, then we plan to have them communicate, thus creating a departmental PACS [2,3]. Selection of pediatric radiology was based on the nature of activities within the section as outlined below. The section is responsible for about 23,000 examinations per year; it has its own conventional radiology, ultrasonography and computed radiography. It also is responsible for performance and interpretation of all pediatric MRI and CT examinations. Therefore, the pediatric radiology section is very similar to a small radiology department and if PACS implementation is successful, then the same model could apply to the rest of the Department. Initially we began our experiment by placing a six monitor 512 x 512 viewing station in the pediatric radiology reading room. This viewing station was utilized for daily conferences with the pediatric house staff. All images were formally interpreted from hard copy. We compared the total length of the conference and length of the discussion and teaching per case between the viewing station and the hard copy system. **The conclusion was that although the total length of the**

conference decreased using the viewing station, the length of the discussion for each case increased, thus improving teaching and probably patient care.

1.2 Laboratory experimental evaluation of accuracy for diagnosis was performed using a humanoid chest phantom and simulated lung nodules. In comparison to hard copy films, **the 512 system was clearly inferior** and insufficient for primary diagnosis.

1.3 *1024 Line System:* We have now implemented a six monitor 1024 x 1024 system in the pediatric radiology reading room which allows higher quality image display. The system is well accepted by pediatric radiologists and the pediatric house staff for conferences and review. Our comparison of this system with hard copy also indicates that the **1K system is inferior to the hard copy** system.

We are now evaluating a two monitor 2048 line system. Based on our preliminary work, it appears that **a 2K system will be comparable to the hard copy** images in evaluating pediatric diseases.

2. *Remote Sites*

2.1 *Coronary Care Unit* (CCU): A three monitor remote viewing station originally 512 x 512 and now upgraded to 1024 x 1024 is placed in the coronary care unit. Images from a computed radiography system [5] are directed to the CCU and is kept in local storage at the CCU. We perform an experiment with the 512 system in the CCU. Hard copy images and soft copy images were available for all patients simultaneously to the clinicians. In 91.4% of the cases clinicians used only soft copy, in 5.4% hard copy and 3.2% using both. Various interpretations can be drawn from this study, one is that **they preferred the viewing station to the hard copy** [4].

2.2 *The Ambulatory Care Center* (Medical Plaza): Across the street from the hospital we now have an out-patient care center (the Medical Plaza) which will house all of our out-patient activities. Radiology is over 30,000 square feet of space, has two MRI's (1.5T Signa), 2 CT scanners, three nuclear medicine cameras and 25 radiograph rooms, including fluoroscopy, mammography, ultrasound and conventional radiology. Development of a new out-patient care center was extremely exciting because of additional space and a newly added capacity, particularly for MRI and CT. Because of the lengthy approval process and the expense of remodeling, it was not possible for us to place these MR units in the hospital in a timely fashion. In addition, a new 3,000 square feet of space of research facility for the MRI has been developed which houses a 1.5T unit. The Department has an additional two CT scanners in the hospital and within the next 12 to 24 months, there will be an additional two CT scanners and two MRI units. Thus, there will be a total of five MRI scanners and six CT scanners in various sites. It is clear that management of these scanners is difficult and very costly. The faculty will have a great deal of difficulty in covering these scanners and providing timely consultation for referring physicians. Alternatively of course, one could hire more faculty which financially is not possible to do. Because of these new concerns, our strategy changed from sectional PACS. We decided it is best if we develop a PACS system for all CT and MRI units which clearly was an urgent problem to solve. Networking and hardware are developed in a way that all images from CT and MRI scanners are stored centrally and viewed either at Medical Plaza or the Hospital. A 1024 line and 2048 line multiple viewing display station are being evaluated in clinical sites.

3. ***Video Monitors:*** All scanners are connected by a broadband fiber optic system to a number of single monitor viewing stations which allow monitoring of the examinations while they are being performed. A channel selector allows selection of a specific scanner. These video monitors are placed in the reading rooms and faculty work areas, allowing real time review of patient's CT/MRI.

4. ***Analysis Station***

Practice of radiology requires providing comprehensive consultation to referring physicians. For example, rather than MRI interpretation of a brain tumor, one could also indicate how surgeons should approach the lesion, so that adjacent organs are less affected. This kind of analytical consultation can become available utilizing the PACS environment as outlined by the following examples. We evaluated a group of children with bone marrow disease. Computer analysis of signal intensity was more accurate compared to subjective analysis available by film and correlated better with the results of the bone marrow biopsy. UCLA is a center for the treatment of arterio-venous malformation of the brain under the leadership of Dr. Fernando Vinuela. We have now developed a computer analysis system to evaluate the amount of blood flow change prior and after embolization therapy. In another project, MRI and PET results are superimposed, creating a four dimensional image of anatomy and function. Yet in another project we are developing a computer algorithm which allows computerized assessment of the bone age [7].

4.1 *Teaching:* The process of teaching seems much more effective if one gets the student excited about the "unknown" rather than the "known". In an attempt to develop an interactive teaching environment, we have undertaken two projects. The first deals with conventional radiology. Here, a student is given an image and is allowed to change KVP and mAs settings and observe the effect on the image. In the second project, an MRI image is provided and a student is allowed to change TE and TR and observe the effect of pulse sequence selection on the resulting image. We hope that this interactive teaching environment allows a better understanding of parameters affecting the creation of an image as well as a disease process.

5. *Problems*

5.1 *Expense:* Development of PACS systems are expensive and at this time it is not clear who is responsible for this expense. Hospital directors in general have not been eager to pay for the cost of these developments. The manufacturer needs to participate and develop cost effective systems.

5.2 *Quality Assurance:* Our experience in the CCU could also be interpreted as **clinicians use a low quality image (512) even though a higher quality hard copy is available, thus adversely affecting patient care.** It is essential then that the interpretation of the radiologists accompany the images for all remote sites.

5.3 *Turf:* Issues related to who would bill for what study needs to be solved. PACS could potentially create conflict between radiology and other departments if not managed appropriately.

5.4 Other technical problems to resolve are related to implementation, training and service. These technical problems require a dedicated group of scientists interested in PACS research and development.

Conclusion: Our experience with PACS has been very positive. We owe this mainly to a large group of dedicated scientists in our Department. We believe PACS is essential for the future practice of radiology in patient care, teaching and research.

References

1. Huang H.K., Kangarloo H., Cho P.S., Taira R.K., Ho B.K.T., Chan K.K.: Planning a totally digital radiology department. AJR 154:635-639 (1990).

2. Huang H.K.: Three-tier system links diverse PACS components. Diagnostic Imaging (Suppl), pp. 13-14,19, September (1990).

3. Focus on PACS - In: Diagnostic Imaging (Suppl), September (1990).

4. Cho P.S., Huang H.K., Tillisch J., Kangarloo H.: Clinical evaluation of a radiologic picture archiving and communication system for coronary care unit. AJR 151(14):823-827 (1988).

5. Kangarloo H., Boechat M.I., Barbaric Z., Taira R.K., Cho P.S., Mankovich N.J., Ho B.K.T., Eldredge S.L., Huang H.K.: Two year clinical experience with a computed radiography system. AJR 151:605-608 (1988).

6. Boechat M.I., Kangarloo H., Dietrich R., Hall T., Taira R.K., Mankovich N.J., Huang H.K.: A picture archiving communications system (PACS) for pediatric radiology. Current status. In: Planning considerations in Diagnostic Imaging and Radiation Therapy. Proceedings of the Fifth International Symposium on the Planning of Radiological Departments, pp.247-249 (1988).

7. Huang H.K., Aberle D.R., Lufkin R., Grant E.G., Hanafee W., Kangarloo H.: Advances in medical imaging. Annals of Internal Medicine 112:113 (1990).

Clinical Evaluation of PACS at the University of Pennsylvania

Harold L. Kundel, Sridhar B. Seshadri, and Ronald L. Arenson

Department of Radiology, University of Pennsylvania Medical Center
Philadelphia, PA 19104, USA

Abstract: The interdisciplinary Medical Informatics Group at the University of Pennsylvania is responsible for designing and evaluating a digital PACS system for the University Hospital. The first clinical sites for PACS implementation and evaluation were the Intensive Care Units. In a controlled study, the clinical efficacy of the PACS was shown by a decrease in the time required to take certain clinical actions after imaging. Changes in image viewing behavior and consultation patterns were also observed. The clinical evaluation is continuing with a shift in emphasis toward measuring parameters that describe the diagnostic value of the system rather than measuring surrogates of patient outcome.

Key Words: PACS Clinical Evaluation, Clinical Efficacy, Diagnostic Value

The Medical Informatics Group

The Medical Informatics Group (MIG) is an applied research group in the Department of Radiology. Since 1983 the Group has been involved in the design and evaluation of Picture Archiving and Communication Systems (PACS). It also provides day-to-day PACS services to selected clinical areas. In its research and development role, the Group conducts *diagnostic efficacy, clinical efficacy* and *cost-identification* analyses. Evaluations of diagnostic efficacy are typically performed in the laboratory under strictly controlled conditions using the Receiver Operating Characteristic (ROC) methodology. Clinical efficacy where the question to be answered concerns the effectiveness of the PACS technology is measured in a *clinical* setting. Even if PACS were shown to be diagnostically efficacious there is no guarantee that the technology would actually improve the physician's patient management capabilities. Additionally clinical efficacy evaluations study the modifications in behavioral patterns when a new technology is introduced into the healthcare environment. Cost-identification analyses are important because even if the PACS technology were proven to be both diagnostically and clinically efficacious, it may still turn out to be so expensive that it would be unrealistic to install PACS on an institution-wide basis.

NATO ASI Series, Vol. F 74
Picture Archiving and Communication Systems (PACS)
in Medicine. Edited by H. K. Huang et al.

In its clinical role, the Group has developed a 12-phase plan to take the Department from its present film-based operation to a largely digital department by the next decade. The strategic plan has three streams of activity. First, image archival and retrieval capabilities will be provided for the digital modalities (MRI, CT, Nuclear Medicine, Ultrasound and PET). Second, image management services will be provided for remote units like the intensive care units (ICUs), emergency rooms (ERs), etc. Third, PACS will be introduced for the conventional modalities within the Department.

In order to meet the Group's research and clinical responsibilities, the MIG comprises professionals with varied backgrounds. Two radiologists are a permanent medical resource while other radiologists and physicians serve as consultants. They also take an active part during the analysis phases. A physicist plays an active role in the design and development of efficacy analysis protocols and a research psychologist interacts with the Group in the area of workstation design and optimization. Software design and development is carried out by a team of six full-time engineers. Their backgrounds include electrical engineering, computer science, physics and biomedical engineering. Finally, a research coordinator acts as a liaison between the Group and the consumers of PACS. The research coordinator's background includes medical technology, radiology, computers and economics, and is an invaluable asset in the various activities of the Group.

PACS at Penn

Hardware technology, in most cases, is developing more rapidly than the required software. There are two reasons for this. First, the vendors do not have a clear understanding of what is required. Second, software development is expensive and does not produce immediately tangible benefits. With this rationale in mind, the Group has concentrated on in-house software development with 'off-the-shelf' hardware. The major strength of the Group is the availability of several software engineers actively writing software *within the clinical environment.* This provides quick, in-depth feedback to the engineers and permits rapid problem resolution for the users. It is very easy for a potential user to loose interest in PACS due to an inadequate software module. Once interest is lost, it is very difficult to regain that user's support.

Figure 1 *(on the next page)* is a schematic of the currently operating PACS. The network backbone is an 80 Mbit/second Pronet fiber-optic token-ring (Proteon Inc., Westborough, MA). Only the high network traffic devices reside directly on the back-bone. These include the image archive, the gateway that serves the satellite link and the film-digitizer. All other

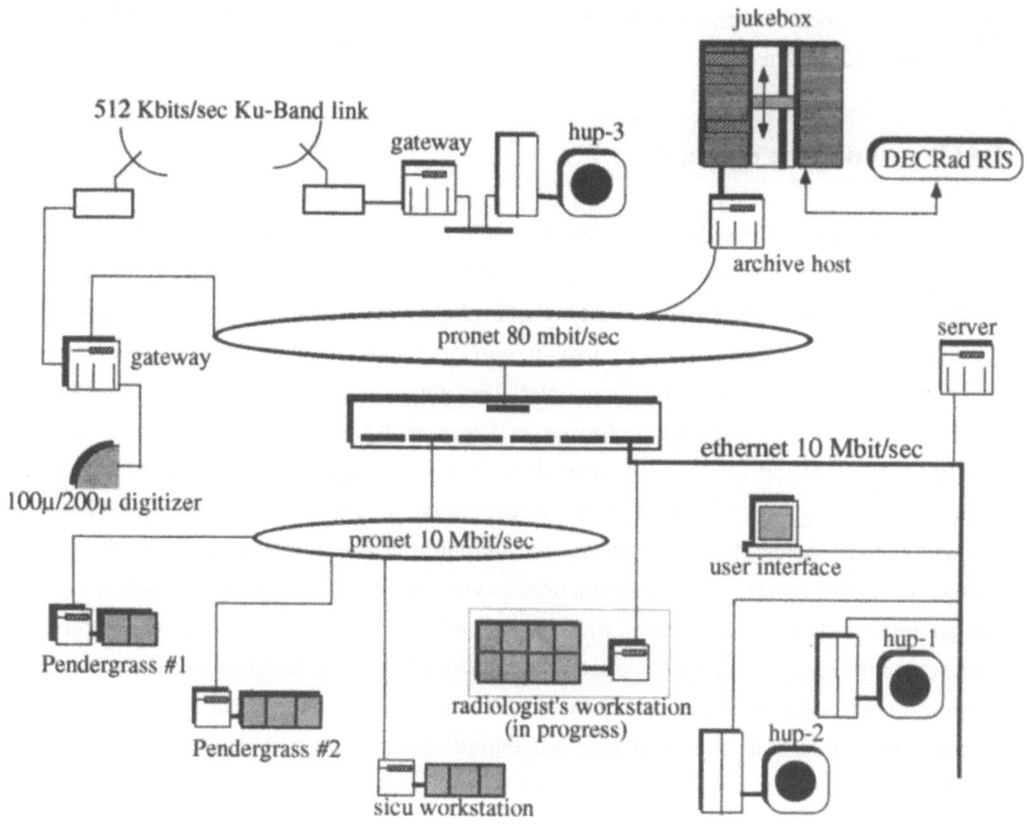

Figure 1. Current PACS Topology

devices communicate with the back-bone through the inter-net router. Images are input into the system either with a 100μ/200μ film digitizer (duPont de Nemours, Wilmington, DE) or through an Ethernet link to the MRI scanners. One sub-net is exclusively dedicated to provide service for the two MRI scanners (General Electric, Milwaukee, WI). Another sub-net provides image access capabilities for the workstations in the Surgical and Medical Intensive Care Units (SICU and MICU respectively). The SICU workstation has a 512x512 spatial resolution and a 16 MByte image memory (Virtual Imaging, Sunnyvale, CA). The MICU workstation has a 2,560x2,048 spatial resolution (Megascan, Hopkinton, MA). The image archive is a hybrid system comprising both magnetic media as well as write-once-read-many (WORM) optical media (Vortech Data Inc., Reston, VA). Currently, it can store 125 GBytes on-line, but is capable of being expanded to 7.6 TeraBytes. Details concerning the PACS have been published in [7].

PACS evaluation: efficacy

Much of the effort put into the evaluation of PACS has been on diagnostic accuracy. The rationale for such studies is simple. A digital PACS is being substituted for a film-based PACS and the crucial common element is the display medium. If the system is going to be clinically efficacious then the digital display must be at least as good as film. Goodness must be determined in a clinically relevant manner. This means using clinical case material and an unbiased estimator of diagnostic performance. The area under the ROC curve is currently popular [6,8]. This approach springs from the efficacy paradigm which asserts that the most valid figure-of-merit for a medical technology is patient outcome [3,4]. Is the patient better off with this technology than with one of the alternatives? The immense difficulty of measuring small changes in patient outcome is overcome by asserting that efficacy is a nested hierarchy consisting, at a minimum, of diagnostic efficacy, clinical efficacy, and outcome efficacy where each level depends upon the preceding level. It then follows that a measure of diagnostic performance or clinical action can be used safely as a surrogate for outcome. This also implies that in order to be efficacious, a digital display must be at least as good as film in the performance arena. The problem with the efficacy paradigm is that the value of a diagnostic technology may depend upon factors other than outcome, such as cost or convenience. Physicians may be willing to trade accuracy for efficiency just as they have been willing to trade accuracy in x-ray diagnosis for dose reduction. Additionally, diagnostic performance studies conducted in the laboratory using highly selected case material and expert observers, who deal only with the images may not be representative of what actually happens in the clinic where images are interpreted by a variety of people in a milieu that is rich in clinical information.

The measurement of diagnostic accuracy although necessary for defining workstation standards is not sufficient for the clinical evaluation of a PACS. We believe that at a minimum a full image management system must be in place when the evaluation is conducted. This is difficult to achieve on a large scale with experimental systems and our approach has been to narrow the scope of the PACS to a self contained clinical entity such as an intensive care unit [1]. Additionally, we have concentrated on measures of clinical efficacy such as the time required to take some specific action that is triggered by an imaging examination.

PACS evaluation: a case study

Between 1985 and 1988, the Group conducted an evaluation of the clinical efficacy of a PACS linking the Radiology Department and the Medical Intensive Care Unit (MICU). The actual evaluation took place over a 12-month period. Eight-week periods during which only films were available (CONTROL PERIODS) were alternated with eight-week periods during which all chest radiographs taken in the MICU were digitized after being developed (TEST PERIODS). Once digitized, the images were available immediately for viewing on an image display console in the MICU. During the test periods, both films and digital images were available. Physicians were given the opportunity to view films, digital images, or both, since we felt it would not be in the best interest of the patient to withhold films from the physician. Sixteen residents and 42 interns who rotated through the MICU as part of their normal course of training participated in the study. The MICU was an 11-bed critical care facility to which patients with both cardiology and pulmonary problems were admitted. Approximately 3,000 chest radiographs were performed and tracked in the MICU during the course of the study. Five forms were completed for each examination by the requesting physician, the radiographer, the film librarian, the radiologist, and the MICU physician who took the action. Fifteen specific clinical actions that could be taken as a result of findings on a chest radiograph were identified. The radiologist/physician methods of communication were also studied. Details of the study are available in [1,2].

There were three significant findings: (1) The image viewing patterns of the physicians were modified by the introduction of the PACS, (2) the time interval between when the examination was completed and when an action was taken (based on image information) decreased in certain cases, and (3) the communication between the radiologist and the physician decreased. The third finding concerning the decreased communication was the most interesting. During the control period, overall consultation with a radiologist, whether by telephone, through written report, or in person, occurred for 95% of the films. During the test period, the overall consultation occurred only for 60% of the digital images. This study showed that there were times when interns and residents acted on information derived from the digital images without consulting with an attending or a radiologist.

PACS evaluation: clinical value

Our experience with PACS evaluation has shown effects occurring in the clinic that cannot be observed in the laboratory. Furthermore, these effects, such as shifts in image viewing times and changes in consultation patterns, may not influence patient outcome, yet they may have profound impact on the process of patient care. Therefore, studies aimed at demonstrating efficacy may not be as revealing as studies of the communication process itself. This does not mean that outcome should be neglected. Indeed, our plans for the intensive care unit studies include measurements of the effect of the imaging medium on the change in the clinical severity of disease and on the occurrence of adverse clinical events. We will, however, focus on modeling the diagnostic process.

We assume that four factors largely determine the clinical value of a medical image management system. They are: (1) the distribution of imaging diagnoses; (2) the importance of the diagnostic information to the clinician; (3) the accuracy of each diagnosis; and (4) the efficiency of communication of the image information. We have proposed a model [5] that combines these factors and provides a single figure-of-merit called diagnostic value. We plan to collect clinical data in the intensive care units and to test the output of the model against measures of efficacy and physician utilization. If this venture is successful, we plan to extend the analysis to other clinical units.

Bibliography

1 Arenson RL, Seshadri S, Kundel H, DeSimone D, van der Voorde, F, Gefter W, Epstein D,Miller W, Aronchik JM, Simson, MB, Lanken PN, Khalsa SS, Brikman I, Davey M, Brisbon N: 'Clinical Evaluation of a Medical Image Management System for Chest Images', *American Journal of Roentgenology* 150:55-59, 1988.

2 DeSimone D, Kundel HL, Arenson RL, Seshadri SB, Brikman I, Khalsa SS, Davey MJ, Brisbon BA : 'Effect of a Digital Imaging Network on Physician Behavior in an Intensive Care Unit', *Radiology* 169:41-44, 1988.

3 Epstein AM. The Outcomes Movement - Will it get us where we want to go? NEJM 1990;323:266-269.

4 Guyatt G, Drummond M, Feeny D, Tugwell P, et al: 'Guidelines for the Clinical and Economic Evaluation of Healthcare Technologies'. *Soc. Sci. Med.* 22:393-408, 1986.

5 Kundel HL, Seshadri SB, Arenson RL. 'Clinical Evaluation of PACS: Modeling Diagnostic Value' *Proc. of SPIE: Medical Imaging IV*, Newport Beach, CA, February 1990.

6. Metz CE.: 'ROC methodology in Radiological Imaging', *Investigative Radiology* 21:720-733, 1986.

7. Seshadri SB, Arenson RL, Sprague DF. 'The architecture of an optical jukebox image archive' in *Proc. of SPIE: Medical Imaging IV*, Newport Beach, CA, February 1990.

8. Swets JA and Pickett RM: 'Evaluation of Diagnostic Systems: Methods from Signal Detection Theory'. New York: Academic Press 1982.

NIH SUPPORTED PACS RELATED RESEARCH [1]

Matti Al-Aish

Radiation Research Program, National Cancer Institute, 9000 Rockville Pike,
Bethesda, MD 20892, USA

Abstract: The diagnostic imaging research program at the National Cancer Institute took a leading role in the development and utilization of electronics and other imaging technology. Research in areas of digital electronic communication technology, picture archiving and communication systems (PACS), image acquisition and sorting, and image enhancement has been actively pursued by our program. The following three research centers were established to bring electronic imaging technology to clinics and to advance early cancer diagnosis.
Keywords: PACS, Image Acquisition.

INTRODUCTION

The Diagnostic Imaging Research Branch (DIRB), Radiation Research Program, DCT, NCI, support research leading to the application of engineering, physical, chemical, biomedical and physiological sciences to improve diagnostic imaging. This entails research designed for the development of innovative instruments and supportive methodologies in the area of medical imaging relating to patient care.

The research objectives are five-fold. The first objective is to develop new and improved non-invasive imaging devices, utilizing both ionizing and non-ionizing radiation technologies to enhance diagnostic imaging for cancer and other diseases, and to reduce the exposure of patients to ionizing radiation. The second major objective is to synthesize new radiopharmaceuticals and other compounds in order to advance research in nuclear medicine. The third aim is to support research in areas of digital electronic communication technology, PACS, image acquisition, sorting and enhancement. The fourth goal is to develop methodologies for the use of important radiolabeled macromolecules in biological, physiological and clinical research, especially relating to imaging of both normal and malignant tissues. The fifth and most significant aim is the clinical use of scientific advances in clinical diagnosis and the monitoring of therapy. Overall, the Diagnostic

---

[1] The opinions or assertion contained herein are the private ones of the author and are not to be construed as official or reflecting the views of NIH, National Cancer Institute or the U.S. Department of Health and Human Services.

NATO ASI Series, Vol. F 74
Picture Archiving and Communication Systems (PACS) in Medicine. Edited by H. K. Huang et al.

Imaging Program has a budget of about 40 million dollars, supporting research using grant-in-aid and contract mechanisms. There are approximately 200 grants and contracts currently in our Program. The ultimate aims of this Program are to develop instruments and techniques characterized by enhanced imaging resolution, reduced ionizing radiation and time of exposure, to lower cost for patients and to enable the physician to diagnose safely and treat cancer and other diseases.

1. PACS IN RADIOLOGY

Dr. H.K. Huang, Principal Investigator
University of California at Los Angeles, Los Angeles, CA, U.S.A.

At the University of California in Los Angeles (UCLA), under the direction of Dr. H.K. Huang, research is in progress on many aspects of picture archiving and communication systems (PACS) in radiology. A digital viewing station with six 1K by 1K monitors was designed and developed several years ago for use in pediatric radiology. The prototype station was successfully utilized for case review, radiology conferences and consultations. The success of the pediatric review station led to the development of a more sophisticated 2K by 2K monitor display station for primary clinical diagnosis. Preliminary results of the evaluation of this station were encouraging.

Meanwhile, hardware modules are being constructed to compare images produced by 2K x 2K, 1K x 1K, and 512 x 512 monitors. These modules will be evaluated in the clinic. The most significant undertaking, however, is the development of PACS modules for chest radiography, neuroradiology, and the intensive care unit (ICU). The ICU module focuses on neonatal, pediatric, and coronary diagnoses. Needless to say, the final results of the PACS research of this team at UCLA will determine whether or not the newly designed digital-based systems can replace the conventional film-based operation for use in radiology departments. Using the valuable experiences gained in the past few years in the development of pediatric radiology and coronary care units stations, Dr. Huang and his staff are expanding their investigations to include other areas for radiological examination. These experiments are planned to prove the hypothesis that PACS are a suitable replacement for conventional film-based radiological examinations in the clinical environment.

The UCLA group is hard at work implementing the required hardware and software for the proposed new systems and evaluating the capability of constructed systems for digital acquisition, processing and display methods in clinical environments as well as user acceptance and system efficiency. They aim to prove that the newly developed PACS imaging technology should equal or surpass the level of diagnostic information provided by the conventional system. Needless to say, the department of radiology at UCLA is taking a major step towards the establishment of a successful and workable electronic imaging system in their hospital.

An important research objective of this program project is to continue the investigations on image compassion. This entails 1) the development of a compression module capable of compressing images from various imaging modalities, 2) the assessment of compressed images in a clinical setting, 3) the extension of a database to include images from all modalities for compression studies, and 4) evaluation and improvement of image quality.

In conclusion, Dr. Huang's team is striving to design and develop an accurate diagnostic imaging system with rapid throughput of patients, timely reporting and efficient retrieval of patients' radiological examinations. They are investing six years of experience in PACS research in this major undertaking. So far, they have success fully implemented two PACS prototype systems in pediatric radiology and coronary care units. These systems have been used for the past two years in the daily routine of patients' examinations. This success encourages them to plan five additional PACS modules in intensive care units, neuroradiology, and chest radiology. It is heartening to say that upon completion of this project, 35% of the total examination procedures at the department of radiology at UCLA will be converted to PACS digital imaging systems.

The anticipated success of this research project will usher in a new era in radiological examination, improved patient care, an health care costs.

2. MEDICAL IMAGE PERCEPTION

Dr. Stephen Pizer, Principal Investigator
University of North Carolina, Chapel Hill, NC, U.S.A.

Professor Stephen Pizer and his colleagues at the University of North Carolina are hard at work developing improved high-quality image display systems. They are utilizing data obtained from conventional x-ray, computed tomography (CT). magnetic resonance imaging (MRI), and single photon emission computed tomography (SPECT) imaging systems. Among the activities of this group is the development of software displays, hardware, stereo, kinetic depth effect, and hard-mounted displays. They are also involved in the assessment of the clinical application of the developed systems in diagnostic medical imaging and planning for radiation therapy.

A major effort of this group is directed towards to development of the appropriate software to compute automatically image description in a rapid manner for both 2D and 3D. Images produced by this system will be evaluated by experienced radiologists in a clinical setting. In addition, improved software is being developed for direct dynamic display and immediate clinical evaluation.

Another significant aim of this research is the development of algorithms for optimal image quality. Various approaches are being employed to enhance image quality. One example is extensive research to improve the gray scale and enhance image contrast. Other studies are directed towards the reduction of noise to signal ratio and to improve spatial resolution.

The North Carolina University group is also engaged in research leading to the development of parallel hardware for image processing and graphic development of display systems. These systems are designed to real-time 3D presentation of dynamic processes and real-time interactions. It is expected that when the system is completely constructed and evaluated, it will have an impact on the electronic image use in clinical examination in both radiology and oncology departments.

In addition to the on-going research into display systems which this team is known for, medical workstations are being designed, developed, and assessed for effective presentation of clinical images for radiologists. The stations are placed in a clinical setting to enable the radiologists and oncologists to evaluate the quality of images produced by the display systems in a clinical environment.

In conclusion, the primary objective of this research program is to develop improved means of medical image processing, utilizing data obtained from various imaging modalities, as well as radiation therapy port films. Efforts are being made a) to produce better gray scale 2D images by using regionally adaptive contrast enhancement techniques and targeting context-sensitive human vision, b) to develop dynamic presentation of 3D images for both anatomical and functional information, and c) to define anatomic objects for fast 2D/3D object definition. This team is in the process of developing a state-of-the-art imaging facility far ahead of many other institutions.

3. EVALUATION OF PACS-RELATED DIAGNOSTIC RADIOLOGY CONCEPTS

Dr. David Gur, Principal Investigator
University of Pittsburgh, PA U.S.A.

Dr. David Gur and his experienced staff at the University of Pittsburgh are involved in the development and assessment of Picture Archiving and Communication System (PACS). Their goal is to develop a PACS system capable of being the primary technological imaging system for clinical imaging in radiology departments. Normal and abnormal cases will be analyzed using the Receiver Operating Characteristics (ROC) Curve. These studies will assess the quality of images if the newly developed PACS systems for clinical imaging in daily routine examinations at the department of radiology.

They plan to observe, examine and evaluate the technical aspect of the system and assess the electronic imaging performance in a traditional setting at the hospital. In addition to evaluating image quality, they will determine the cost effectiveness of the system. The extensive and continuous evaluation and improvement of electronic imaging is essential for developing acceptable medical imaging equal to or better in quality than conventional imaging. A series of workstations were assembled by Dr. Gur and his colleagues for clinical use in their department. The display system has the following advantages over other display systems: 1) it is extremely user-friendly (all function are controlled by a trackball and three push buttons and it does not require typing), 2) it has very high image quality (uses state-of-the-art display), and 3) it is very fast (it displays up to 400 images at full resolution in 1/3 of a second).

They are able to use this system to perform large-scale clinical studies under conditions which simulate the environment of a contemporary radiology reading room. Based on a series of multi-observer studies involving 22 radiologists and the examination of over 25,000 images, the results are summarized below:

1) Sophisticated digital equipment (soft display) can be used for primary diagnosis in a clinical environment and as efficient as conventional systems.

2) The system is capable of detecting nodules or masses with an accuracy level comparable to that achieved by conventional films.

3) The availability of concise computerized clinical patient histories during the image reading does not increase the accuracy level for the detection of nodules and interstitial disease.

4) The availability of image processing options increases the accuracy level in the detection of the above-mentioned abnormalities.

5) In subspecialty examinations, e.g., computed tomography (CT), digital technology is more efficient and has equal or better diagnostic accuracy than conventional techniques.

Another major objective of this project is to evaluate the cost effectiveness of the digital imaging system. Preliminary multi-institutional financial studies were conducted to assess the potential impact of PACS on health care delivery and cost effectiveness. All institutions had a similar clinical environment and followed the same methodology of system evaluation. The results were encouraging. Preliminary results showed electronic imaging is competition for conventional imaging. With refinement and improvement in technology, electronic imaging will surpass conventional imaging in quality of examination and reduce patient care costs.

Digital Imaging Network Systems in the U.S. Military: Past, Present, Future

Jerry A. Thomas, Medical Service Corps, United States Navy

Department of Radiology/Nuclear Medicine
Uniformed Services University of the Health Sciences
4301 Jones Bridge Road
Bethesda, Maryland 20889-4799

Abstract: The United States Military (Army, Navy and Air Force) has been active in Picture Archiving and Communications (PACS) for over ten years. Early work includes three teleradiology studies and a full scale prototype PACS implementation at two civilian universities. An Educational PACS and Computer Aided Instructional (CAI) programs have been developed by the Uniformed Services University of the Health Sciences (USUHS) to support Gross Anatomy, Neuroanatomy, Pathology and Radiology training of military medical students. Current activities include PACS implementation into military hospitals and clinics under the Medical Diagnostic Imaging Support Systems (MDIS) program and developing additional digital image based CAI programs to supplement first year (GME-2) medical, surgical and radiology residency training. Future plans include implementation of PACS in all military treatment facilities, operational units and combat medical support units.

Keywords: PACS, U.S. Military PACS, Educational PACS, MDIS, Teleradiology

I. Introduction

Past, present and future PACS activities in the U.S. Military medical department will be described. In the late 1970's the U.S. Military along with the U. S. Public Health Service became interested in Tele-Medicine as a means to more effectively utilize available personnel, time and financial resources in meeting expanding health care delivery responsibilities and to improve the combat care capabilities of the military medical department. The availability of cheap, compact and portable computers in the early 1980's fueled Teleradiology and PACS research interests and efforts. Early research focused on determining clinical efficacy, cost effectiveness, technological capabilities, and potential usefulness of teleradiology in combat casualty care. Three teleradiology studies in 1978,

NATO ASI Series, Vol. F 74
Picture Archiving and Communication Systems (PACS) in Medicine. Edited by H. K. Huang et al.

1982 and 1985 and two PACS prototype implementations at civilian universities have been completed. Since 1985, Uniformed Services University of the Health Sciences (USUHS) has developed an Educational PACS and three digital image based CAI programs for use by first, second and fourth year medical students and first year radiology residents.

Current programs include implementation of PACS into three military clinical centers and a Teleradiology hub and spoke at a fourth. Additional digital image based CAI programs are being developed for Graduate Medical Education in Radiology, Surgery and Medicine.

Future PACS implementations are anticipated to impact significantly on health care delivery methods and on the utilization of personnel, financial and time resources in times of peace and war. PACS should allow more efficient management of these resources and consequently should provide for increased efficiencies in health care delivery.

II. Past Research Efforts

A. 1978 Tele-medicine. In 1978 a Teleradiology study using a Colorado Video analog teleradiology system was evaluated by the Navy for Ship-to-Shore transmission of radiographic images. Transmitted image quality was not that required for primary diagnosis. Image transmission times were to long and were unacceptable. No further efforts in Ship-to-Shore image transmission have been made.

B. 1982 Teleradiology Field Trial. In 1982, MITRE Corporation under a contract from the U.S. Public Health Service and with cooperation from the U.S. Military, built a Teleradiology system (512 x 512 x 8 bit image) to support a six month field trial. The study objectives included determining the technical feasibility, clinical effectiveness, professional acceptance and operational costs of transmitting digital radiographic images over leased telephone lines to a central site for interpretation. Four remote sites -- Army, Navy, Air Force and Public Health Service clinics -- digitized and transmitted images to a central site where, daily, they were "read out" by a radiologist. Films were interpreted in the usual manner. Reported findings from film interpretation and digital image "read-out" were compared. Any difference in findings between the two reports were evaluated by a consensus panel. ROC analysis were performed. The results showed the overall accuracy (sum of correct findings divided by the total of possible findings) to be 96.7% for findings and 95.0% for impressions.

C. 1985 Field Trial. A follow-on study to the 1982 teleradiology field trial was conducted in 1985. MITRE Corporation under contract from USUHS purchased commercially available teleradiology hardware with 10:1 image compression capability. The four Army, Navy, Air Force and Public Health Service remote sites used in the 1982 study were used in this study. The central reading site was changed to USUHS. Images were digitized into a 1024 x 1024 x 8 bit format and transmitted at 9600 baud over non-treated telephone lines. Ten percent of the images were compressed 10:1 using a hardware cosine transformation. Films and digital images were evaluated in the same manner as the 1982 study. ROC analysis of the data gave results similar to the 1982 study.

D. 1988 - 1990 Hospital Based PACS. Success from the two Teleradiology Field Trials lead to a two year study designed to determine the requirements for a fully functional hospital based PACS. Georgetown University (Washington, D.C.) and University of Washington (Seattle, Washington) were selected to serve as the clinical evaluation sites. Fiber optic networks and digital imaging equipment (phosphor plate readers, video frame grabbers, film digitizers and various sized reading stations) were installed at each University. Time-in-motion studies, data transmission rates, clinical acceptance, and a cost analysis of PACS were made. The results from this study were used to prepare specifications for Hospital based PACS systems and Teleradiology hubs and nodes. Army and Air Force medical facilities will be the first to receive PACS systems. Navy medical facilities will begin to procure PACS systems after 1994.

III. Current work

A. Educational PACS.

During the 1985 Teleradiology Field Trial, the need to acquaint medical students, graduate students, and physicians in graduate medical training programs with the digital image and the usefulness of digital images in medical education was recognized. Gross Anatomy, as taught at USUHS, includes weekly lectures from radiologists on normal radiographic anatomy. Due to film costs and concerns about lost and damaged films students were given limited access to the radiographic images used in each lecture. A CAI program utilizing digital images could make an increased number of radiographic images available to the students. Additionally, students would have 24 hour access to the radiographic images without having to have a radiologist available. Since 1985, three CAI programs using digital images on an Educational PACS system have been developed at USUHS. The Educational PACS is integrated into the University Local Area Network, a fiber-optic backbone with Ethernet co-axial drops throughout the University. Students and faculty members may access over 1500 digital images and CAI programs from any office, laboratory or study area. Simple image manipulation to include zoom, pan and scroll and image inverse is available at all locations were images may be viewed.

The three CAI programs developed for student use and an on-line radiographic image data base available for faculty use are used extensively throughout the academic year. Two programs, "Normal Radiographic Anatomy" and "Neuroanatomy" are used by first year medical students in the Gross Anatomy and Neuroanatomy courses. A third program, "Rad/Path Correlation" supplements the second year medical student Pathology course. Faculty members utilize the Educational PACS images for self-study and for developing examinations, teaching and lecture aids. A "Digital Case of the Week" and answers to the weekly Radiology Quiz are available to second year medical students on the Educational PACS. A new program "How to Read a Chest X-ray" for use in the second year Radiology course is being developed.

1.) "Normal Radiographic Anatomy". The first CAI program developed, "Normal Radiographic Anatomy", provides over 300 radiographic images grouped according to major body section -- Back and Upper Extremities, Chest, Abdomen, Pelvis, Lower Extremities and Head and Neck. All modalities used to image the body section are represented. Anatomic structures on the digital image are labeled. Test and continuous display modes are available.

The viewing station is a PC-AT based computer with an Imaging Technologies PC Vision Plus card. Two screens are used, one for the digital image and the other for text.

Since its introduction in 1985, this program has become an integral portion of the Gross Anatomy course. Both faculty and students have found the program to be invaluable in teaching and learning Gross Anatomy and normal radiographic anatomy.

2.) "Neuroanatomy". A follow-on to "Normal Radiographic Anatomy", "Neuroanatomy" uses two digital image display screens and one text screen and contains 325 digital images. Digital Images of gross brain sections and the corresponding CT and MRI are simultaneously displayed. Structures in each image are labeled with the key being displayed on the text screen.

3.) "Rad/Path Correlation". A new program this year, "Rad/Path Correlation" supplements the second year Pathology course. Digital and analog images are used in conjunction with each other. A background discussion of the case material is presented and is followed by abnormal digital radiographic images and an analog color image of the gross specimen with a corresponding discussion. Low power and high power microscopic images of the disease are stored on an analogue video disc. Normal radiographic, gross and microscopic images are available for comparison.

4.) Digital Image Data Base. Over 1500 digital images are available for faculty use. The images are cataloged according to imaging modality, view, anatomic structure and radiographic finding. Faculty members use the images to supplement their teaching materials. Weekly quizzes, supplemental study materials, and lecture aids are made from the digital image data base.

B. Medical Diagnostic Imaging Support Systems (MDIS).

The MDIS project was created by the Surgeon's General of the Army, Navy and Air Force to exploit the results of the extensive research projects of the past ten years. This project is designed to procure film-less radiographic systems through a contracting approach for several military health care facilities. The project is described in detail in an accompanying paper "Medical Diagnostic Imaging Support Systems for Military Medicine" by Lieutenant Colonel Fred Goeringer.

IV. Current Resources Impacted by PACS

Implementing PACS into military medical treatment facilities will impact personnel, time and financial resources and will provide the military manager with management options not previously available..

Personnel impacts will range from better utilization of reserve officers to increased retention of radiologists. Currently, radiologists serving in reserve units do not utilize their radiology skills during required monthly weekend drills. PACS and digital imaging will allow the distribution of radiology workload from clinics to large hospitals, from large hospitals to small clinics and from either large or small hospitals to reserve units thereby allowing workload rather than personnel to be shifted between facilities. Staffing of military hospitals in geographical regions containing hospitals from more than one service may be augmented or reduced as necessary by using PACS to redistribute radiology workloads among the different service hospitals. Increased job satisfaction, immediate access to imaging experts and the opportunity to work with state-of-the art imaging equipment should result in increased retention of active duty radiologists.

PACS will eliminate time wasted looking for lost films and locating current studies. Referring physicians (internists and surgeons) will have immediate access from their work place to the radiological images and to the radiologists report. Second opinions may be obtained without having to track the expert down.

Contracts for radiology services will be able to be eliminated or competitively bid, resulting in significant savings to the U.S. Military for these services. Film costs for repeated and lost films will be eliminated.

References

1. Cerva, J.: Technical Specifications for a Teleradiology System, The MITRE Corporation, MTR-85W239, November 1985

2. Harrington, M.B., Miller, K.D. and Wilkinson, M.: The Results of an Image Quality Evaluation Using the 1982 Field Trial Digital Radiographic Imaging System. The MITRE Corporation, MTR-84W255, December 1984

3. Harrington, M.B. and Miller, K.D.: Results of the Image Quality Evaluation Conducted with the 1984-85 Teleradiology Field Trial. The MITRE Corporation, MTR-85W236, November 1985

4. Imaging Systems Group: Technical Specifications for a Hospital Based digital Imaging Network, The MITRE Corporation, MTR-85W242, January 1986

5. MDIS Technical Development Team: MDIS Medical Diagnostic Imaging Support System Acquisition Document, 15 October 1990

Medical Diagnostic Imaging Support Systems for Military Medicine

Fred Goeringer

Office of the Surgeon General, United States Army
5109 Leesburg Pike, Falls Church, Virginia 22041-3258, USA

Abstract
The Medical Diagnostic Imaging Support (MDIS) project will achieve the objective of implementing filmless medical imaging systems at several military medical treatment facilities over the next 4 years. The Surgeon's General of the military services created the MDIS project to exploit the results of extensive imaging research efforts over the past ten years. Filmless medical imaging systems are a superior health care delivery alternative to film-based image support systems which are inherently limited by film as a hard copy media. Four enabling core technologies make it possible to system integrate an effective filmless system for military medicine. These filmless MDIS systems will be acquired from industry through a contracting approach that (1) functionally describes subsystem and system performance for acceptable clinical operations, (2) validates proposed systems in performance evaluation, and (3) makes a system selection and contract award on the basis of best value to the government.

Key Words
Filmless Medical Imaging, Medical Diagnostic Imaging Support, MDIS, Military Medicine, Teleradiology, Total PACS, System Integration.

MDIS Background and Objectives

The Defense Medical establishment has been actively involved in filmless medical imaging research and development (R&D) throughout the decade of the 80's. Thirty million dollars has been invested in R&D with the intent of exploring incarnations of

NATO ASI Series, Vol. F 74
Picture Archiving and Communication Systems (PACS) in Medicine. Edited by H. K. Huang et al.

filmless medical imaging technology–in the areas of image acquisition, display, storage, and communications technology–that are prudent for implementation for military medicine. These efforts have included such projects as the Teleradiology Field Trials sponsored by the Uniformed Services University of the Health Sciences and the DINS or Digital Imaging Network Systems project sponsored by the U.S. Army Medical Research & Development Command in the late 1980's.

As the decade of the 90's begins, the military medical community is now at a stage where it will achieve a major return on investment in these filmless imaging research efforts. In January of 1990 a Medical Diagnostic Imaging Support (MDIS) System Project was jointly established by the United States Army, Air Force, and Navy Surgeon's General with the explicit goal of pro-actively achieving filmless medical imaging excellence early-on in this decade throughout the military health care delivery system. The MDIS project will achieve three major objectives for improving military health care:

Objective 1. MDIS will establish a contract with industry that makes clinically effective filmless medical imaging technology available to the Defense Medical establishment and other Federal agencies.

Objective 2. MDIS will exploit the economies of scale achieved in this contract to set new standards of cost effectiveness and quality for filmless radiologic practice above those that were previously available.

Objective 3. The MDIS contract will be the primary vehicle for all filmless medical imaging applications in the Defense Medical establishment over the next four years.

The discussion of the MDIS project is organized into three parts: (1) Technology Assessment and Contract Design, (2) Major Functional Specifications, and (3) Conclusions.

Part 1. Technology Assessment and Contract Design

1.1 The MDIS Technology Curve. Figure 1 depicts our macro-assessment of the state of filmless medical imaging technology through a learning curve paradigm. We are at a point now at the knee of this curve. The MDIS contract specification gets us beyond the knee of the curve in directly addressing the final major issues of system and image database integration issues–all of which build to achieving clinical acceptance. Our view is in the mainstream with most filmless medical imaging technology assessments today which typically report (1) that 'all the components are there' and (2) that an increasingly-informed radiology and referring-physician community fairly well know 'what now must be done' for a clinically acceptable filmless implementation–all with the aim of improving productivity and quality of medical diagnostic imaging support in a health care setting.

Figure 1 - The MDIS Technology Curve

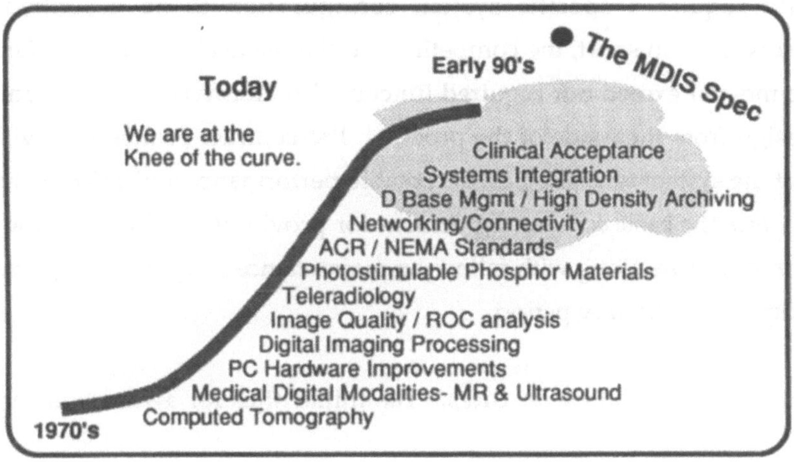

1.2 MDIS Enabling Core Technologies. Our basic conclusion regarding the technology itself is that there are four enabling core technologies that make MDIS feasible for active system integration and implementation. Each of these four areas now provide a depth of technical performance and clinical capability that did not exist as recently as 24-36 months ago. These four areas are: (1) emerging cost-effective computed radiography technology with a variable range of performance; (2) high performance, customizeable imaging workstation technology with nimble user interfaces; (3) high bandwidth throughput and connectivity in local and wide area networking; and (4) high density digital storage technology with now-effective writing and retrieval speeds for supporting robust medical image data sets.

1.3 MDIS Contract Design Overview. The nature of our technology assessment was the main determinant of our basic contracting approach. The MDIS contract will be an indefinite delivery requirements contract that will seek a competitively-awarded solution from industry under provisions of the Federal Acquisition

Regulation. In the specification, various MDIS system requirements are described by functionally and clinically acceptable performance–

(1) at the device/subsystems level,
(2) at the integrated system level, and
(3) at the clinical operations level in various 'clinical scenarios.'

This basic functional approach to achieve a contract is contrary to a design approach that would require a specific system configuration to be described on the buyers/users side. In short, the competition in the medical imaging marketplace, in seeking to meet or exceed our required functional performance, will determine final system design from their side of the process. Pre-contract evaluations will look to selection of the system with the best acceptable performance and price/value. After contract award, the basic contract agreement for providing MDIS technology will be in place for four years along with follow-on maintenance support for eight years after the basic first year warranty period.

1.4 Start-up and Follow-on Sites. The initial sites for award will include the following:

- Madigan Army Medical Center at Fort Lewis, Washington
- US Air Force Medical Center, Wright-Patterson at Dayton, Ohio
- Brooke Army Medical Center at Fort Sam Houston, Texas
- Teleradiology hub at Luke AFB Hospital at Phoenix, Arizona linked to a spoke at Davis-Monthan AFB Hospital at Tuscon, Arizona

Additionally, there are contract options to issue follow-on delivery orders for nine other government intra-hospital MDIS (PACS) sites as well as twelve teleradiology hubs and 96 teleradiology supported spokes over the four year period.

1.5 Project Timing and Vendor Evaluation. The MDIS contract bid specification was issued to industry in June 1990. Evaluations of vendor proposals will begin in late 1990 with benchmark testing scheduled for early 1991. The vendors' systems will be scored and a contract will be awarded on a "best value" basis to the government considering both technical and managerial assessments as well as price.

"Best value" will be determined through two phases of evaluation prior to award: (1) scoring vendor's written proposals of their systems, and (2) benchmark testing at both the component and integrated system level of all vendors that are judged competitive after the phase 1 written evaluation. Vendor selection and contract award will be complete by mid-1991 with first MDIS installations and site acceptance inspections in late 1991. Installations at the larger sites will be implemented over a twenty-for to thirty-six month period using a phased approach as discussed below.

1.6 Embedding Defense Against Obsolescence. A key element of the contract is the defense-against-obsolescence clauses. Any filmless medical imaging effort must embed technology upgradeability into the original basic design. MDIS makes use of 'engineering change proposal' (ECP) contract clauses which *require* the vendor to periodically propose technology upgrades for the system over the life of the contract. Prices for these upgrades are cost-negotiated based on the overall performance improvement and anticipated value to the government. Additionally, the contract provides for "value engineering" in which new items that clinically and technically improve the productivity and efficiency of the system shall be included.

1.7 The MDIS Phased Implementation Approach. At larger sites–those facilities with over 1,000 plain radiographic exposures per day and about 2000 digital exposures per day (i.e., CT, MR, Ultrasound, Nuclear Medicine, and others)–MDIS will be implemented in discrete phases over a two or three year period. The figure below depicts the implementation approach for a large medical treatment facility. After installation, phase one begins with implementation of the nucleus of the system–primarily in the radiology department. Initially all plain radiographic images are acquired by computed radiography (CR). To build clinical acceptance, about 40% of exams are read from soft copy displays and 60% are read from single-emulsion film. Hard copy non-accountable "reference images" are printed for many referring clinical areas outside of the radiology department. Phase two involves proliferation of the system to high volume referring activities such as orthopedics. In this phase there will be more reliance on soft copy interpretation and image delivery and the RIS-MDIS interface will become more sophisticated. Finally, phase three will involve total proliferation of the system to all areas of the hospital to complete the implementation. A mature MDIS system will make the facility about 90% filmless.

Figure 2 - The MDIS Phased Implementation Approach

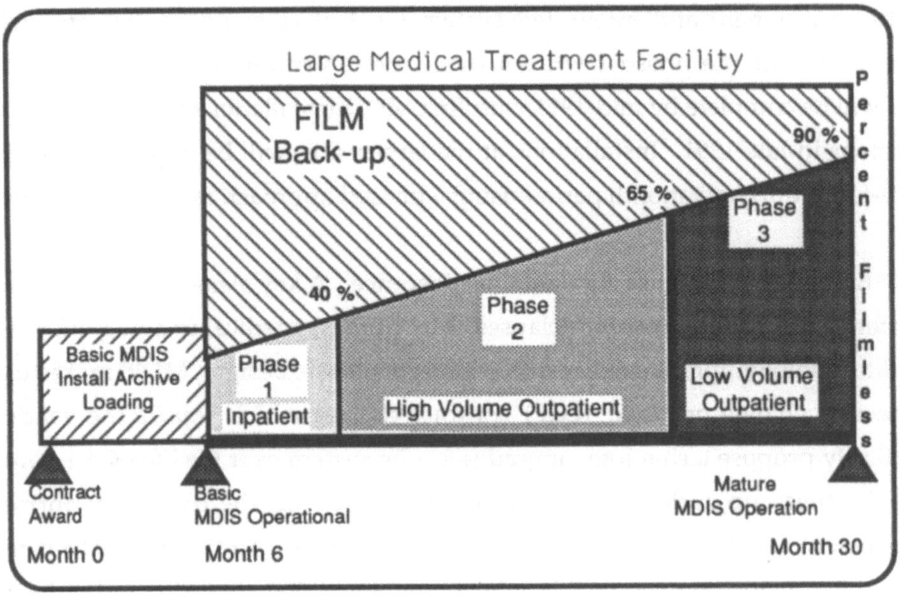

Part 2. Major Functional Specifications

2.1 Overview. Functional technical specifications are described in four major sub-system areas in the MDIS contract solicitation. These are:

(1) Image Acquisition
(2) Image Display and Output
(3) Image Database and Storage
(4) Communications and Network

Additionally, System Integration performance and Systems Installation, Maintenance, and Support functions are also specified in order to acquire a total system implementation. Specification highlights of each of these areas are discussed below.

2.2 Image Acquisition

2.2.1 Plain Radiography. Primarily because of operational issues associated with patient identification (in addition to accruing substantial double-emulsion film dollar savings), we require maximum use of computed radiography (CR) as our sole acquisition technique for new plain radiographic images from the outset. This is in lieu of relying on digitized hard copy analog film for new exams. Our view is that hard copy film digitizers will be used only in three supporting roles: (1) to introduce previously-existing hard copy analog images into the system, (2) to support small teleradiology spoke sites where computed radiography is not cost-feasible, and (3) to serve as an emergency system fail-safe to computed radiography at sites where there may be only one CR device. Hard copy film digitizers will not be used routinely for new plain radiographic image acquisition; these devices are more 'operator intensive' than CR on a per image basis for new images. Figure 3 below summarizes our main computed radiography requirements in which three levels of functional performance are specified.

Figure 3 - Three Levels of Computed Radiography

Computed Radiography (CR) Attribute:	Performance Requirements*:		
	High	Medium	Low
Throughput (plates / hour)	60	40	20
Multiple Plate Size Processing **	Yes	Yes	No
Input Stacking Capacity (Plates)	30	10	No
Spatial Resolution- 14" X 17" Plate Size	2.5 lp/mm	2.5 lp/mm	2.5 lp/mm
Spatial Resolution- 8" X 10" Plate Size	5.0 lp/mm	5.0 lp/mm	5.0 lp/mm
Dynamic Resolution- Minimum: 10 Bits/pixel	Yes	Yes	Yes
Image Processing	Yes	Yes	Yes
Max Footprint	NA	NA	12 sq ft Desktop

NOTES: * Numerical entries represent minimum acceptable performance levels, except where indicated otherwise.

** Varying plate sizes typical of those used in conventional radiography.

2.2.2 Image Acquisition Highlights, Other Imaging Modalities. Computed tomography (CT) and magnetic resonanace (MR) images will be introduced into the MDIS system at their full 12 bits of dynamic range along with the image examination parameters (e.g., MR signal intensities) using ACR-NEMA standards. Most CT and MR medical imaging vendors have announced this capability for filmless support. In situations where there is no ACR-NEMA interface–such as with a GE 8800 series CT scanner–a custom interface is acceptable. Because of the relatively limited dynamic range of ultrasound, these images will be acquired by MDIS using video frame-grabbing techniques. Nuclear medicine images will be acquired by individual customized solutions because of the ad-hoc nature of most nuclear medicine imaging system configurations. MDIS does not require the digitization of real-time fluoroscopy–only the image that would be recorded as a photospot film is required for entry into the system. If the fluoroscopic system has digital output, this will be exploited for acquisition; otherwise video frame-grabbing from the image intensifier will be used for image acquisition. The acquisition of conventional spot films that use the spot film device will utilize CR plates for digital acquisition.

2.3 Image Display and Output

2.3.1 Soft Copy Image Presentation. For MDIS, we will retain a range of options for soft copy image display in support of a variety of clinical environments. There are contract options for specialized workstations to handle unique situations such as cine and 3D rendering of volumetric images from CT and MR as well as on-call portable teleradiology. To handle the majority of our soft copy display requirements, there will be two primary categories of image workstations, referred to as 'SCIDs'–or soft copy image displays. These are: (1) the standardized workstation or SCID-S and (2) the optimized workstation or SCID-O. The SCID-S is intended to support primary interpretation clinical support situations, while the SCID-O is intended for the referring physician setting where ease of use in a simple interface and throughput is 'optimized.' For the SCID-S, our minimum requirements are a '1K window' with the full 2K dataset in the frame buffer. The SCID-S workstation can be re-configured to a full '2K window' if warranted by the clinical situation. In addition to these monitor configurations, the full range of commonly accepted image manipulation, image enhancement, and image supplement features (e.g., roam, pan, and window and level functions) are also required. Figures 4 and 5 below, summarize the major functional features associated with these workstations.

Figure 4 - Soft Copy Image Display (SCID) Functions

Workstation Functions	Standardized Workstation	Optimized Workstation
Pictorial Image Directory	Yes	Yes
Worklist / Patientlist	Yes- Worklist	Yes- Patientlist
Display on Single Monitor	Multi-images	Multi-images
Image Rearrangement within & between Monitors	Yes	Yes
Paging within Exam	Yes	Yes
Edge Enhancement	Process/Display	Display
Window/Level Adjustment	Yes	Yes
Inverse Video	Yes	Yes
Image Roam	Yes	Yes

Figure 5 - Soft Copy Image Display (SCID) Requirements

Attribute:	Standardized Workstation	Optimized Workstation
Number of Displays	1-8 monitors	1-2 monitors
Monitor Type	Class A, B, or Both	Class C
Monitor Type Spatial Resolution (pixels)	A ≥1536 x 2048 B≥1024 x 1280- Portrait	C ≥ 880 x 1150 or C ≥1150 X 880
Viewable Raster Diagonal	17"to 23 "	17" to 23"
Data set- Rad Images	Full Data Set	Downsampled Data Set
Brightness (luminescence)	A>40 / B>60 FtLamberts	C> 60 ft Lamberts
Viewable Greyscale Display	8 bits	8 bits
Refresh Rate	Flicker Free	Flicker Free
Monitor Calibration	Yes	Yes
*Brightness Uniformity	<15%	<15%
*Linear Distortion	<3%	<3%
*Geometric Distortion	< 3%	< 3%
*Spot Size Variation	<50%	<50%
Frame Buffer	12 bits	8 bits or greater
Auxiliary Output for Slave Monitors	Yes	Yes

Note: All * items are measured concurrently

2.3.2 Image Output & Display Highlights, Hard Copy Output. MDIS requires that conventional laser printers operate as assets of the entire network for supporting hard copy medical image output from the full range of image acquisition modalities. If laser printers are currently components of existing CT or MR devices at each site, then these printers will be retrofit into the MDIS network. The laser printers will produce hard copy single emulsion images during the phase-in period–perhaps as much as 60% of the image volume–when an MDIS system is first implemented. This will gradually draw down to less than 10% of the image volume as soft copy reading and SCIDs proliferate throughout the facility. With the increasing availability of SCIDs throughout the facility, a key operational parameter is controlling access to film printing privileges from workstations; not all clinicians should have this privilege to foster building confidence in using the system 'filmlessly.' Additionally, to support residency and medical education programs at larger teaching facilities, 35mm hard copy slide capability is also required as a specialized hard copy device.

2.4 Image Database & Storage

2.4.1 Database Overview. The MDIS database shall serve as patient image exam manager. As such, it shall; (1) accept and store image exams through the inter-MTF or intra-MTF network that are acquired by image acquisition devices; (2) communicate stored and newly acquired images across the network on demand or by a logical autorouting algorithm to individual workstations or hard copy devices either for reference with the accompanying previous radiologic interpretation or for initial interpretation; (3) automatically restore exams to workstations according to a predetermined procedure; and (4) function through a useful interface with the radiology information system (RIS) or with the DoD Composite Health Care System (CHCS) as applicable. The image database shall make use of design strategies to optimize cost and performance tradeoffs between image archiving efficiency and image delivery times. Specifically, logical designs are required that feature short-term storage for high demand imaging exams for fast retrieval and long-term archiving for low demand imaging exams for slower retrieval.

2.4.2 Image Database, Short Term Storage. Images and related information that shall be in the short-term storage are (1) exams that are newly acquired in the past 48 hours, (2) exams awaiting primary interpretation, (3) exams acquired in a period equal to the the facility's average length-of-stay for in-patients, (4) selected historical exams for autorouting to a clinical area according to a daily clinic appointment schedule, and (5) selected supporting historical exams of patients who have had new image exams. Short-term storage may be centralized or distributed throughout the system. Design strategies such as activity-based caching may be acceptable approaches, provided that short term storage throughput performance can be maintained or exceeded, for the image categories specified here.

2.4.3 Image Database, Long Term Archive. This archive shall contain all images and related information necessary to support (3), (4), and (5) in the sub-paragraph directly above. The long-term archive shall be capable of storing the current year plus 4 additional years of imaging exams for a total of 5 years of imaging exams; 2 years will be 'on-line' and 3 years will be 'off-line.' Additionally, there are instances where exams must be retained for longer periods–e.g., pediatric images must be retained until a patient's 21st birthday or digitized post-interpreted mammography exams must be retained for the life of the patient. The particular archiving periods shall be site-specific, depending on the clinical situation. The archive shall also reflect the site-specific compression strategy as discussed below.

2.4.4 Image Database Compression Strategy, Digital Microfilm. The data compression strategy shall be site-specific and offer the option of both bit-preserving and non bit-preserving compression techniques for exam management. Reasonable compression ratio options range from 2:1 to 10:1. A site-specific scenario where image exams can be displayed at full acquisition device resolution prior to primary interpretation followed by post interpretation long-term archiving at compression ratios of up to 10:1 shall be provided. Moreover, this post interpretation long-term archiving compression operation shall also be site-specific and capable of being varied by type of exam and by acquisition modality. This operation is termed "digital microfilm."

2.4.5 Image Database, Key Operational Features. The database shall use the metaphor of patient and teaching folders that hold previous images and examination reports. Moreover, these folders shall be capable of being 'opened' at the workstation to display relevant images and reports. This feature is termed "Pictorial Image Directory." The database also will make maximum use of image autorouting in background to deliver relevant clinical images to support clinical activities. For example, once each day the hospital information system will communicate a list of daily appointments for the orthopedic clinic through an interface to the MDIS system; this will trigger routing of the last previous orthopedic exam from the archive to the workstation in the orthopedic clinic. This autorouting will occur at a time of day–for example at 1 am–when there is minimal traffic on the network.

2.4.6 Image Database, Interface to the Radiology Information System (RIS). In a larger sense, medical images can be regarded as just another form of patient information; any filmless medical imaging solution that does not recognize this fact and drive for total integration of hospital information systems and the image network will be limited. Accordingly, the MDIS system addresses this integration by initially requiring a one way interface between the image database and the RIS in which there is a master-slave relationship with the RIS as the information master. Over the course of the implementation phases, this interface will evolve to a true two-way interface between the RIS and the image database. The interface will achieve and maintain information integrity of the two systems by observing three primary data management requirements. These are:

(1) Using a consistent exam and reporting approach between the two databases
(2) Using common data elements between the two databases
(3) Requiring ACR-NEMA compatibility

2.5 Communications and Networking

2.5.1 Intra-Facility Highlights. Intra-hospital PACS networks will make use of high speed communication protocols such as FDDI for support to high volume image areas such at the radiology department and other selected patient areas such as the emergency room. Other areas of the facility will be serviced by a lower speed communications protocol to support image delivery for consultation purposes to SCID-O review workstations. It should be noted that image traffic over this secondary portion of the network may be downsampled 1Kx1K size eight bit deep reference images that are for image consultation by referring physicians.

2.5.2 Teleradiology Highlights. Figures 6 and 7 below summarize MDIS teleradiology requirements for receiving teleradiology hubs and for sending teleradiology spokes.

Figure 6 - Teleradiology Hubs

Required Performance				
	Spokes Supported	Output & Display	Storage	Images & Reports of Record
Small Hub*	4 small or 1 large Spoke w/out added capability	2 Screen Diagnostic Workstation	Magnetic or Erasable Optical	Hardcopy Film & Reports at Spoke
Medium Hub*	4 small or 2 large Spokes w/out added capability	2 Screen Diagnostic WS	Single Drive Optical Disk	On Disk at Hub No RIS Support
Large Hub*/**	8 small or 4 large Spokes w/out added capability	two each 4 Screen Diagnostic WS	Optical Disk Juke Box	On Disk at Hub RIS Support

Notes: * Added capability from individual CLINS will allow upgreadability to 8 spokes
**must support upgradeability to intra MTF capability, must be capable of transmitting images (e.g., CT) to spoke

Figure 7 - Teleradiology Spokes

Required Performance				
	Communications Workload*	Input	Display & Output	Communication**
Small Spoke	50 / day	Small Film Digitizer and DeskTop CR	Single Screen Clinical Workstation	19.2 Kbps or 56 Kbps
Large Spoke****	150 /day	Large Film Digitizer*** and Mid-Performance CR***	2 Screen Diagnostic WS Compact Laser Printer	56 Kbps or T-1 Grade

Notes: * 14'by17" images or equivalents
** Line determined during installation site survey
Line must also support report Transmission to/from hub
*** Includes image review monitor
**** Shall support Receipt of images (e.g., CT) from Hub

2.6 System Integration

2.6.1 Functional Perspective. The specific MDIS system integration solution is one that meets the generic system solution indicated in the basic figure below. The emphasis is on efficient functional image and patient data flow without regard to specifying any particular system topology. Specifically, the diagram does not imply that a star configuration is required. A system integration solution is acceptable as long as it meets the overall system performance requirements described below. The figure indicates a generic operational configuration from a functional perspective.

Figure 8 - System Integration, Functional Perspective

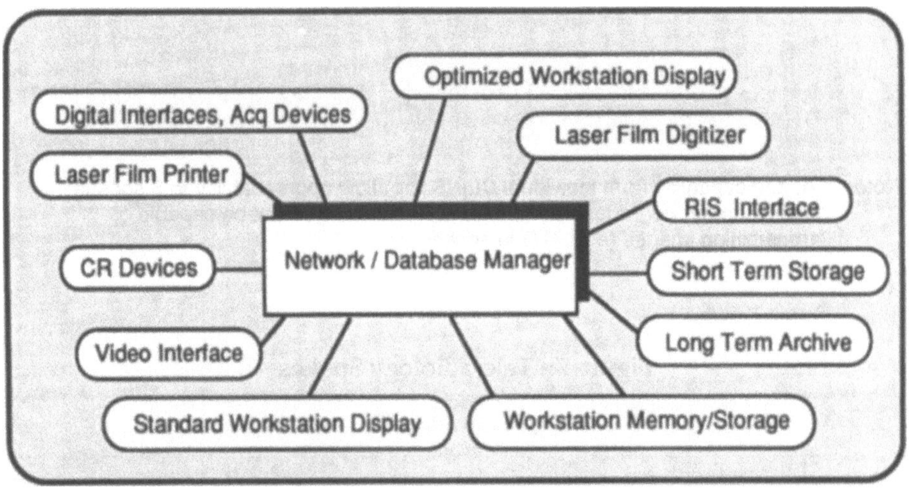

2.6.2 System Integration, Performance Measurement. Beyond performance at the device level, the nucleus of a successful system lies in its overall system performance from various operational perspectives. MDIS takes the approach of describing throughput performance along with image quality as perhaps the key system integration performance issues. Throughput is assessed by a specified standard image exam set under "loaded network conditions"–i.e., the busiest two hours of the clinical day. Benchmark testing prior to award and acceptance testing after award and installation at each phase will look at and validate performance that meets or exceeds these described standards. The three figures below summarize key MDIS areas for benchmarking and performance validation parameters that the system must meet or exceed.

Figure 9 - Key Performance Parameters in Each System Area

Subfactor	Key Element
Systems Integration	Throughput Performance
Image DB & Storage	RIS Interface Integrity
Image Acquisition	Cost Effective CR
Image Output & Display	Standard WS Operations
Commo & Network	Teleradiology Reliability
Install, Maint, & Training	Maintenance Plan

Figure 10 - Standard Image Set for Benchmarking

Image Type	Data Set per Image	Compression Type	Images Study
New Chest	2K x 2.5K x 10 bits	bit preserving	2
Previous Chest Image	"	10:1	4
New CT Study	512 x 512 x 12 bits	bit preserving	50
Previous CT Study	"	bit preserving	50

Figure 11 - Required Throughput, Standard Image Set

System Performance at SCID-S (4 screen)	Exam Thruput	Interval 1- first 3 min after thruput	Interval 2- >3 min after thruput	Text Info
from Acq Device, Routine	≤5 min	≤5 sec/screen	≤2 sec/screen	≤1 sec after Images
from Acq Device, Emer	≤2 min	"	"	
from Short Term Storage	≤2.5 min	"	"	
from Long Term Archive	≤5 min	"	"	
System Performance at SCID-O (2 screen)				
from Acq Device, Routine	≤15 min	≤5 sec/screen	≤2 sec/screen	≤1 sec after Images
from Acq Device, Emer	≤3 min	"	"	
from Short Term Storage	≤5 min	"	"	
from Long Term Archive	≤10 min	"	"	
Image Acq Device	Device Unimpeded by System	—	—	—

2.7 System Installation and Training Support Highlights. MDIS implementation must cover the gamut of system support issues. Installation shall be provided on a full "turn key" basis–i.e., basic site preparation and installation support details are bundled in the overall system acquisition. Additionally, training shall routinely be provided to the full hospital staff–including referring physicians, radiologists, ancillary professional personnel as well as supporting staff such as radiology technicians and ward masters.

2.8 System Maintenance

2.8.1 System Performance Requirements. During warranty and maintenance periods, the MDIS system shall maintain a monthly system uptime of 99%. For the purpose of this calculation, the "system" includes the MDIS database, storage devices, the CR and workstation functionality, the network, other network peripherals as well as the MDIS interfaces to the imaging equipment and to the RIS. For intra-MTF teleradiology it also includes the wide area telecommunications network between the hub and spoke. It does not include the imaging devices themselves (e.g., radiography units or CT scanners). "Uptime" means that all system elements/activities (e.g., CR capability, digitizer capability, workstation capability, or database management) comprising the system can perform their designated functions. If this is not the case, then the system is incurring "downtime". Specifically, for the image acquisition and display functions the system is considered "down": (1) if more than 20 % of standardized workstations are down, (2) if more than 10 % of optimized workstations are down, (3) if two or more film digitizers are down, (4) if two or more CR devices are down, or (5) if the MDIS portion of the interface to a CT or MR device is down. Planned system downtime consists of approved scheduled maintenance or installation of systems-wide software, firmware, or hardware upgrades or changes. Planned system downtime shall be less than 1/4 of 1% for the most recent previous 12 months total available system hours. Component uptime requirements are 95% by individual device.

2.8.2 System Maintenance Support Personnel. Large MDIS systems will have the following support personnel:

- **System Engineer**– responsible for optimal operation of all the computer components of the MDIS system, including the database, image transmission, interfaces, and electronic image archives.

• **Database/Archive Manager**– responsible for all aspects of image archiving, both electronic and hard copy; assures that images are entered into the archive in the correct format and are readily available for physician review and diagnosis; assures integrity of the database; shall demonstrate competence in the use of computers for data management, and in operation of a medical center image archive system.

• **Computer Technician/Trainer**– responsible for maintenance and training on the system; shall demonstrate competence in all aspects of computer systems maintenance and have a thorough understanding of trouble shooting techniques; shall be proficient in the operation of all components of the system, and fully capable of training others in the proper operation of the equipment.

Part 3. Conclusions

From our perspective as a large not-for-profit corporate health care setting, we advocate that filmless Medical Diagnostic Imaging Support systems can be developed that are clinically and economically superior alternatives to conventional film-based medical imaging support systems which are inherently limited by film as a hard copy media. Moreover, filmless MDIS systems can be acquired, installed, and maintained using an approach that functionally describes full scale filmless systems in detail through a contract specification that industry can respond to competitively.

With this approach, the medical imaging industry will be provided with a substantial incentive to system integrate a cost effective, clinically responsive system from enabling core technologies such as computed radiography. Such an approach presumes that there is a technology benchmark validation step in the process and that "best value" systems can be acquired in a wholesome, competitive process. It also presupposes that this process is administered by a group of informed medical users who can functionally describe and validate what is required in various health care settings in terms of filmless medical imaging technology. In short, the whole must exceed the sum of the parts.

In the final analysis, filmless medical imaging systems make sense today for implementation in the not-for-profit health care environment. Properly implemented filmless systems provide a fresh approach in achieving new levels of productivity and quality in medical imaging beyond what is presently available in the mainstream of military medicine and U.S. medicine today.

Acknowledgements

The author wishes to acknowledge the substantial contributions of the entire Medical Diagnostic Imaging Support System Technical Development Team. These are a group of individuals who have totally dedicated themselves to achieving success in making filmless medical imaging a reality for military medicine. They are a credit to their country and to the military health care tradition of total commitment to the patient. It is a genuine honor to be on their team.

The author also wishes to acknowledge the support and guidance from Lieutenant General Quinn E. Becker, Lieutenant General Frank F. Ledford, Jr., Major General Philip K. Russell, and Major General Richard T. Travis and their vision of a filmless future for military medicine in launching this initiative.

References

1. ________________: **Medical Diagnostic Imaging Support System, Specification:** issued by U.S. Army Engineering Division, Hunstville, Alabama, Solicitation # DACA87-90-R-0058, June 22, 1990.

2. Goeringer, F., Mun, S.K., and Kerlin, B.D.: **Digital Medical Imaging– Implementation Strategy for the Defense Medical Establishment**, Medical Imaging III , SPIE Proceedings, Newport Beach, Ca., January-February 1989.

3. Glass, Harold, Northwest Thames Regional Health Authority, London, United Kingdom: **Personal Communications** in 1988, 1989, & 1990.

4. Huang, H.K., Medical Imaging Division, Department of Radiology, University of California at Los Angeles Medical Center: **Personal Communications** in 1987, 1988, 1989, and 1990.

5. Kim, Yongmin, Electrical Engineering Department, University of Washington: **Personal Communications** in 1987, 1988, 1989, and 1990.

6. Ledford, Jr., F. F. LTG: **Filmless Medical Imaging for Military Medicine**: an invited presentation at the Image Management and Communication (IMAC 89) Conference, June 8 1989.

7. Mitre Corporation: **Digital Imaging Network System, Final Report**: prepared under U.S. Army Medical Research and Development Command Contract # DAMD 17-86-C-6145, June 1990.

8. Mun, S.K., IMAC Division, Department of Radiology, Georgetown University Hospital: **Personal Communications** in 1987, 1988, 1989, and 1990.

9. Mun, S.K., Goeringer, F., Saarinen, A., et. al.: **Total Digital Department- Implementation Strategy**, Medical Imaging III, SPIE Proceedings, Newport Beach, Ca., January-February 1989.

10. Zimble, James A., VADM: **Navy Medicine and Digital Imaging**: an invited presentation at the Image Management and Communication (IMAC 89) Conference, June 8 1989.

CommView® - A System from AT&T and Philips. A Technical Status and Evolution Plan

Anthony V. Lando
Ruttger Heu
Philips Medical Systems

Abstract

Philips Medical Systems and AT&T have been involved with various PACS projects for over ten years. These activities have encompassed small application driven projects and large filmless diagnostic clusters. This presentation will provide an overview of these activities and address in detail the major elements of our PACS architecture and its planned evolution.

Introduction

Philips and AT&T have provided to the worldwide market PACS products in a range of configurations to meet various applications. These products have been based on technology and design requirements which centered on a common core architecture and network configuration. As various projects moved into clinical use, product limitations appeared and required sustaining engineering programs to tailor equipment for clinical operation. The number of PACS sites and their geographic position worldwide demanded a technical support network to interpret the clinical feedback to our development organization. This process provided significant direction for future development.

CommView Systems

Major CommView installations are based on a distributed processing architecture with a central data management system (DMS) supporting various size magnetic storage sub-systems, optical archives, network management sub-systems, and external interfaces to information systems. Both acquisition and display components are interfaced to the DMS via a fiber link utilizing an optimized transport protocol. Acquisition devices (AM) are capable of supporting up to 5 modalities via an analog interface. The AM can also support an ACR/NEMA logical, DR11W physical interface used for digital image transfers from a computed radiology device, laser film digitizers, or digital modality interface adaptor.

Viewing stations represent both PC-based products and high end diagnostic consoles.

Networking is based on a zone concept with the most critical areas being supported with fiber runs, remote display areas (ICU, CCU, etc.) utilizing a coaxial base band network, and areas outside the facility supported with public networks.

NATO ASI Series, Vol. F 74
Picture Archiving and Communication Systems (PACS) in Medicine. Edited by H. K. Huang et al.

Overview of Projects

CommView has been involved with various projects ranging from stand alone computed radiography systems to full functional filmless applications.

Objectives for each project varied. Some were sponsored by government agencies to evaluate the technology and implementation scenarios. Others were required for specific clinical applications or to enhance scientific research. In addressing these objectives, we gained experience in the following areas:

- Site Support
- User Derived Application
- Product Limitations
- User Interface Requirements
- Modality Integration
- Archiving Requirements

Lessons Learned

Site Support

Projects require, for successful implementation, the following elements:

- 24 hour central technical assistance
- Local project teams composed of engineering, applications, and technical support
- On site operation specialists hospital based must be identified and trained to facilitate uptime.
- Access to scientific development activities and rational mechanism to exchange image data with external workstations.

User Denial Implementations/Benefits

By observation of use and monitoring operational scenarios, there are clear clinical benefits to application driven configurations. An example is digital transmission of portable examinations to an ICU viewing station. These applications require modification to user interfaces and system operations.

- Auto-Routing: Through the use of a site configurable decision table, images can be auto-routed to specific Network stations.
- Hard Copy Capabilities: Central server to print from any workstation the appropriate image.

Product Limitations

When fully utilized in a clinical environment product performance envelopes are delivered. Where this performance fell short of clinical expectations, efforts are made to resolve them in both current and future products. Major limitations which have been defined include:

- Viewing Platforms
- Archiving Strategy
- Digital Interfaces
- Open Networking

Future Direction

Based on our experience there is a need to migrate our current architecture and equipment portfolio to the following elements:

- Networks
  - FDDI Dual Token Ring supporting up to 100 mb/sec.
  - PC Networking through Brouter providing 10 mb/sec. on standard TCP/IP Ethernet
  - Network manager to support multi mode configurations (Unified network manager)
- Database
  - Single database, not distributed
  - RS232 support to extend information system
- Viewing
  - $2K^2$ graphic display systems
  - Parallel transfer disk options
  - Multi-monitor (up to 8) display configurations
  - Improved hardware reliability
- Archive
  - Migration to new media (5 1/4, MOD's)
  - Archive/Dearchive protocol with integration to an external information system
- Interfaces
  - Archiving will support ACR/NEMA over DECNET or ETHERNET
  - Open architecture with standard interfaces for foreign workstations

- User Interfaces

    - Window environment for user interface
    - Pre-window data display

Additional product scenario will include:

- Shared hard copy networks
- Scientific workstations
- Provider workstations

GE PACS STRATEGY AND DEVELOPMENT

Patricia LAYZELL, Olivier LIS, Emmanuel PAVAGEAU

Multimodality Department, GE-CGR, 283, rue de la Minière
78530 BUC - FRANCE

Abstracts: GENERAL ELECTRIC MEDICAL SYSTEMS invested in PACS as early as 1983 in the worldwide Research and Development Centers - in the USA (MILWAUKEE), Europe (BUC-FRANCE, GE-CGR Company previously owned by THOMSON), and at the YMS Headquarters of TOKYO, JAPAN.
The work started with 1K workstation, network and standard interface. A first generation of systems were installed in 1987. Clinical feedback showed the necessity of 2K display for diagnostic display, and better integration at users level. Having defined the Development and Marketing strategy of the three poles, GEMS is now ready to launch the ID-System concept, handling Diagnostic Workstations (2-4 screens, 2048x1536) fully integrated in PACS environment with archiving and communication access, including HIS/RIM integration .
GE PACS R and D effort is committed on both product integration (standard ACR-NEMA, standard Modalities CPU, peripherals, user interface) and Multimodality workstation application such as CT/MR 3D visualisation, and Digital X-Ray diagnostic/therapy applications such as Cardiac and Angiography workstations.

Keywords: Diagnostic Workstation, Network, Multimodality, standard

NATO ASI Series, Vol. F 74
Picture Archiving and Communication Systems (PACS)
in Medicine. Edited by H. K. Huang et al.

BACKGROUND

Sophisticated Imaging Acquisition devices such as CT and MR scanners, requires powerful image processing information treatment, based upon computer techniques. The rationalisation of hardware configuration linked to price/performance ratio , and the necessity to access images from at least 2 modalities on the same workstation for diagnosis and review purpose, lead to the first Multimodality approach at the Technical Imaging level : the "MINIPACS" approach : one or two workstations with local image archiving connected through a local network to 3 or 4 CT or MR's devices. In the meantime, Management Information System were becoming an accepted administrative tool for every aspects of data management directly related to patient records and Hospital data (logistic, finance).

On one side, a Computer Image data base. On the other side, a Medical Image data base. The convergence of the two computer technologies, the rapid increase of computer performance coupled with the increasing demand of Hospital for Computer Aided techniques and information communication, led to emerging PACS technology and data implementation.

To develop and assess the GEMS Multimodality approach, a PACS department was created in MILWAUKEE. In 1984, the engineering work started on a 2 or 4 screen 1K workstation with network, standard interfaces plus review station for clinicians. An agreement was signed with IBM, to include the development of an archive system with full database management and access to HIS/RIM system. The first clinical system was installed at Milwaukee Medical Centre in 1987 and more followed in 1988 and 1989.

Clinical feedback from some 20 workstation installations in the USA lead to development of higher resolution display with better performance.

In JAPAN, recognising the need for "MINIPACS" implementation, a product was launched in 1989 at the technical imaging level for CT and MR applications: 3 imaging acquisition devices, (CT-PACE, MR-MAX lines) linked to 3 viewing stations with capabilities similar to the acquisition stations : 512x512 resolution, 1 or 2 screens - proprietary network through fiber optical cable - and local archiving capabilities. Such systems are installed in JAPAN and are in progress in EUROPE (around 25 sites operational in 1990).

GE STRATEGY AND COMMITMENT TO PACS

Marketing studies prove the rising interest and emerging market for PACS. GE is fully committed to PACS and GE is taking every action to fulfil PACS commitment, for SYSTEM INTEGRATION,and for DELIVERY OF PACS PRODUCTS as part of its core business of Medical Imaging Acquisition and Treatment systems.

For **SYSTEM INTEGRATION**, GE is directly involved in the ACR-NEMA standard development. As a direct consequence of this involvement, every major piece of Digital Imaging Equipment delivered by GE will be compatible with the ACR-NEMA standard.

System integration implies Industry standard. A major investment occurs in the USA with the CSE (Computer Service Engineering) Department, responsible of common computer platform development and standardization for the whole company : modalities Imaging acquisition or image treatment, and multimodality/PACS equipment.

For **PACS PRODUCT DEVELOPMENT**, GE has a strong investment in Multimodality Engineering and Marketing teams based in all 3 poles, synchronising major Company development at worldwide level: in the USA, the focus is overall PACS product integration in the ID-System (Integrated Diagnostic System) concept in liaison with the Modalities (CSE team), the standard Network development, called ID-NET II, ACR-NEMA implementation, Archiving and database Management linked to HIS-RIM integration - this work being performed in cooperation with IBM. And peripherals interfaces such as Cameras and digitizers.

In EUROPE, the major development is linked to Multimodality workstations: High performance 2 or 4 screen 2K Diagnostic Workstation (clinical evaluation in 1990), Review station, and dedicated stations such as the TARGET II Radiotherapy dosimetry and treatment planning station. The Diagnostic Workstation can also handle a powerful 3D-visualisation option, based on OCTREE technique. A graphical accelerator allows to manipulate a full 3D model in real time, with sectioning, thresholding and "electronic scalpel" capabilities. Mid term development is a 1K Workstation with various software options such as Multimodality treatment; X-Ray angiography and Cardiac applications; total compatibility with the ID-system for archiving and Database Management.

On the application level, the European Multimodality team is working on MINIPACS system integration around the Diagnostic Workstation, including local archiving and HIS-RIM interface to HIS-RIM systems based upon DEC Hardware. The "MINIPACS" systems handle Wide Area Network capability, through European telecommunication standards such as ISDN for teleradiology.

In JAPAN, the Company is focusing on the CT/MR MINIPACS concept, to include standard interface with the other GE product lines. CR technology with technical interfaces is under investigation.

MULTIMODALITY RESEARCH : ACCESS TO IMAGE DATA BASE, ON STANDARD PLATFORM

In order to satisfy the needs of Application Research in a Medical environment, and gather feedback on Multimodality clinicals, GE is working with key University Hospital and R and D oriented institutions to enable access to proprietary image database, for CT systems (CT 9800 lines and CE line from GE-CGR), MR systems (MR-SIGNA), and DSA systems (DG/DX). The images are available on a totally standard computer platform with currently available 1K colour screens, using TCP-IP/ETHERNET communication.

Some 50 installations of this kind are now in place in EUROPE and the USA, providing valuable feedback on Multimodality environment, and on image treatment for future diagnostic and therapeutic applications.

A European example : the VAL DE GRACE Military Hospital in PARIS, where MR-SIGNA and CT-9800 images are transfered on a standard SUN 4 station through fibre optical cable, to perform stereotactic preparation.

SHORT TERM ACTIONS AND LONGER TERM DEVELOPMENT

1990 and 1991 are busy years for the GEMS PACS teams : the MINIPACS installations are a reality today. First clinical tests for HIS-RIM integration are conducted in MILWAUKEE. The first production line of the 2K Diagnostic Workstation is presently launched in BUC - FRANCE, for distribution in 1991. Later on in 1991, the Multimodality stations will be expanded to handle Digital X-Ray data base, with data communication either from angiography/cardiac modalities or from CR devices. Many clinical data are being collected to get feedback on these PACS installations, to understand the actual usage and finetune the GEMS offering.

In the future, we will witness with the ID-System approach an overwhelming spread of computer techniques in the Hospital world, where PACS concept is of primary importance in merging patient record and image data base. Then, Computer Aided Techniques will be used for Diagnostic and Therapic application, with Artificial intelligence technique and worldwide communication procedure.

Introduction to Hitachi PACS

Tetsuo Okabe, Kazuhiro Satoh, Shun Kabata

R & D Center, Hitachi Medical Corp.
2-1, Shin-toyofuta, Kashiwa-shi, Chiba-ken. 277 Japan

Abstract: A brief introduction of the present status of medical instrument and system business of Hitachi group is presented first. Hitachi Medical Corp. is responsible for Picture Archiving and Communication System (PACS) under close cooperation of several R & D institutes of Hitachi group. The strategy of our company about open architecture and system integration with the modalities from other manufacturers is introduced by showing several implementations. Our technical achievements and PACS related products available are presented.

Keywords: PACS, Image management, LAN, Workstation

Introduction

Hitachi Medical Corp.is one of subsidiary companies of Hitachi Ltd. and has been responsible for the business of medical imaging modalities and systems in Hitachi group, as our R & D activity has been supported by several R & D institutes of the group. They have provided us high technology components of PACS such as optical disk , optical fiber cable, and many other kinds of devices. We have assembled them with our original components into the systems for hospitals. Hitachi Ltd.himself has been strongly concerned with Hospital Information System with main frame computer.

Hitachi Medical Corp. has had the joint projects with Kyoto Univ. Hospital for development and evaluation of prototype PACS and with Kobe Univ. Hospital for evaluation of digital chest X-ray images for these five years repectively. With these achievements, we have already shipped ten systems so far.

Strategy on PACS

1) Systematization of medical imagings and image management using computer will be on the future trend in the hospital. As the medical imaging modalities manufacturer, Hitachi Medical Corp. has been actively responding with these situations under R & D supports by Hitachi group. We have installed PACS based on X-CT, MRI and DSA image management and based on the system installed inside radiology department so far. The hospital wide PACS has been also developing, where optical fiber network will be distributed in the clinic rooms and wards in addition to the radiology department.

2) We have many experiences on the coupling of other manufacturer's imaging modalit-

NATO ASI Series, Vol. F 74
Picture Archiving and Communication Systems (PACS) in Medicine. Edited by H. K. Huang et al.

ies to our system. In Japan, we have already had MIPS standard as the digital interface standard,and we are now developing its hardware and software. However, it needs both improvement and time for being widely implemented. At present, they would often want us to use the interface such that they have already experienced.

3) So far as the data format on the system is concerned, we have provided "Converter" as an acquisition console."Converter" has a microcomputer and buffer memory for converting data format to that of our system, and it can manage two imaging modalities.

4) Integration with Hospital Information System will,at present status, be performed through the image database management system in uni-directional way, using same patient ID etc.. But we have not received this kind of order from customers as yet.

Technical Achievement

Hitachi Medical Corp.has piled up the experiences with joint projects with Kyoto Univ. Hospital and Kobe Univ. Hospital so far. With these achievements, we have already distributed ten systems in the hospitals. Some of these systems will be presented at the meetings. The largest system is composed of star and ring network where three workstations, three X-CTs, two laser printers, MRI, DSA, FCR, film digitizer, two optical disk libraries and image database management system. The smallest one is composed of two X-CTs and one workstation with the optical disk unit.

We have installed the workstation in the clinic room of radiation therapy in Niigata Univ. Hospital, which is coupled with X-CT, CR and the image database through the network system. Manual measurements of size, angle, distance, density etc on the images need complicated procedures which have been desired to be improved for long years. And calculations of them by computer will be done quite easily. Radiologists have certainly used the system everyday for estimating and/or calculating tumor size on the images in comparison with previous images on the workstation.

We have introduced the workstation in the Lecture Hall of Kyoto Univ. Hospital coupled with the image database through optical fiber network. Before introducing workstation, lecturers must sometimes stopped their lectures to call students to the viewbox in order to interprete interesting focus on the images. Nowadays, they don't meet such a break anymore, and furthermore, they could show dynamic display such as DSA and 3-D display on the CRTs to the students.

We have also experienced teleradiology system between Fukui Univ. Hospital and local hospital, which is aiming at MRI image diagnosis via dedicated telephone line. Three or four patient's MRI images have usually been sent to the University Hospital in deep night. Radiologists have diagnosed them in the next morning and sent the reports by TeleFax. They seem glad to use the system because of time saving.

PACS-related Products

We have provided the commercial products of PACS-related components of 1) image workstation, 2) archiving system, 3) optical LAN, 4) acquisition and 5) others.
The specifications of these components are listed bellow.

1) Image Workstation for Diagnosis:

Four kinds of workstations are now available;that is, 2-CRTs workstation, 3-CRTs workstation, 6-CRTs workstation and 2-CRTs slave station. These workstations have dual buses, high speed image processor, 1.2 GB magnetic disk, 64 MB display memory, and I/O port for image transmission. They can process the images of 2048 x 2400 pixels at 12 bits density with the operation of mouse and can display images on 20" CRT with 1024 x 1200 pixels at 8 bits density in non-interlace mode. Several kinds of application software for the workstation have been provided so far.

* Frequency Enhancement
* ROI Presentation
* Dynamic Display of Angiographs
* Editing Images on CRTs
* Zooming & Reduction of Image Size
* 3-Dimentional Display
* Scrolling of Images
* Subtraction & Superposition of Two Images
* Measurement of Distance, Angle and Density of ROI

2) Image Archiving System:

We have marketed several kinds of image archiving system: 1) 12" (300mm) Optical disk and its library, 2) 5.25" (130mm) Optical disk and its library, 3) 5.25"(130mm) Magneto-optic disk and its library, 4) Image databse management system, 5) Digital VTR for medical application. All these components have been provided by Hitachi Ltd. and Hitachi Maxell Ltd.. And we have given them additive values for their medical applications when shipping to the hospitals.

3) Optical LAN:

We have provided the optical fiber network with the combination of star and ring topology. Ring mode is operated at 100 M bps with circuit switching, and its logical structure is composed of 10 M bps by 8-channels and low speed 121-channels. The network is controlled by Loop Control Station, which can manage up to 14 terminals. The distances between nodes is designed to be shorter than 500 meters.

4) Acquisition Console:

We have "Converter" as an acquisition console,which can receive images from modalities, convert their data format into that of our system and give them buffer for transmitting. The converter can manage up to two modalities on the network. We have many experiences with the converter for coupling modalities of GE, Siemens, Toshiba, and Fuji Photo-Film.

5) Others:

Other PACS-related components such as laser image printer,film digitizer, and MODEM for teleradiology have been coupled to the network owing to the customers requirements. In these cases, digital interfaces have been designed by each situations.

R & D Activity for Future Plan

R & D activities for future plan of PACS have been performing under close cooperation with Hitachi group. Some of them are introduced as follows:

1) Optical Disk :

The development of optical disk in WORM mode has been rapidly going. Hitachi Ltd. will introduce the optical disk of 7 GB memory capacity and its library system of up to 448 GB memory capacity in the market early '91. The transfer rate, access time and mount/demount time have been improved to 2 MB/s, 150 ms and 7 sec. respectively, those which must contribute to faster throughput of the archiving system.

2) IC Memory :

The development and evaluation of 64Mbit DRAM was announced by Hitachi Ltd.early this year. One chip can memory one entire CR image without any data comression. Although the introduction of commercial products in the market will be on the year '93-'94, they will make display memory larger and will make access of more images faster on the workstation, which will be expected to contribute to the ease of diagnosis on the workstation.

3) Personal Health Data (PHD) Recording System :

PHD Recording System has been developed on the idea that every citizen should have his health care card which files all examination data including typical diagnostic images. And it will become the utmost strategy for PACS. Dr.H.K.Huang of UCLA, USA, and Hitachi Maxell Ltd. have had the joint project for PHD for these two years. 3.5 inches optical disk of 256 MB memory, archival station and related software have been developed for this purpose, which can be used with workstations such as IBM PC/AT, Apple Macintosh II and Sun Microsystems SUN. At the clinical test, about 1000 disks will be given to the patients in order to watch what will be happened and to investigate what will be most important to PHD system.

4) The Diagnostic Reporting System :

The reporting system for image diagnosis has been extensively required by radiologists in order to fulfill PACS utility. With Kyoto Univ.Hospital, we have tried to couple the word processor of the lap top personal computer with the workstation. Radiologists can write diagnostic reports with the word processor and see them on the workstation's CRT display. The reports can be either sent to the optical disk library through the optical fiber network or filed in the dedicated workstation's magnetic disk.

5) I.I. Digital Imaging :

The digital X-ray imagings, which can be fully accepted by Dept. of Radiology, are said to be most important key point for implementation of PACS in the hospital. The conventional R & F images, as well as chest images, have been required to be digitized. The digital R & F table with the images of 2k x 2k pixels at 12 bits density and 7.5 frames/sec. has been studying. Its clinical evaluations on gastro-intestinal examinations will be presented at RSNA '90 held in Chicago, USA,.

6) Hospital-wide PACS installation :

We have been trying to install hospital-wide PACS in Tokyo Hitachi Hospital located in Tokyo. The hospital has 150 beds and 300 out patients. The system will be distributed in the Nurse Centers on every wards and five clinic rooms as well as in the Radiology Department. It will couple with X-CT, CR, digitizer, laser image printer, image database and ten workstations through optical fiber network. We are installing this system in order to know what are most important for the workstations installed in the nurse stations and clinic rooms in the hospital-wide PACS.

Concluding Remarks

We have been concerned with PACS development for these several years. Although both radiologists and administrators have not always been satisfied with the existing PACS implementation, we could believe to overcome these technical difficulties in near future, and we believe that the age of PACS will come to the hospitals for better medical environment.

References

1. Minato,K. et al; "PACS development at Kyoto University Hospital" Proc. of SPIE: vol.1093, p20-23. (Medical Imaging 111. Newport Beach, CA. USA. Feb. '89)
2. Yamasaki,K. et al; "Diagnostic Capability of Digital Chest Images" Proc. of SPIE: vol.914, p577-580. (Medical Imaging 11, Newport Beach, CA. USA. Feb. '88)
3. Okabe,T. ; "PACS : Future Perspective of Technology" Proc. of 1st Int'l Conference on IMAC (IEEE publication): p191-199. (Washington, USA. June. '89)

Contact to :

Tetsuo Okabe FAX: 81-471-33-2543, TEL: 81-471-31-4159 (Japan)

The Vienna SMZO Project

Hans Mosser, Alfred Mandl, Michael Urban, Helmut Hradil, Walter Hruby

SMZO Hospital, Betriebsorganisation, Langobardenstraße 122, A-1220 Vienna, Austria

Abstract: This paper presents the project of the Integrated Digital Radiology System (IDRS), i.e. the integration of PACS, HIS and RIS into a totally digital environment as planned at the SMZO hospital in Vienna, Austria. The hospital -a 1400-bed teaching institution- will start its operation in March 1992. Since 1988 a team consisting of computer engineers and radiologists has been planning the design and the operational structure of this integrated digital radiology service. The goal of this project is to optimize the radiology service by improving communication within the radiology department and with other departments and to increase the accessability of radiologic images, the basis of our daily work. The improvements in these areas certainly will result in better patient care.

In this paper the major considerations in the planning and development phases are outlined as far as the radiologist is concerned and the estimated maximum data volume to be handled - based on the analysis of a hospital of comparable size and functions - is discussed. An overview of the physical structure of the system will be given according to the current plans, which still are open to changes if necessary as planning progresses.

Keywords: SMZO, Integrated Digital Radiology Systems (IDRS), PACS, HIS, RIS

Introduction

The SMZO is a hospital under construction, which will start its operation in early 1992. SMZO stands for "Sozialmedizinisches Zentrum Ost", meaning "Social and Medical Center East". It is a complex patient care system including a nurse training center, a geriatric center for social activities and communication, and the 1400-bed teaching hospital. The whole complex is located in the eastern part of Vienna, Austria, covering a resident population of more than 100.000 people.

NATO ASI Series, Vol. F 74
Picture Archiving and Communication Systems (PACS) in Medicine. Edited by H. K. Huang et al.

Planning for a new hospital and in particular for its radiology department has to take into account the emerging technologies in DR, PACS, HIS and RIS [1, 5, 6].

In the decision-making process three alternatives had to be considered: planning (1) a conventional film-based radiology department, (2) a mixed digital/conventional system still using films, and (3) a fully digital department in the sense of an integrated digital radiology system (IDRS), covering a HIS, RIS and PACS with the final goal of a filmless radiology department.

At the turn of the century the larger part of radiologic images is assumed to be operated with digital modalities and manipulated within a PACS, in particular as soon as digitally generated images will account for more than 40 to 50 percent of all imaging [1, 3, 4, 6]. In the light of these considerations and with an optimistic view to the further rapid development of digital technology we decided to plan and equip the radiology department of the SMZO with digital radiography and, as its logical consequence, a total PACS integrated to a RIS and HIS.

During initial planning a team of radiologists and engineers was formed which, after vendor selection on a tender basis in November 1989, has started to cooperate within an extended project team including the vendor´s development representative. This collaboration proved to be of vital importance and is regarded as a prerequisite for the successful planning and implementation of an IDRS [1, 2, 6].

Description of the planning process

Involving the radiologist in the strategic planning of an IDRS is a must if the system is to be in line with the real needs of daily radiology practice. Several phases have to be systematically run through from the initial decision to the working installation. The approach to this project is primarily a top-down-strategy but with inductive elements as well.

(1) Goals and objectives: The objective of our project is to overcome some inherent problems of the film based system: the loss of films in studies where a comparison is necessary e.g., is a major problem in conventional radiology departments. This happens at least in 10 percent of the cases in the Rudolfstiftung hospital. We hope to minimize this problem in the SMZO hospital because one of the goals of an IDRS is improved accessability to radiological images. Another goal is to improve communication links within the hospital and making patient data faster and more easily accessible for the radiology department. Intensive communication within the radiology department with its numerous subsections as well as to the clinical departments is mandatory to improve the diagnostic capabilities of both radiological staff and modalities.

(2) Problem analysis: The basis of this task is to study and define the current operating system, its (a) organizational structure, (b) data production and (c) communication links. The outcome of this analysis results in the current status regarding the estimated number of patients and exams, as well as data volume, speed and direction of data transmission, frequency of access and related issues. Since the SMZO-hospital is under construction and therefore does not rely on old structures, problem analysis was carried out in a hospital comparable in size and functions: our planning was based on the data analysis performed in the Rudolfstiftung hospital in Vienna.

(3) Macro planning: Based on these analyses the concept of an IDRS is worked out by the computer science department of the hospital and the development group of the selected PACS vendor.

(4) Micro planning: The logical planning concept and its implications for the physical design and the impact on the operational structure of the radiology department is discussed by the project team and brought in line with the radiologist´s requirements. This leads to further modifications of the concept, until the plan for the IDRS meets the needs of the final users.

System architecture of the SMZO-IDRS (current planning status)

A token ring topology (100 MBit/s FDDI-ring) connects seven modules for acquisition and/or display of images. Each module has a local storage capacity of 4 GByte (Parallel Transfer Disk). A fault tolerant central image managment system for storage and archiving includes 2 DOD-Jukeboxes with on-line access to 2x179 GByte. Interfaces, databases and networks are based on the ACR-NEMA standard and SPI-specifications. For diagnostic and consulting purposes many workstations (2 to 8 screens per console) with interactive manipulation possibilities will be installed in the radiology department. Additionally numerous remote viewing stations with limited display capabilities on a PC basis will be located in specific areas of the hospital. The necessary number of diagnostic reporting consoles and viewing stations is still to be determined. It becomes evident only now that in the initial planning stages the number of reporting consoles and viewing stations was grossly underestimated.

The PACS will be interfaced to a RIS and HIS, providing effective communication links within the radiology department as well as to the administration and the clinical departments. Text- and data information, communication and patient documentation are keywords in this field.

The daily data volume for the first two years of operation is estimated at 9 GByte, which certainly will place high demands upon the system´s performance. Table 1 shows the estimated figures with regard to the different matrix sizes.

Perspective

(1) Installation-phase: The system will be installed in 1991.

(2) Preopening-phase: The radiology department will start its operation in September 1991, that is 6 months before the official opening of the hospital. This phase consists of (a) the technical evaluation of the installed devices and (b) the clinical assessment. While the former will be performed with phantom studies, the latter will be done with phantoms and with patients of the geriatric part (400 beds) of the hospital, which already has been in operation since 1987.

(3) Opening phase: In March 1992 the hospital will start its operation and the radiology department its routine work. Problem analysis and finding solutions will continue throughout this phase and modifications will be made where necessary.

(4) Five-year upgrading and updating: This phase involves continuous modification as well as expansion of the system to meet the needs as they arise in daily routine work, in line with state-of-the-art technological advances in PACS and related fields.

(5) Documentation: The SMZO project team continues to document and report progress and implementation phases of this project.

Table 1: *Estimated workload for the first two years of operation with regard to the different matrix sizes:*

MATRIX	*IMAG./ DAY*	*MBYTE*
512	1.730	907
1024	617	1.483
2048	800	6.712
TOTAL		**9.102**

References

1. Huang H.K., Kangarloo H., Cho P.S., Taira R.K., Ho B.K., Chan K.K.: Planning a totally digital radiology department. AJR 154,635-639 (1990)
2. Huang H.K., Mankovich N.J., Cho P.S., Taira R., Stewart B.K., Ho B.K. Picture archiving and communication systems in Japan. AJR 148,427-429 (1987)
3. Jost R.G., Mankovich N.J.: Digital archiving requirements and technology. Invest. Radiol. 23,803-809 (1988)
4. Lemke H.U.: Computer-assisted radiology. Radiologe 28,189-194 (1988) (german)
5. Mezrich R.S.: The implications of PACS for radiology practice. AJR 151,828 (1988)
6. Mun S.K., Benson H.K., Elliot L.P.: Total digital department: implementation strategy. Proc SPIE 1093,133- 139 (1989)

PACS and PACS-related research in Belgium

R.Mattheus, M.Osteaux

Pluridisciplinary Research Institute for Medical Imaging Sciences (PRIMIS), University hospital of Brussels,Laarbeeklaan 101, B-1090 Brussels,BELGIUM

Abstract: This paper gives a brief overview of PACS and PACS-related research in Belgium. Three institutes are active in this field: KUL-UCL and VUB / PRIMIS. At the University of Leuven (KUL) the PACS project's main application is to support research activities in acquisition techniques and image processing methods. One paper covers this item [1]. At the University Hospital of Brussels (ULB) a project mainly for medical evaluation in the field of tele-radiology is going on. At the University Hospital of Brussels (VUB) together with the PRIMIS (Pluridisciplinary Research Institute for Medical Imaging Science) institute, started a project in 1986 and, has one of the few multi-vendor installations world-wide.

Keywords: PACS, networking, medical application, decentralized, imaging computers, network management, tele-radiology.

Introduction

Belgium has played a very active role in Medical Informatics, more specially in PACS developments, image processing, standardization coordinating initiatives and EuroPACS.
The project at the VUB is one of the oldest, and concentrates on communication topics: communication between systems and between man and machine. Three main issues are studied: imaging computers, network architectures and clinical integration. These are also the main components of the phased approach project. At the ULB, the French speaking colleague of the VUB, a project is being set up to asses the clinical feasibility of diagnostic image transmission between the University Degli Studi di Firenze (Italy) and the ULB[2].

PACS at the University Hospital of Brussels - VUB

PRIMIS (Pluridisciplinary Research Institute for Medical Imaging Sciences).
This institute was created by the V.U.B.(Vrije Universiteit Brussels) in 1987 in answer to a call from the Ministry for Science and Technology to create a centre of excellence concerning one specific subject in each Belgian university. The PRIMIS group is a pluridisciplinary centre for the study of the application of digital imaging techniques and related computer techniques; one of the PRIMIS projects is the PACS project. The university hospital itself is a 700 bed teaching hospital, with a large medical imaging department which has the latest imaging equipment serving also the outpatient facilities, resulting in 150.000 examinations a year.

NATO ASI Series, Vol. F 74
Picture Archiving and Communication Systems (PACS) in Medicine. Edited by H. K. Huang et al.

Phased build-up approach

The VUB - PACS / IMACS project is executed in distributed upgradeable phases. In a first step, a PACS system on a limited scale was installed: a local network with two CTs as acquisition units, two 512x512 screen reporting console and storage possibilities. In the beginning of 1989 we decided to shift towards a total distributed approach resulting in small PACS islands for each medical sub-section.

The CT-PACS consists of two CT units (Siemens), a prototype three 1kx1k screen reporting station (Siemens), local storage and archiving possibilities (2 Gbyte). A hardcopy unit (3M) is shared by the 2 CTs and the reporting console.

The installation of the 3M laser hardcopy unit at the CT-PACS is an example of a full usable image management environment with low down time (Fig 1).
The MMU (Multi Modality Unit) is used for the connection of the two CT's and the PACS network (server of the reporting console).
operation:

CT1,CT2 ——> Laser Hardcopy

PACS reporting console —— > Laser Hardcopy

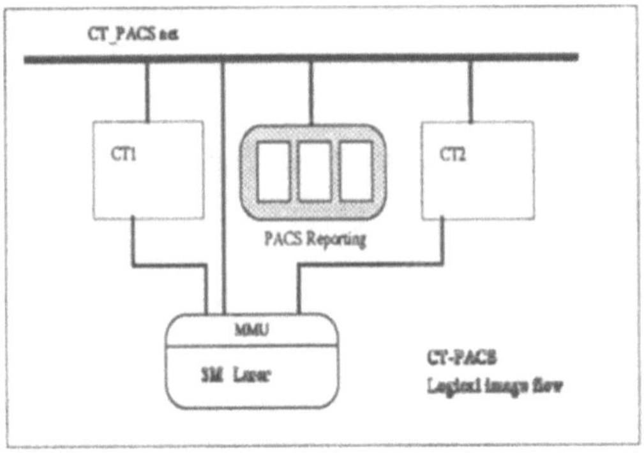

Fig 1.
CT-PACS section at University hospital of Brussels
(VUB)

A component may break down, without having problems making hardcopies. On the PACS reporting console instead of using the keypad (3M) to produce hardcopies, selections on the screen can be done by mouse. Options like "format "and "print" have been implemented. A unique item is the possibility to place images of different examinations of one patient on one film, which can be useful during operation in the operation room. A patient history film is a combination of the important images of that patient. These documents are more easy to handle.

The MR-PACS section consists of two MR units (Siemens), a powerful image-processing station, local storage, shared hardcopy (3M) and archiving possibilities (3.2 Gbyte). At the emergency department the PACS section consists of a laser-digitizer (Dupont), a two 1kx1k screen reporting console (Dupont) and a prototype DR (II) (CGR-GE) system. Evaluation studies are going-on for the PACS emergency, with promising medical results but technical problems at the level of integration of the patient identification on the laser scanner.

The DSA (Philips) system is integrated in our network and a logical step for next year will be the connection of the DSA (Siemens) which was installed in the beginning of 1990 and will result in the creation of our DSA-PACS section.

At the end of 1990, a test at the orthopedics department with a 1kx1k viewing station (Imlogix)will be set up with a personal computer based image server (Macintosh). Also a project has been set up to manage ultrasound images, all interfaces are digital except thse from US.

These subsystems are linked via Ethernet to an acquisition backbone network in the radiology department. This backbone network consists of a buffer storage (MOD), optical archive , a multi-modality reporting console and a hardcopy unit (3M). The network currently in use (MR, Emergency and radiology) has a length of 300 metres. At the moment, an optical fibre network with a length of 1,5 km is being completed. This network consisting of 20 parallel fibres will not only be used for PACS, but also for a number of other hospital informatics projects. It will allow for further extension of the PACS system towards the nursing units. On this fibre we will still use ethernet and TCP/IP until other protocols and interfaces become more cost effective. Experiments are also being considered which will extend outside the hospital walls (e.g. towards a radiologist on duty, who can then, during the week-ends, give advice, based on images sent to his home, to less experienced radiologists inside the hospital). A project for inter hospital communication based on ISDN is under study.

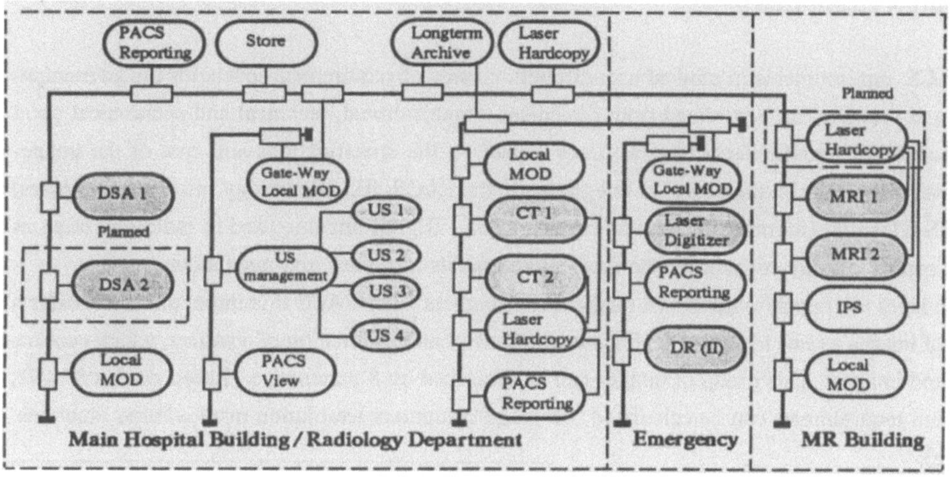

Fig 2. PACS installation at the University Hospital of Brussels (VUB) 1990.

Images are grouped in folders (examinations), one patient can have different folders. Images can be retrieved by number, name, date, but also by ACR (American Collage of Radiology) code number. All examinations (for the connected modalities), for which no report has been dictated yet, remain automatically available for direct viewing. Images on the long term archive are stored under a compressed form and selected images are stored under their original form. A radiologist can build his own collection of selected cases, stored together with personal comments and review possibilities based on a personal computer (Macintosh). During the night, an elementary link between RIS (Radiology Information System) and PACS allows automatically search of all images on the archives of those patients scheduled for control or new examinations the following day. Also the necessary information is downloaded to ensure consistent HIS/RIS/ PACS data-bases. The SPI (Standard Product Interconnected) proposed standard (ACR-NEMA super-set) is used for image format and network commands. The knowledge based network manager shell insures that the traffic on the network is optimal. Buffering, decentralization and prefetching are techniques which are used to allow faster access of examinations (image transport).

The PACS at the CT, MRI and emergency department is in real clinical use, which has the advantage of:

- less missing films
- multiple copies and mixed films
- feedback for assistant radiologist
- more images for research
- sharing hardcopy units
- full redundant system

The PACS installation can be divided in several interconnected sub-sections connected to a backbone network (Fig 2). With the multi-vendor, distributed and network management approach this project is often used as reference site, and we have welcomed visitors from all over the world. Our goal is to try to have a clinical integrated system; to realize this, medical knowledge is added to the technical possibilities. Major research is going on to realize this, in the direction of Imaging-Computers, PACS architectures and Networking.

Imaging - Computers

For PACS environments in clinical use, different classes of requirement-levels for image manipulation and viewing can be defined from a medical, organizational, technical and economical point of view. These requirements have a direct impact on the specifications and cost of the image-computers as well as on their level of integration in the PACS, RIS (Radiology Information System) and HIS (Hospital Information System) environments. Digital imaging used in radiology requires a high quality display console for the medical specialists to review and manipulate images.

Images are central objects in a medical environment. For PACS it is important to consider a group of images as one identity. A clinician will look at an examination of a patient, which consists of several images. This group of images can be described by 8 parameters. Based on this 8-tuple, minimum requirements can be calculated for image-computers (resolution manipulation functions, storage).

Definition of an 8-tuple (M,E,N,S,R,D,T,C) :
Examination type

M: {CT,MR,US,DR,XR,DSA}
modality source

E: {body, head, neuro, cardiac, abdominal, vasc,bone} examination

N: maximum number of images

S: maximum number of marked images

R: {64,128,256,512,1024,2048,4096}
matrix n.n

D: {2,3,4} display technique
(depends on acquisition)
2 : 2D image (x,y)
3 : image of a 3D set(x,y,z)
4 : image in a dynamic 3D set (x,y,z,t)

T: average examination time

C: {8,12} contrast resolution in grey level

M	E	N	S	R	D	T	C	class
CT	Body	34	21	512	2	30	12	A
	Head	20	15	512	2	30	12	A
MR	Body	80	13	256	2	40	12	A
	Neuro	128	13	256	3	40	12	E
	Bone	173	25	256	3	55	12	E
	Cardio	19	19	256	4	35	12	C
US	Body	8	4	512	2	20	8	A
DR	Body	8	8	4096	2	15	12	B
DSA	Cardio	400	20	512	4	20	12	C
	Cardio	100	20	1024	4	30	12	C
	Vasc	400	20	512	4	30	12	C
PET	Neuro	20	5	256	2	40	8	D
	Cardio	20	5	256	2	40	8	D

Table 1.
Parameters for 8-tuple of the Radiology dep. of the University of Brussels (VUB) 1989.

Five main display classes can be defined (Table 1). Each class has it own hardware and software needs for displaying the examinations depending on (r,d) and the image-computer classes. Various image-computers are required to obtain an effective PACS. Two main types can be considered:

- Image/data acquisition (Input)	- Image display stations (Output)
- Film input	- Specialized image stations (SIS)
- Data input	- Images processing station (IPS)
	- Multi modality reporting console (MRC)
	- Multi modality viewing station (MVS)
	- Remote viewing Station (RVS)
	- Multi modality film output (MF)

Specialized image stations (SIS)

A specialized image station forms the interface between an acquisition unit and the medical and technical users. The hardware and software of these stations is dedicated to each type of acquisition modality (CT, MRI). This console belongs to the acquisition unit and allows processing of the raw data and image manipulations. It also has to provide the possibility to make the image available in the PACS environment. Three main problems appear for integration in a PACS environment. First by images are in the format of the modality (company and modality dependent). Hopefully a standard format will soon be commercially available. Second, the patient information (patient id, patient name) is added on this level to the images, with the necessary problems of inconsistent PACS and HIS/RIS database (miss-typing). Thirdly, the acquisition unit has not yet been developed towards integration in a network structure (user-interface, security, system organisation).

sc_{sis}: storage capacity SIS for one modality type (disk)

$4n(r^2/10^6)*exa\_day = sc_{sis}$ Mbyte (1)   exa_day:examinations per day

Image Processing Station (IPS)

The IPS should have multiple levels of user interaction, so that it can be used by specialist with different backgrounds (engineers, specialists, clinicians, programmers).

	Requirements for integration IPS
- Specific selection of programs:	- flexible and easy to use
3D reconstruction	- dedicated programs
CAD/CAM prosthesis	- extensive image processing tools
flow manipulation	- large RAM
3D view	- fast array processor
- Interactive image processing	- easy to integrate in a network structure
- Programming environment	- fast image transfer
- Array processor programming	- integration with hardcopy unit
	- different display possibilities

Multimodality reporting console (MRC)

Function : interpreting and reporting of examinations

Two types of stations, "multiple" (MRCM) and "virtual" (MRCV) display stations can be considered. A multiple display station has to be capable of displaying several images simultaneously on different image monitors. Instant transfer of images from monitor to monitor allows convenient side-by-side comparisons and some simple low-level image operations.

A virtual display unit uses panning and scrolling in a large image memory to obtain analogous capabilities with a single monitor. The idea of the virtual image station is to roam the images on the screen the same way you do with a microfilm, but instead of moving the head to the left and right, you move the image memory mapping and look at one monitor. This type of reporting can speed up the examination because there is no waiting time after selecting a patient. Both implementations mentioned above are in fact electronic simulations of the arrangement of the conventional films.

msc_{mrc} : minimum storage capacity MRC for one examination type (disk)

exa_day (n + s) $4r^2/10^6 = msc_{mrc}$ Mbyte (2)

Requirements for integration MRC.
- modular design (1 to n high resolution monitors)
- display and individual image processing on each screen
- simultaneous display facilities of digital system
- text annotation
- compression/decompression
- fast access to the archive
- networking capabilities
- graphic editing (ROI; annotations)
- reporting tools
- contrast modification
- spatial frequency
- user friendly interface

Multimodality Viewing Station - MVS

Functions: Review of reported images

This type of workstation offers any authorized clinician the possibility of having a quick review of a patient study. It should be a cheap entry to the PACS environment, with a reduced image set and good spatial resolution. Only a few image manipulation functions are required. At this user-friendly station it is necessary to have the possibility of displaying images and reports

Requirements for integration MVS.

- simple to use
- not expensive
- image manipulation tools
  window centre
- report (RIS) / image integration
- fast access to reduced image set
- good spatial resolution

To overcome the display technology barrier, the review console was designed to keep larger images in memory and to display them at a lower resolution. In this unit, no image buffer is available (3). An image server is implemented on a Macintosh computer. Pan and zoom functions were built onto the console to allow the user to look at certain sections of the image at full resolution.

msc_{mvs} : minimum storage capacity MVS for one examination type (disk)

exa_day (s) $4r^2/10^6 = msc_{mvs}$ Mbyte (3)

Remote Viewing Station - RVS

Functions: Review of reported images at remote sites and scientific archive images.

Requirements for integration RVS.
- teler-radiology inside / outside the hospital
- tele-medecine
- teaching rooms
- simple to use
- not expensive
- image manipulation tools
  window centre
  width
- fast access to reduced image set
- good spatial resolution

msc_{rvs} : minimum storage capacity RVS for one examination type (disk)

exa_day (s) $2r^2/10^6 = msc_{rvs}$ Mbyte (4)

Multimodality Hardcopy Output - MH

Functions : Printout of images : different mixed modalities (CT,MR,US,...)
time mode (old and new images)
patient history (selected images)

In a radiology department, hardcopy units are connected to the modality by video or digital connection. In a PACS environment it is very useful to connect a hardcopy unit to the network in a digital way, so they can also be used as back-up unit for each other. Some vendors offer a multi-modality option, which allows the connection of a group of inputs by means of a multiplexer, which is important in a full function PACS.

Requirements for integration MH.
- workstation driven
- network driven
- all images
- selected images
- easy and intuitive access to images and patient records
- concurrent view of exams from different modalities
- support for the report generation process

Distributed sub-networks

The distributed architecture and network management techniques were introduced at the first IMAC (Image Management And Communication)world congress in Washington in June 1989 .
Four main PACS topologies can be defined: closed line, central line, distributed network and distributed subnetwork [3].

Closed line.

A closed line is modality-dependent and only aimed at connecting devices of the same modality (CT, MR,...) together, possibly, with a second console. A typical construction will contain two modalities of the same type connected with an image console.

Central line.

Added to the first topology is a central computer system with a local store, data base, archiving possibilities, display and research software for the reporting node. It is possible to interconnect devices of different modalities.
Figure 3 gives a view of the image traffic on the network on which two CT's and one MRI were connected and measured for one month. Every 15 minutes the total of the transmitted images is represented including images which have been generated, re-archived and sent to the hardcopy unit.

Distributed network.

Added to the previous topology is the possibility to make use of distributed processing on the different computers in the network.

Distributed subnetworks.

Each sub-department (MR Department, CT Department, Emergency Department,...) has its own small PACS environment with local store, reporting consoles, hardcopy units and a gateway or bridge to the PACS-backbone network which interconnects all the sub-networks. This approach allows distribution of the image-flow over the network, making traffic on a single subnetwork possible without interfering with the traffic on the other subnetworks. Figure 4 gives a view of the same units connected in a sub-network. The maximum activity on the network is reduced for more than 50% compared with the central line installation. Activity on the back-bone is due to the re-archiving and multi-modality viewing.

The practical throughput can still be higher with the introduction of the network management technology, and result in a faster display time of images on the screens.

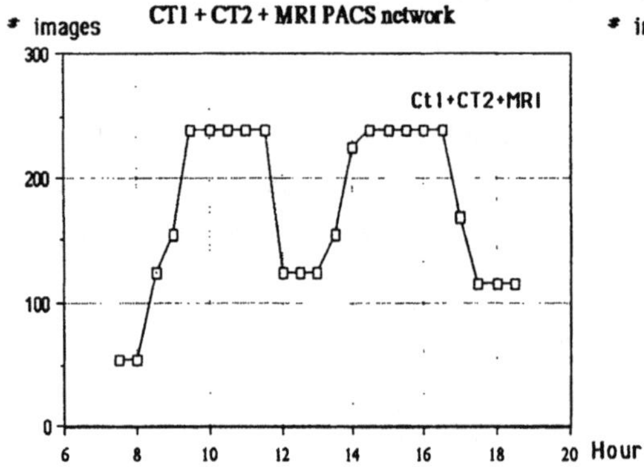

Fig 3.
Central line PACS with 2 CT and 1 MRI at the University of Brussel(VUB)(1989).

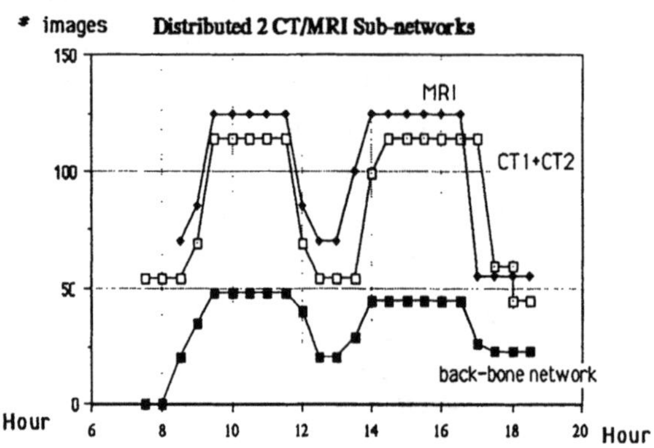

Fig 4.
Distributed sub network PACS:
CT_PACS consist of 2 CT and
MR_PACS consist of 1 MRI at the
University of Brussels (VUB) (1990)

Network manager

The network manager shell insures that traffic on the network is optimal. Buffering, decentralization and pre-fetching are techniques used to allow faster access of examinations (possible delays are due to image transport).

A program data-base in KEE is used to maintain, change and further develop these programs. KEE is a commercially available knowledge representation system embedded in Lisp. The installation has a hierarchical architecture introducing several levels for image transport, control, security and management[4].

Conclusion

Even in a small country such as Belgium, a lot of research activities are taking place in the PACS field. Most of the projects are driven from a medically oriented point of view, which is a possible next step in the PACS developments after the technically driven approach.

References

1. E.Bellon,P.Suetens,G.Marchal:*PACS at the University Hospital of Leuven*; Nato ASI; Springer Verlag. 1990.
2. D.Baleriaux;G.Dal Pozzo; *Magnetic resonance image transmission in neuroradiology*, AIM reports, PACS-IMACS project EC 1990.
3. R.Mattheus,R.Luypaert,M.Osteaux.:*PACS Topology*, Proceedings, IMAC 89;Image management and communication in patient care: implementation and impact. Washington,DC
4. R. Mattheus,F. Moyson,Y. Temmerman,M. Osteaux.:*Hospital Integrated Picture Archiving and Communication Systems; A European Project*;pp.396-404; SCAR,90 Symposia Foundation 1990.

PACS at the University Hospital Leuven

E. Bellon, P. Suetens, G. Marchal

Interdisciplinary Research Unit for Radiological Imaging (ESAT/MI2 & Radiology),
K.U.Leuven, Herestraat 49, 3000 Leuven, Belgium

Abstract: The configuration and the use of the PACS in the university hospital are described. Currently, this configuration is used mainly for supporting our research activities on image acquisition techniques and on image processing methods. This research as well as our own PACS developments are situated. Actual problems and future plans are discussed.

Keywords: PACS, Image Processing.

Introduction

The PACS development at the university hospital of Leuven is performed in the Interdisciplinary Research Unit for Radiological Imaging. This research group has been established jointly by the Department of Radiology and the ESAT division of the Department of Electrical Engineering. There is a close collaboration with other departments of the faculty of Medicine, such as Cardiology, Nuclear Medicine, Neurology and Neurosurgery. Currently, the PACS installed in the university hospital 'Gasthuisberg' is used mainly for supporting our research activities on acquisition techniques and image processing and analysis methods.

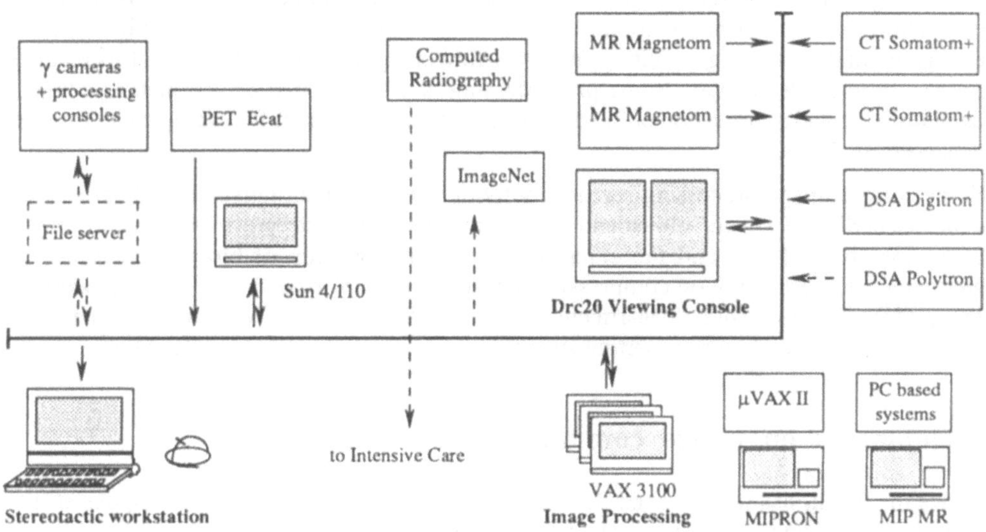

Figure 1 : Configuration of the PACS in the university hospital Gasthuisberg

NATO ASI Series, Vol. F 74
Picture Archiving and Communication Systems (PACS) in Medicine. Edited by H. K. Huang et al.

Current configuration and use

Figure 1 shows the current configuration and the extensions planned for the near future. The different PACS components are interconnected using Ethernet. The configuration can conceptually be divided into 4 units :

- The first unit is centered around a Siemens DRC20 PACS console and associated database. 2 CT scanners (Somatom+) and 2 MR scanners (Magnetom) from the department of Radiology and a DSA modality (Digitron) from the department of Cardiology are connected to the net. The connection of an additional DSA (Polytron) will be provided in the near future. A prototype Computed Radiography system (ADC, Agfa) is being evaluated.
- In the department of Nuclear Medicine, a PET scanner (ECAT 931, Siemens) and a Sun 4 workstation used for display and image processing are connected to the network. The plans are to make this department totally digital.
- The Image Processing subunit is equipped with a cluster of VAX/VMS workstations and display computers.
- A 3D graphics workstation (IRIS 2400 T, Silicon Graphics) is turned into a workstation for stereotactic neurosurgery planning [4]. This workstation can simultaneously display images from different modalities, allowing the surgeon to get an integrated view of CT, MR, and DSA data. This allows for an accurate planning and intraoperative control of the trajectory of the biopsy needle. Essential for the clinical application is a fast and easy access to the image data available on the network.

The PACS is configured mainly to support the clinical evaluation of image acquisition methods and the fundamental research on medical image processing and analysis :

- While routine MR and CT examinations are processed and stored locally on the modality, some image series acquired for specific studies are transferred to the PACS viewing unit for evaluation at an appropriate time. This viewing console is particularly suited for the evaluation of 3D data (e.g. resulting from MR flow imaging studies) because of its abilities for fast 'cine mode' display. Some preprocessing is often applied on the images, ranging from reslicing and reprojection of 3D data sets, to the application of for example blood vessel enhancement filters.
- Some current image processing research activities are the quantitative analysis of the left ventricle in SPECT [1] and ultrasound images, the automated interpretation of blood vessels in angiographies [2,3], and the enhancement and recognition of blood vessels in MR angiograms [5]. Research on compression algorithms is performed in the image processing group of the ESAT division (Department of Electrical Engineering). We plan to start a project on developing a radiological reporting system based on the IBM RISC/6000 system.

Developments on image communication

Because the PACS has mainly been developed to support our research activities, our efforts have been focused on making an easy transfer of image data possible.

The connections of the CT, MR and DSA modalities and the integration of the DRC20 console and the image processing unit have either been provided by Siemens or have been realized with the aid of the Siemens 'PACSNET Research Kit', a software package developed for assisting in the implementation of communication programs between PACS components.

The availability of such software packages is of particular importance in our situation, where 'turnkey' commercial software solutions are unlikely to be easily integrated in our research software environment. However, this software package can only be used on VAX/VMS computers using the DECnet network protocol for communication.

As no commercial solutions were available for the connection of several other modalities, the software for the integration of the PET scanner, the Nuclear Medicine viewing console and the stereotactic workstation has been developed in-house. These modalities are based on computers running the UNIX operating system, and we have used TCP/IP as the lower-layer network protocol.

In our own image communication developments, we have tried to adhere to the philosophy of exchanging all data according to the ACR-NEMA protocol (with SPI extensions where necessary). By ensuring that all data on the net are in the same format, it becomes easier to add new modalities to the network.

Present problems and future plans

In the near future, we plan to extend the use of PACS towards more clinically oriented applications and to evaluate on a small scale the potentials as well as the organizational problems associated with the use of PACS concepts in our clinical routine.

One possible project, now in a preliminary phase, is the installation of relatively low-cost viewing consoles in the intensive care unit, for evaluating the impact of providing diagnosed computed radiography examinations digitally.

Our experience with the current configuration is that it accomplishes its task in transmitting and storing images for our research applications, but that it imposes constraints for the application in clinical practice. A typical problem is that the PACS database associated with the DRC console cannot be queried by other nodes in the network. In our opinion, the flexibility necessary in a PACS requires very standardized database modules, controlled by a central database management system that relieves the user of much of the practical issues. In our situation, we would like any PACS console to be able to interrogate any of the distributed databases (aside from security-issues) and to retrieve images as well as other patient related information.

References

1. Nuyts J., Mortelmans L., Suetens P., Oosterlinck A., Deroo M.: *Model-based quantification of myocardial perfusion images from SPECT.* The Journal of Nuclear Medicine, Vol. 30, No. 12, December 1989, pp. 1992-2001.
2. Smets C., Van de Werf F., Suetens P., Oosterlinck A.: *An expert system for the labeling and 3D reconstruction of the coronary arteries from two projections.* The International Journal of Cardiac Imaging, Vol. 5, Nos 2-3, pp. 145-154, 1990.
3. Suetens P., Smets C., Van de Werf F., Oosterlinck A.: *Recognition of the coronary blood vessels on angiograms using hierarchical model-based iconic search.* Proceedings IEEE Computer Society Conference on Computer Vision and Pattern Recognition, CVPR'89, San Diego, California, June 4-8, 1989, pp. 576-581.
4. Suetens P., Vandermeulen D., Gybels J., Marchal G., Wilms G.: *Stereotactic arteriography and biopsy simulation with the BRW system.* To be published in "Computers in Stereotactic Neurosurgery" (P. Kelly, ed.), Blackwell Scientific Publications, New York, 1990.
5. Vandermeulen D., Smets C., Suetens P., Marchal G., Gybels J., Oosterlinck A.: *Knowledge-based 3-D segmentation of bloodvessels on a spatial sequence of MRI and ultrasound images.* Proceedings SPIE, Vol. 1092, Medical Imaging III : Image Processing, Newport Beach, California, January 31 - February 3, 1989, pp. 142-148.

PACS activities at the University hospital of Rennes

Bernard Gibaud, Christine Toumoulin, Jean-Marie Scarabin

INSERM U335, Service de Neurochirurgie, Hôpital de Pontchaillou, 35033 Rennes, France.

Abstract: The authors present the major concepts guiding the PACS activities carried out at the University Hospital of Rennes; the past, current and future developments are briefly presented.

1. General approach of PACS problems

1.1 Motivation for PACS

Our laboratory "Signals and images in medicine" has been working on image and biological signals understanding since 1981. In this area, the complementarity of the various investigation modalities and the capability to refer to anatomical or functional models have very exciting applications. Consequently, the abilities to index, retrieve, process and represent in various modes (especially in 3D) the medical data arise with a critical importance for the future.

Even in 1983 PACS seemed able to us to solve some of these problems. Our interest for PACS had two aspects: on the one hand, PACS would facilitate the access to medical images and stimulate the development of computer assisted procedures based on image data, especially in neurosurgery and radiation therapy planning; on the other hand, our objective was to participate to the general effort devoted throughout the world towards integrated solutions for image archiving and distribution [1 to 3]: though there is no doubt for a general evolution towards integrated information systems, nobody can tell whether this evolution will take 5, 10 or 20 years to spread into the hospital world. The development of a PACS prototype in Rennes seemed to us the best way to get a practical experience in order to derive and assess general concepts.

1.2 Major options

• Departmental or hospital wide PACS ?

Ever since 1983 we have been convinced that PACS issues had to be addressed at the hospital level and that integration problems could not be bypassed. We are very aware that significant simplification results in the design process when limited configurations are considered (departmental PACS); nevertheless, we believe that a PACS can't be really integrated and extensible if it was not designed with enough generality.

As a consequence, in our opinion, PACS experiments should associate different kinds of image users (radiologists, clinicians, surgeons, etc...). Furthermore, we believe that the added value of PACS will more easily be demonstrated in medical domains requiring quantitative utilizations of image data; neurosurgery (especially stereotaxy), radiotherapy, reconstructive surgery are domains where images are involved in the therapy planning and require that the data are kept in digital form. Multi-modal confrontations are more and more necessary (in 2D or 3D) which imposes the digital acquisition of image data including the related acquisition parameters (which excludes a priori any solution based on video tapping). In such medical domains PACS are neither a luxury nor a gadget but an essential tool. Another issue to be highlighted is cost: the number of procedures involved in such departments is relatively small, as compared with the volumes of data produced in a radiology department; besides, image resolutions (256 or 512 for CT or MRI, 512 or 1024 for angiography) are small and consequently workstations remain at a reasonable cost. Lastly, therapeutic procedures are mostly achieved in delayed mode which allows a flexible organization of data transfers.

NATO ASI Series, Vol. F 74
Picture Archiving and Communication Systems (PACS) in Medicine. Edited by H. K. Huang et al.

These factors make us believe that PACS should be experimented in these departments because the cost efficiency ratio seems to us very interesting.

- **Integration of the PACS in the hospital**

Our experience is that a successful integration of a PACS can only be obtained if adequate relationships can be established at 4 levels:

- *medical level*, of course, because the final objective is to improve patient care and only physicians can appreciate such improvements,
- *administrative level*, for mainly two reasons: on the one hand the hospital administration is responsible for investments (at least in France) and on the other hand it manages the Information System,
- *research level*, because industrial PACS are not really available today; the integration of the various components (image sources, archives, network, workstations), the setting up and the evaluation of PACS (from medical, organisational, economical viewpoints) need the collaboration of research laboratories,
- *industrial level*, lastly, because the interfacing of industrial products on multi-vendor PACS can hardly be achieved without the collaboration of the manufacturers (although significant advances have been obtained in standardization during the last 5 years). Besides, the PACS market is a new one and the manufacturers themselves seek collaboration with the end users: the expressions of PACS requirements are still un-complete, un-structured and in rapid evolution so that a practical way to understand real needs is to analyze the users reactions at a early stage of the development of prototypes. Another feature has to be pointed out: PACS experiments have highlighted that PACS could not really be used in routine unless they are based on industrial products; the industrialization of prototypes derived from a cooperation between research and manufacturers also needs a strong motivation from the manufacturers.

2. Current developments in Rennes

2.1 Setting up of the SIRENE prototype

The first stage of the SIRENE program (85-89) allowed to set up and evaluate a prototype at the University Hospital of Rennes [4]. Although the early objectives (defined in 83-84) have not been completely reached (reduced configuration, only one image modality could actually be interfaced in digital mode), we consider that the experience was globally positive. The configuration is actually used, mostly in neurosurgery and radiotherapy (clinical activities) and by the researchers of the INSERM U335 laboratory for research activities (multi-modal imaging and 3D).

Some developments, especially related to the network have turned out as technological dead ends, mainly due to the lack of motivation of the concerned manufacturers. Other developments, which were based on good ideas (notion of multi-media patient records) did not lead to a satisfactory utilization because all aspects of the problem could not be properly addressed.

2.2 Current and future activities

From 1989 two action lines were carried out:

• *A second phase of the SIRENE program* has been initiated (fig. 1): it aims at extending the PACS configuration; new imaging modalities will be interfaced and diagnostic consoles will be installed in the radiology department and properly interfaced to the image server; sub-networks will be set up in the Cancer Centre and in the University. These new developments are carried out by the CERIUM[1], in partnership with GE-CGR and Copernique (image server manufacturer). The new PACS configuration will allow interworking with the newly installed RIS. This extension will link together several sub-networks by means of two networks: a low-rate, wide-spreading network, compatible with public networks, and a high rate backbone (based on Ethernet firstly, on FDDI, in a second stage) to support image transfers.

• *The second aspect of our action* deals with the design of multi-modal image databases. This component of a PACS seems strategic to us because it will determine the final capabilities of the PACS: intelligent access to images, processing and re-processing of images, various representation possibilities (3D, multi-modal, reference to anatomical atlasses). These topics are being intensively discussed within the NRV-PACS group[2] in order to elaborate a comprehensive model able to represent in a suitable way the whole set of data involved in medical imaging. The latest advances of this work are reported in Dr Bizais's paper (same issue).

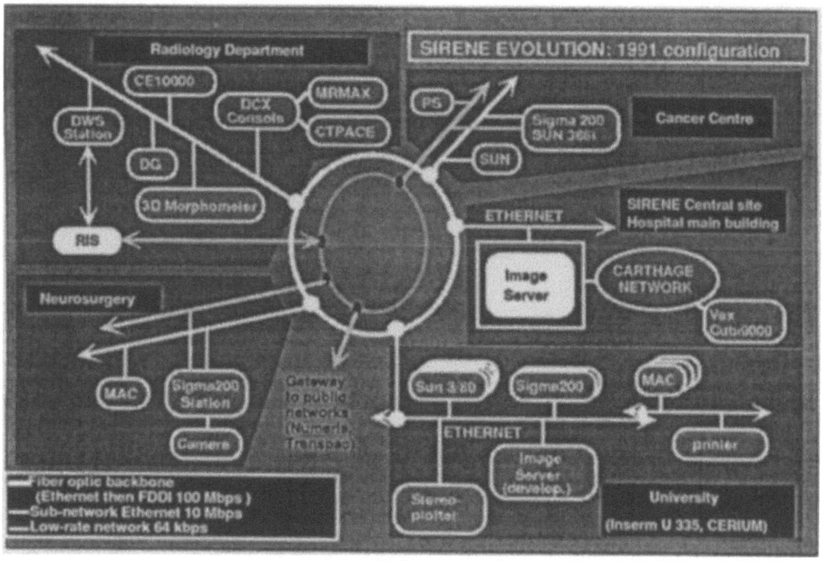

figure 1: the SIRENE configuration expected in 1991

References

1. Zeleznik M.P., Maguire G.Q., Baxter B.: PACS data base design, SPIE PACS II, Vol 418, pp 287-295, 1983 .
2. Meyer-Ebrecht, D.:On the architecture for pictorial information systems.NATO ASI series, Pictorial information systems in medicine, HK Höhne Ed, Springer Verlag, Berlin Heidelberg New-York Tokyo, 1986.
3. Jost, R.G.; Hill, R.L.; Blaine, G.L.; Cox, J.R.:PACS experience as a motivation for a campus-wide picture network. SPIE PACS pp 549-556, 1986.
4. Gibaud B., Picand F., Benslimane A., Grassin P., Urbano L., Toumoulin C., Fresne F., Rouvière M., Scarabin J.M.:SIRENE:Desin and evaluation of a PACS prototype, Proc. CAR 89, Springer Verlag Ed, pp 484-488, 1989.

[1] Centre Européen de Recherche en Imagerie à Usage Médical

[2] Collaboration on PACS activities between Nantes, Rennes and Villejuif (France)

PACS and related research in France

J. Vermont, P. Cinquin, J. Demongeot

TIMB-TIM3-IMAG, Faculty of Medicine, University J. Fourier of Grenoble
38 700 La Tronche France

Keywords : PACS, hospital information system, medical image workstation

Introduction

In a first part, we present the state of the art in research about PACS in France, with an overview on the main french PACS projects (Grenoble, Lille, Montpellier, Nantes, Rennes and Villejuif).

We will take afterwards the representative example of Grenoble in order to analyse :
- in which way and by which actors the PACS has been specified
- what financial supports has been found
- what type of multipartnership has been built (university, hospital, industry)
- what obstacles have been encountered : political (at the local, regional, national and european level), legal and socio-medical
- how the project could be integrated in the european research on PACS.

Finally, we will propose a perspective of the PACS as a subsystem of an integrated hospital information system and we present an example of methodology for specifying a PACS workstation devoted to the manipulation of images by a surgeon. We will give an example of functionalities and multimedia facilities that such a workstation has to bring in helping the pre-, intra- and post-operative strategies.

PACS research in France

PACS research in France is active and has some reasons to see "the light at the end of the tunnel" [1]. Seven experiments are in process :
- in Grenoble, a PACS is in building in the university hospital with a special connection to neurosurgery (Grenis project and AIM project Murim) and a private one is designed in the private sector (Race project Multimed)
- in Lille, a PACS is in designing in the university hospital in the frame of a IHIS project (DEC-CHRUL project)
- in Montpellier, a PACS is in building in the university hospital (Race project) [2]
- in Nantes, a PACS has been built from a nuclear medicine network in the university hospital (DIMI project) [3]
- in Rennes, a PACS has been built in the university hospital with a special connection to neurosurgery (Sirène project) [4]
- in Villejuif, a PACS is in building in the Gustave Roussy Institut in the frame of a IHIS project (Simbad project).

The three last Projects are closely connected in the NRV-consortium, in which common concepts and tools are developed.

The french projects have been planned in collaboration with scientific or technologic research institutions (like CCETT, IMAG, INRIA) and industries (like Apsis, CGR-GE, Copernique,

NATO ASI Series, Vol. F 74
Picture Archiving and Communication Systems (PACS) in Medicine. Edited by H. K. Huang et al.

Digital Design, Siemens, Sopha Medical). They participate to the EEC-COMAC-BME project ISCAMI (Integrated System for Computer Assisted Management of Medical Images) and to numerous other EEC projects (like Esprit, AIM, Race). The concept of multimodality display workstation and interventionnal clinical or surgical workstation is often proposed in the french PACS design step, and also the interface with the underlying Integrated Hospital Information system (IHIS).

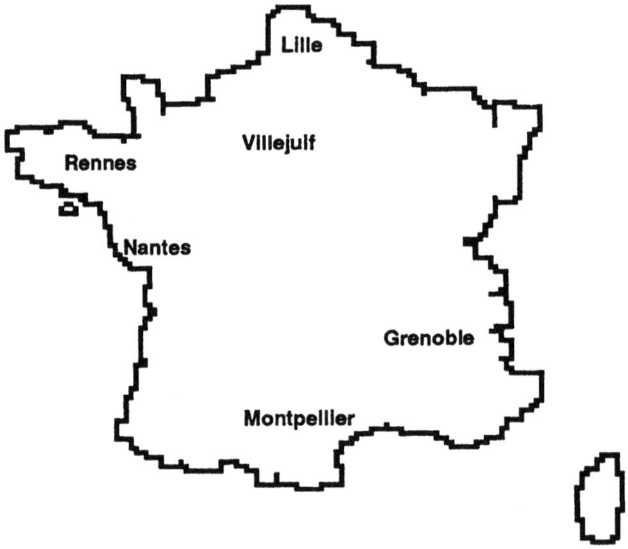

PACS projects in France

Building a PACS in Grenoble

The design phase began in 1988 in the frame of the director scheme for informatizing of the university hospital (Racines method - third phase). The actors were, after deciding to leave separate the public and the private (Multimed) Grenoble projects :
- the medical information department of the university hospital
- all medical imaging services except paediatric radiology (central radiology, neuroradiology, nuclear medicine, NMR Institut)
- the neurosurgery service, interested by the robotization of the stereotaxy
- the radiotherapy service, interested by the problem of the inverse dosimetry and its automatization
- three private companies (DEC, Siemens and Digital Design).

The financial support were manifold : University Hospital (medical material and informatics budgets), Rhône-Alpes region (research commission), Ministry of Health (Direction des Hôpitaux), Ministry of Research and Technology, companies (short contract or long partnership). An official partnership has been for example put in concrete form by a contract between University J. Fourier, University Hospital (CHRUG) and DEC.

The major obstacle has been encountered on the socio-medical side : the hospital sociology, the work habits make very difficult the exchange of images between classical image producers and clinicians : only the whole medical image (the physical one and its medical interpretation) is

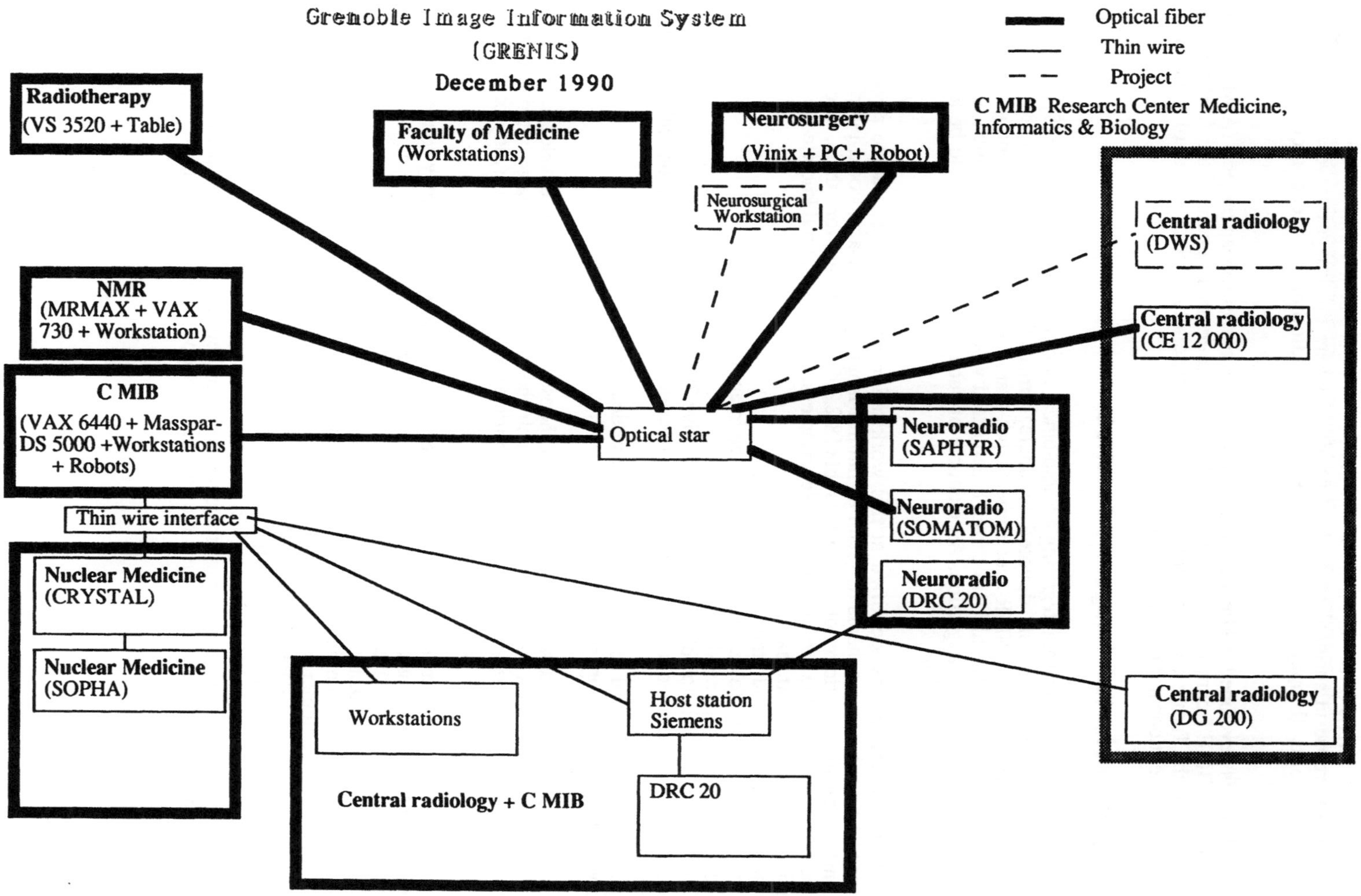
Grenoble Image Information System
(GRENIS)
December 1990
Optical fiber
Thin wire
Project
C MIB Research Center Medicine, Informatics & Biology
Radiotherapy
(VS 3520 + Table)
Faculty of Medicine
(Workstations)
Neurosurgery
(Vinix + PC + Robot)
Neurosurgical Workstation
NMR
(MRMAX + VAX 730 + Workstation)
C MIB
(VAX 6440 + Masspar-DS 5000 +Workstations + Robots)
Optical star
Central radiology
(DWS)
Central radiology
(CE 12 000)
Neuroradio
(SAPHYR)
Neuroradio
(SOMATOM)
Neuroradio
(DRC 20)
Thin wire interface
Nuclear Medicine
(CRYSTAL)
Nuclear Medicine
(SOPHA)
Workstations
Host station Siemens
DRC 20
Central radiology + C MIB
Central radiology
(DG 200)

susceptible to be transferred on the network. That causes obstacles to the diffusion of images for therapy purpose in surgical or medical services, where the images serve as support of certain treatments (biopsy-puncture, curie-therapy, radiotherapy, deep recording with electrodes, ...).

The PACS is developed in Grenoble in connection with two european projects (AIM Murim and COMAC-BME Iscami) and some worktasks of the future main phase of the AIM project correspond to the future development of the PACS, especially in the domain of building a dedicated interventionnal imaging workstation.

A surgeon oriented image workstation

An image information system integrating a Radiological Information System (RIS) [5] and other specific information systems devoted to nuclear medicine, functional neuro-imaging (magnetoencephalography, NMR spectrometry, pet-scan, fluorescent dye mapping,...) has to be interfaced with the general IHIS through dedicated workstations [6,7], which realize the concrete location of the integration. An example of such a workstation could be a surgeon oriented image workstation.

Multimediafunctions identified at this workstation are numerous : some already are supported by existing mechanisms (such as natural language interface, vocal interface, 3D reconstruction stereotactic robot, cf. for example [8,2,3,4,5,6,14]) and for others, mechanisms of the computerized issue are to be found. Nevertheless, today's acquired knowledge makes it possible within the 5 next years.
These functions can be summed up as follows :
- consulting a medical folder enclosing multimedia images subfiles;
- vocal and tactile command for tools;
- intraoperative simulation of the action, by interactive 3D imaging;
- consulting of multimedia databases enclosing operative protocols and anatomical atlas used as reference;
- performing the surgical action and intra-operative decision making;
- monitoring of the patient or of cut-off organs.

This conception approach allows to isolate and choose applications to be developed.
One example is given on Figures 1 and 2, which show a preoperative segmentation of the wrist (using a neural network enhancement technique) and the dynamical study of the 3D-reconstructed articulation.

You will find hereafter some examples of on-going applications which may converge to the realization of a workstation able to help the surgical intervention , such as :
- automatized help to realize surgical interventions :
  . Stereotactic neurosurgery [8-14],
  . Computer-assisted intervertebral puncture [15,16],
  . Pedicular screwing [17],
  . Cranio-facial surgery [18],
  . Wrist surgery [19,20].
- matching of images from various origins.
- definition of an atlas necessary to estimate relationships between normal (with areas to avoid) and pathological objects, and to use the localizing value of images in a diagnostic and therapeutic aim.

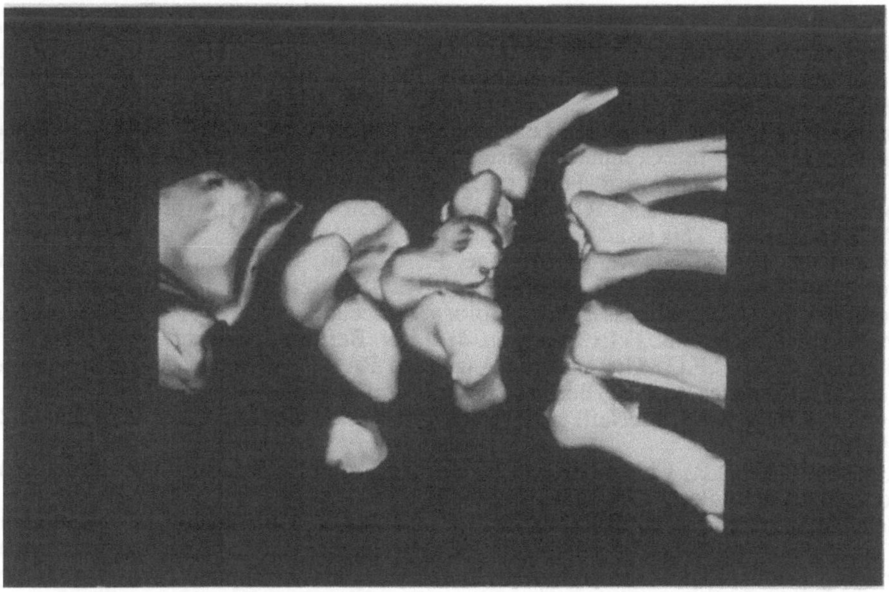

Figure 1 : contrast enhancement of a scanner wrist cut by using a neural network technique

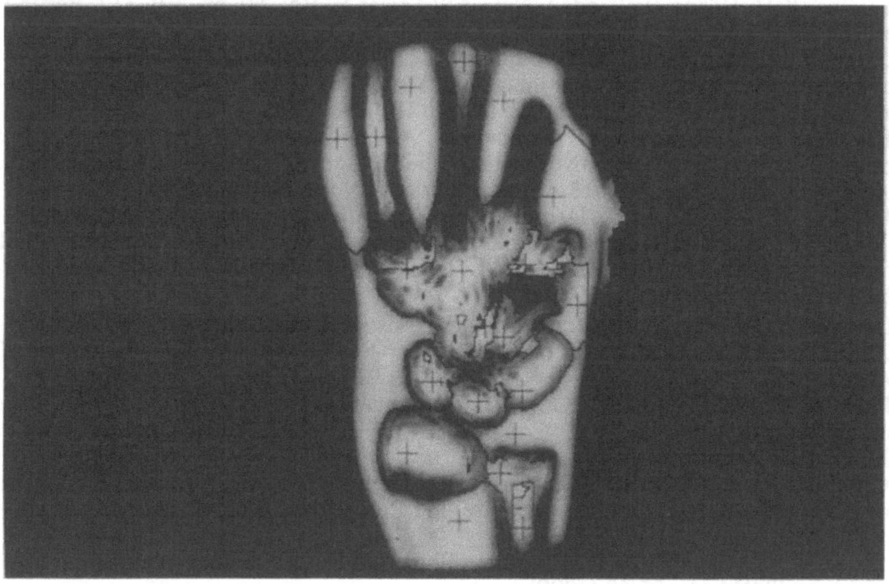

Figure 2 : preoperative dynamical study of the 3D-reconstructed wrist

References

1. Jost, R.G., W., Wessel, G.J., Blaine, J.R., Cox & R.L., Hill : PACS - Is there light at the end of the tunnel. In : Medical imaging III. PACS system design and evaluation. SPIE 1093, 74-84 (1989).
2. Lamarque, J.L., J.M., Bruel, H., Lestienne, P., Lopez & M., Gareil : PACS and medical selection imaging. In : Computer assisted radiology, Proceedings CAR 89 (H.U. Lemke, M.L. Rhodes, C.C. Jaffe & R. Felix, eds.), pp. 483. Berlin : Springer Verlag 1989.
3. Forte, A.M., & Y., Bizais : A mulyimodality medical image interpretation system as a PACS component. This issue.
4. Gibaud, B., F., Picand, A., Benslimane, P., Grassin, L., Urbano, C., Toumoulin, F., Fresne, M., Rouvière & J.M., Scarabin : SIRENE : design and evaluation of a PACS prototype. In : Computer assisted radiology, Proceedings CAR 89 (H.U. Lemke, M.L. Rhodes, C.C. Jaffe & R. Felix, eds.), pp. 484-489. Berlin : Springer Verlag 1989.
5. Rienhoff, O. & C.F.C., Greinacher : A general PACS-RIS interface. Lecture Notes in Medical Informatics, Vol. 37. Berlin : Springer Verlag 1988.
6. Hunter, T.B. & K.M., McNeill : the radiologist workstation. Invest. Radiol. 25, 282-284 (1990).
7. Arenson, R.L., D.P., Chakraborty, S.B., Seshadri & H.L., Kundel : The digital imaging workstation. Radiology 176, 303-315 (1990).
8. Benabid, A., P., Cinquin, S., Lavallée, J.F., Le Bas, J., Demongeot, J., de Rougemont : A computer driven robot for stereotactic surgery connected to cat-scan magnetic resonance imaging. Technological design and preliminary results. J. Applied Neurophysiology 50, 153-154 (1987).
9. Marque, I., S., Lavallée, C., Goret-Lezy & P., Cinquin : Towards a 3-D medical image analysis system based on a continuous modelling. In : SCAR 90 (R.L. Arenson & R. M. Friedenberg, eds.), pp. 191-197. Carlsbad : Symposia Foundation 1990.
10. Lavallée, S. & P., Cinquin : Computer assisted medical interventions. In : 3D-imaging in medicine (K.H. Höhne et al., eds.). Nato ASI Series F, Vol. 60, pp. 301-312. Berlin : Springer Verlag 1990.
11. Leitner F., I., Marque, F., Berthommier, T., Coll, O., Francois, P., Cinquin & J., Demongeot : Neural networks and image processing. In : From pixels to features II (J.C. Simon, ed.), pp. 173-194 . Amsterdam : North Holland 1990.
12. Lavallée, S., P., Cinquin, J., Demongeot, A.L., Benabid, I., Marque & M., Djaïd : Computer assisted interventionist imaging : the instance of stereotactic brain surgery. In : Medinfo 89, Singapore, pp. 613-617. Amsterdam : North Holland 1989.
13. Lavallée, S., P., Cinquin, J., Demongeot, A.L., Benabid, I., Marque & M., Djaid : Computer assisted driving of a needle into the brain. In : Computer assisted radiology, Proceedings CAR 89 (H.U. Lemke, M.L. Rhodes, C.C. Jaffe & R. Felix, eds.), pp. 416-420. Berlin : Springer Verlag 1989.
14. Lavallée, S. : A New System for Computer Assisted Brain Surgery. In : Proceedings 11th IEEE Engineering in Med. and Biol. Conference, Seattle, pp. 926-927. Piscataway : IEEE 1989.
15. Lavallée, S., P., Cinquin, C., Goret, J., Demongeot, G., Crouzet & P., Peltié : Ponction assistée par ordinateur. In : Actes Congrès AFCET RFIA 87, Antibes, pp. 543-550. Paris : AFCET 1987.
16. Cinquin, P., S., Lavallée, C., Goret, J., Demongeot, G., Crouzet & P., Peltié : Computer assisted intervertebral puncture. In : Proceedings MIE 87 (A. Serio et al., eds.), pp. 847-852. Rome : EFMI 1987.
17. Mazier, B., S., Lavallée & P., Cinquin : Chirurgie de la colonne vertébrale assistée par ordinateur. ITBM 11, 55-62 (1990).
18. Cinquin, P., J.C., Saget, G., Plasse & P., Antoine : Chirurgie plastique maxillofaciale assistée par ordinateur. In : Proceedings A. I. Biomed 86, pp. 95-99. Montpellier : CRIM 1986.
19. Moutet, F., A., Chapel, P., Cinquin & L., Rose-Pitet : Imagerie du carpe en 3D. Ann. Radiol. 33, 128-133 (1990).
20. Brunie, L. & S., Miguet : 3D reconstruction of the wrist : a new method of segmentation. DEA Report. Grenoble : IMAG 1989 and (submitted).

PACS Germany

Gerd Witte

Radiology Clinic, Diagnostic Department, University Hospital Eppendorf, Martinistraße 52, 2000 Hamburg 20, F.R.G.

Keywords: PACS Germany, PACS projects, PACS sponsorship

In Germany no PACS project has yet been used under clinical conditions in the daily routine of a hospital.

The most advanced German PACS project is the Berkom project in Berlin (see Lemke's paper) sponsored by the Federal Postal Administration. Other goverment sponsorships were announced in 1989 by the ministry of research and technology in order to evaluate PACS installations in light of medical economics. Until now, however, no further information concerning the extent or nature of support for these projects has been released. Due to the special situation caused by German unification, it seems rather unlikely that PACS projects will be sponsored by the ministry of science in the foreseeable future. Other governmental sponsorships is improbable.

The currently existing projects depend on the cooperation between the medical electronic industry and clinical institutions. In these cases, a cooperation agreement specifies that industry supplies the hardware for the project while the clinical institutions perform the practical application. Such agreements exist between industry and seven radiology departments of university hospitals as well as one private radiology institution. In three university projects computer science departments are working in cooperation with radiologists. The main fields of the projects can be described as follows:

- PACS and cut film (CT, MR) diagnostics (6 projects),
- PACS and computer radiography (1 project),
- PACS integration and existing radiology information systems (3 projects),

NATO ASI Series, Vol. F 74
Picture Archiving and Communication Systems (PACS) in Medicine. Edited by H. K. Huang et al.

– PACS - consequences for medical economics (2 projects).

The various projects are in different states of development, from hardware installation to first pre-clinical experiments.

The project at the Diagnostic Radiology Dept. of the University Hospital Eppendorf in Hamburg includes the integration of existing cut film modalities as well as radiology information systems which are already in use. Medical economic issues and their direct effect on the clinical radiologic routine are being explored in cooperation with the Institute of Economics of the University of Osnabrück. At present the installation of the hardware is being finished. First results are to be expected in about one year.

The Berlin Communication System (BERKOM)

Heinz U. Lemke

Technical University of Berlin, FR 3-3, Franklinstr. 28/29, 1000 Berlin 10, FRG

The BERKOM project was initiated in 1986 by the German PTT and the Senate of Berlin. Management responsibility for the project has been given to Deutsche Telepost Consulting (DETECON). BERKOM serves as a test bed for future developments of broadband communication services, terminal devices and applications. This aim is shared by a spectrum of project partners comprising some 40 scientific institutions, 17 industrial companies and 5 users.

The budget made available for the project totals 110 Mio. DM for services, applications and end-systems, as well as 65 Mio. DM for the test network, gateways and adapter cards. An additional 50% project financing is derived from industry participation.

Communication Services and Interfaces

The basis for the communication services is the existing Berlin optical fibre overlay network, which extends over the urban area of West Berlin. It includes subscriber lines, a variety of

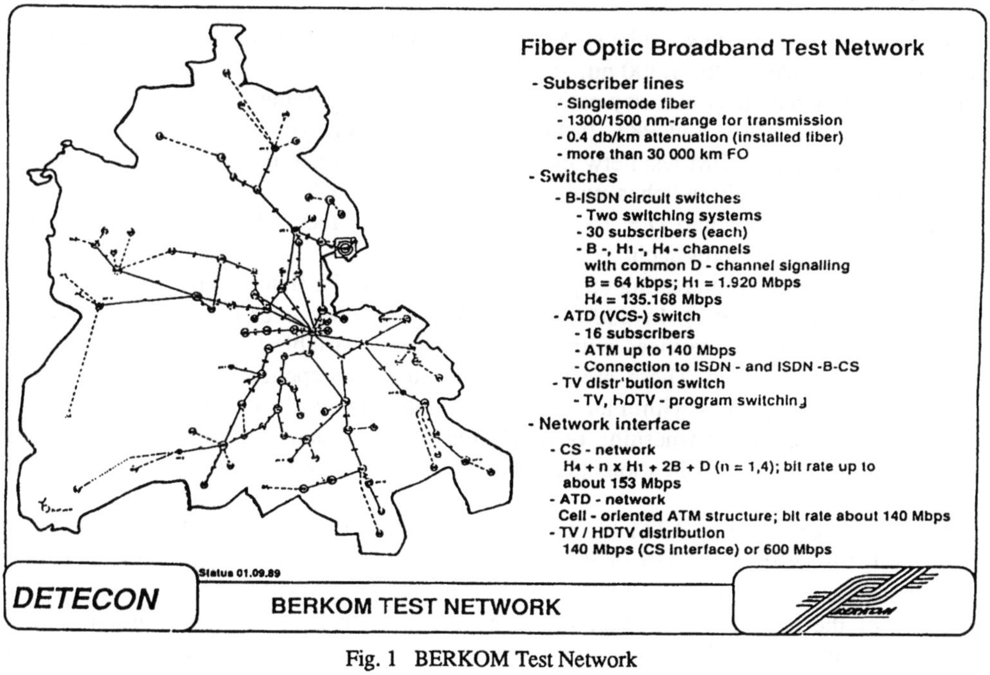

Fig. 1 BERKOM Test Network

NATO ASI Series, Vol. F 74
Picture Archiving and Communication Systems (PACS)
in Medicine. Edited by H. K. Huang et al.

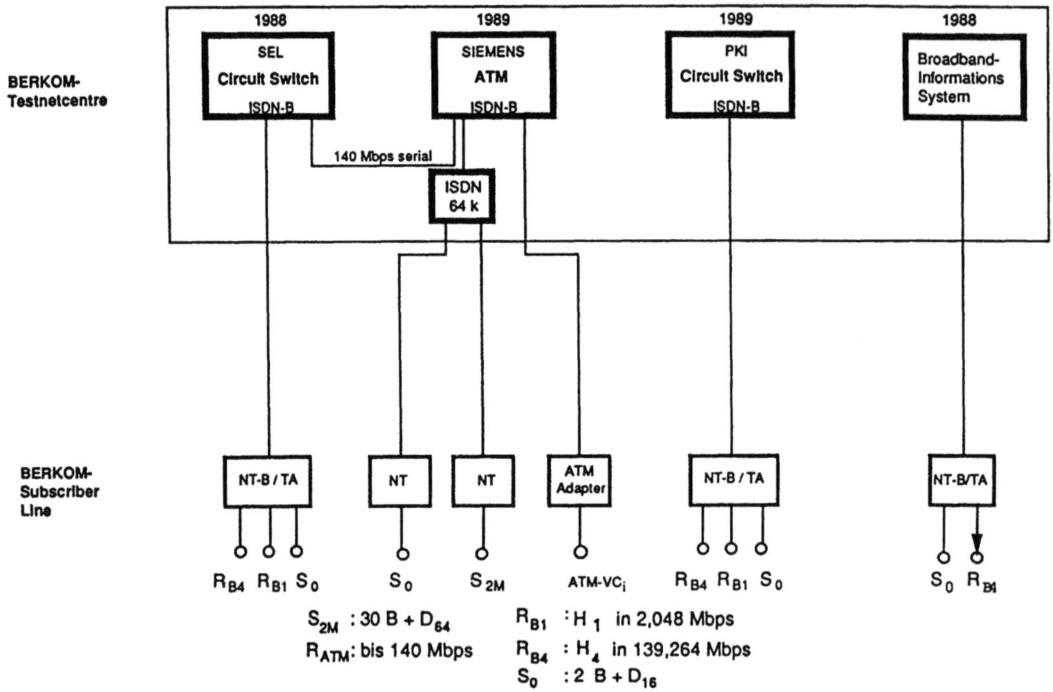

Fig. 2 Switches in the BERKOM Test Network

switches and network interfaces (Fig.1). Monomode fibres are the basis of this test network. Transmission is effected in the 1,300 nm-sector.

The telecommunications industry actively involved in developing prototypes of ISDN-B switches are Siemens, SEL and PKI. In the meantime, several types of switches have been installed within the BERKOM test network and are already in operation (Fig.2). The systems of SEL and PKI are circuit switches with a channel structure of H_4 = 135,168 Mbps, H_1 = 1,920 Mbps, B = 64 Kbps and D = 16 Kbps. Signalling is effected by the D-channel. Both switches serve 30 end-systems each.

Siemens has taken an interesting step in BERKOM, with the first official commissioning of an ATM exchange (Ref.1). In addition to 64-kbit/s services over fibre-optic subscriber lines, several motion-video services with various fixed and variable bit rates are offered via flexible subscriber accesses. These are linked to mutually independent virtual private networks for closed user groups within an "imaginary" public network. This makes possible the coupling of PABXs, the intermeshing of local area networks and the realization of special networks, e.g. for medical applications (such as PACS) and the print media (joint editing), see Fig.3.

The transfer principle of ATM (asynchronous transfer mode) is generally regarded as a possible basis for broadband ISDN (Ref.2). Whereas previous networks were characterized by the simultaneous use of circuit switching and packet switching, ATM essentially unifies the advantages of the two methods. With ATM, no definite physical channel is assigned to a connection. A bit-string to be transported is therefore first split into cells of constant length and then tied up according to actual needs with packeted-like messages; each packet contains a definite address. The cells of several mutually independent communication connections, possibly with different bit rates, can be transmitted over the same line of an ATM-oriented network. Cells of the individual connections are not assigned to any definite places, instead they are transmitted in random order of their arrival ("statistical multiplexing").

In BERKOM, the ATM-exchange will be interconnected to ISDN and ISDN-B switches, in order to test hybrid network concepts for a variety of applications (Ref.3). Individual switching technologies, e.g. packet- or circuit-switching, have characteristic advantages and disadvantages for different traffic modes (continuous or discontinuous bitstreams).

A further network complexity is introduced when different LAN (Local Area Network) and other MAN (Metropolitan Area Network) concepts have to be integrated into an overall communication infrastructure (Ref.4). Several LAN and MAN standards will have to be considered in this context. Connectionless coupling and the expansion of LAN`s into MAN`s is already part of the MAN standard IEEE 802.6 (distributed queue dual bus, DQDB). Bell Communications Research (Bellcore) has specified a connectionless service on the basis of this standard, the "switched multi-megabit data service, SMDS", which can be also operated over public ATM networks.

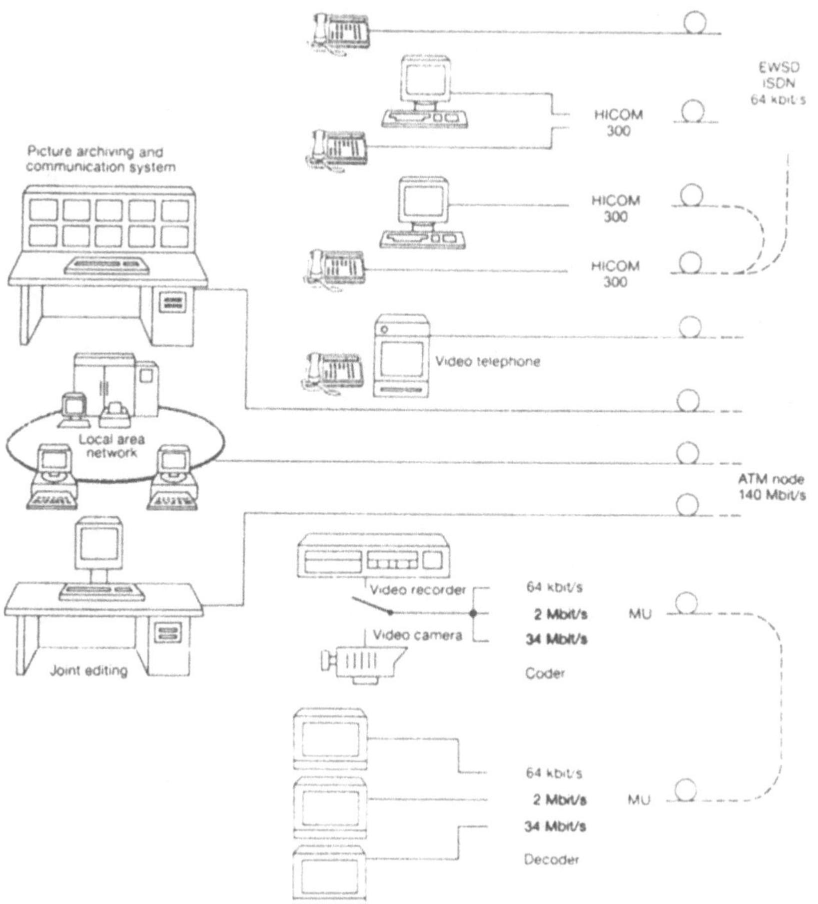

Fig. 3 ATM Applications in BERKOM and Siemens ATM switch

Further investigations are needed to determine which LAN and MAN access methods and WAN switching systems appear to be the most appropriate for the realization of network features. They should satisfy real practical demands for applications, end-systems and users.

The gateways being developed for BERKOM are aimed to permit the connection of different LAN types (Ethernet, Token Ring, FDDI, etc.) to the test network. They are often integrated with adapter cards, connecting computer systems with standard busses (e.g. VME,

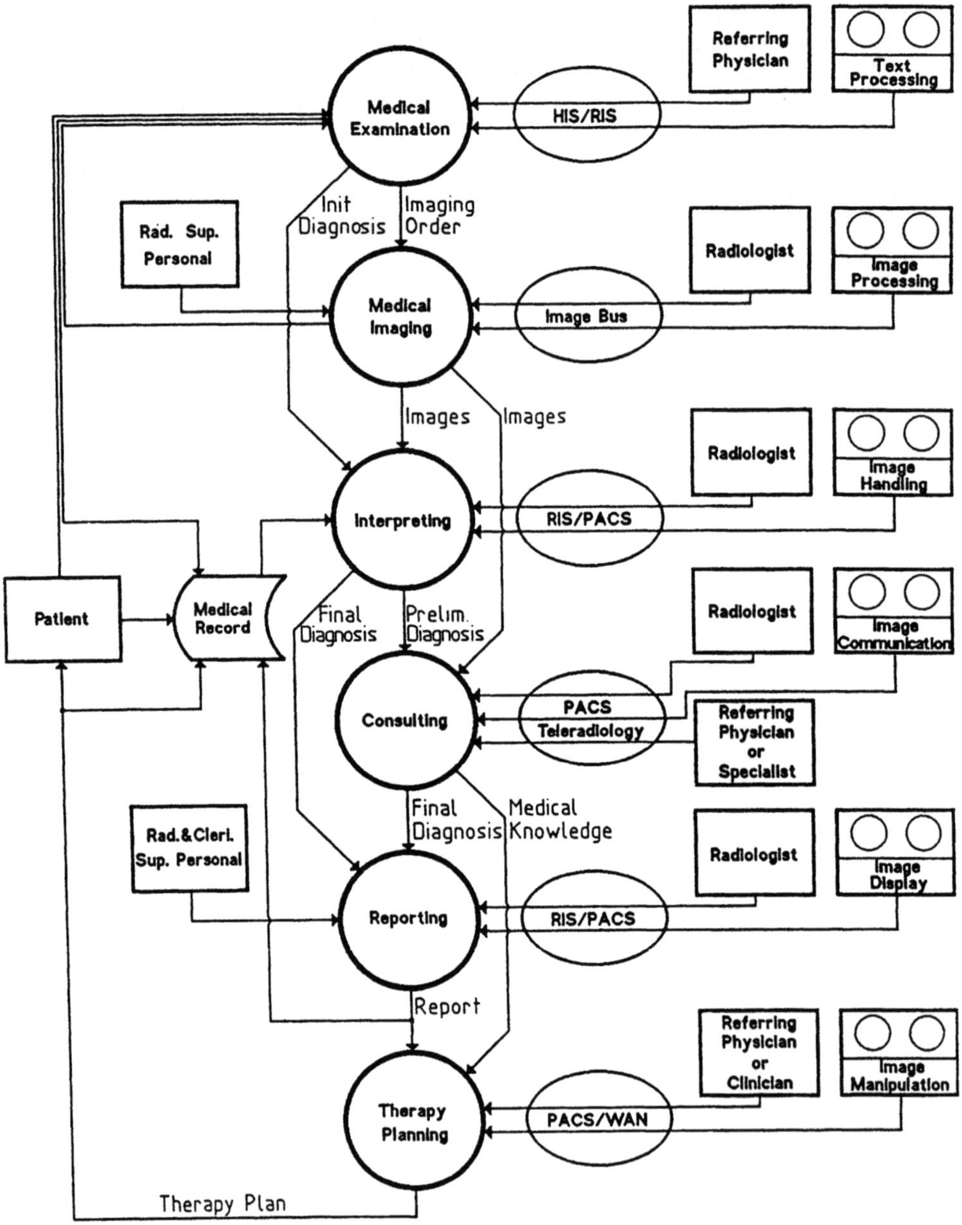

Fig. 4 Functional Types of Workstations in Diagnostic Pathway

AT, Microchannel) to the test network. Although currently these cards are expensive, with projected larger quantities in the future, considerable reduction in price can be expected.

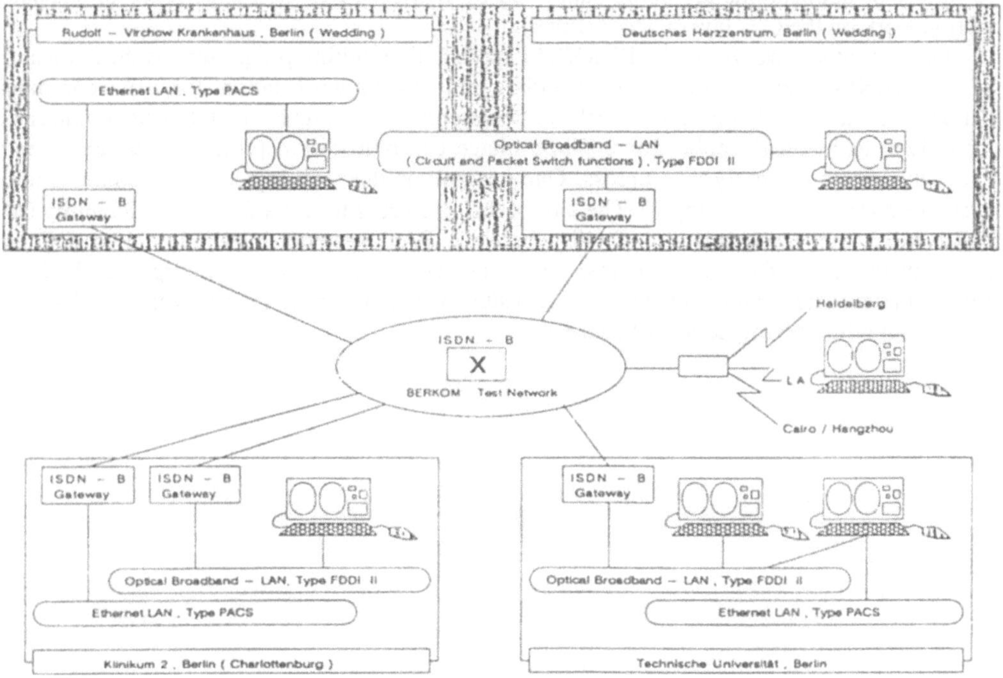

Fig. 5 Teleconferencing Environment for Medical Workstations

End-Systems

Within the BERKOM Project, special attention is being given to end-systems, particularly workstation development with multi-media communication capabilities. Many forms of document communication contain all elements of basic data types, such as text, graphics (vector data), still pictures (pixel data), speech/audio and moving images (cine). In this context, importance is being attached to basic terminal development for mass and individual communication. It is hoped to profit from the advantages of consumer electronics (i.e. low costs made possible through high production rates) when it will be applied to the field of individual communication. With this aim, technical components are developed, which are applicable for TV and video communication terminals or for HDTV and office terminals.

Similar objectives are also followed in a BERKOM project dedicated towards medical workstation development. Based on the diagnostic pathway, different functional types of medical workstations have been identified (Fig.4). They are being interconnected for multimedia communication purposes. The primary working mode considered here is real-time teleconferencing (Fig.5).

Applications

Applications of users are supported by standardized hardware connections of their systems to the test network and the provision of compatible communication protocols (OSI layers 1 to 4). A variety of applications are being supported by BERKOM to gain experience of standardization work on the OSI layers 5 to 7 (Fig.6).

One essential task within the BERKOM application working programme is the realization of the BERKOM service concept by providing multi-media workstations for office communication. The idea of a "paperless office" is often mentioned in this context. Special emphasis is also given to medical applications. Three image communication projects in various health care settings are currently in progress. One example is the radiological data communication system RADKOM (Ref.5, Fig.7). The aim of this project is to test a fast ISDN-B facility (140 MBits/s) as the communication facility between PAC systems installed in two different departments of the Rudolph Virchow University Hospital at Charlottenburg and Wedding. Medical teleconferencing between these two sites, which are more than 10 km apart, has been carried out with varying degree of success.

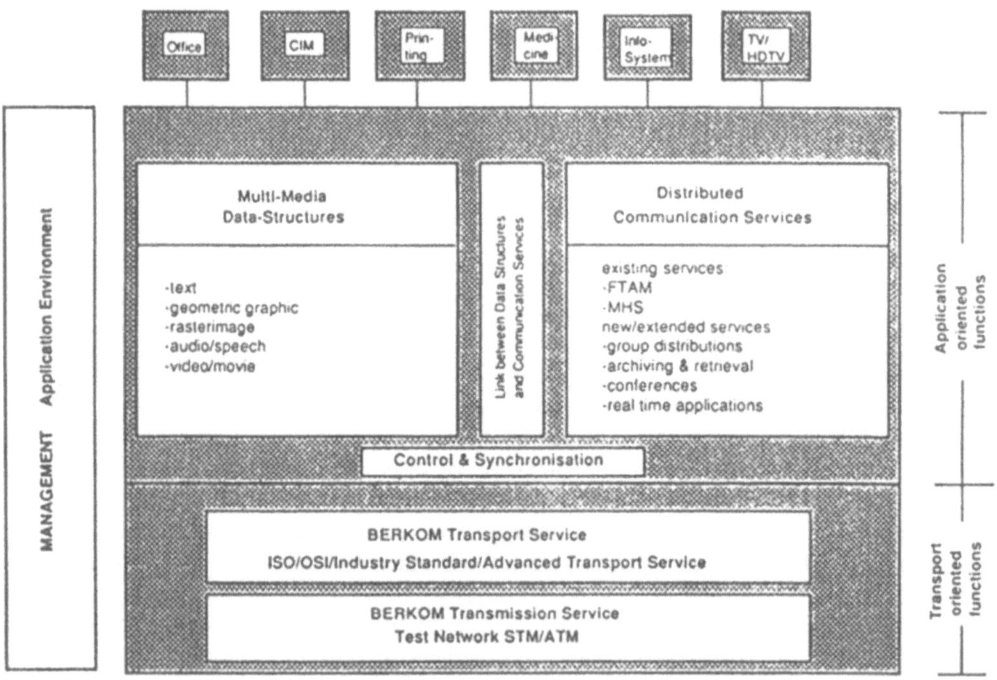

Fig. 6 BERKOM Reference Model Pilot Applications

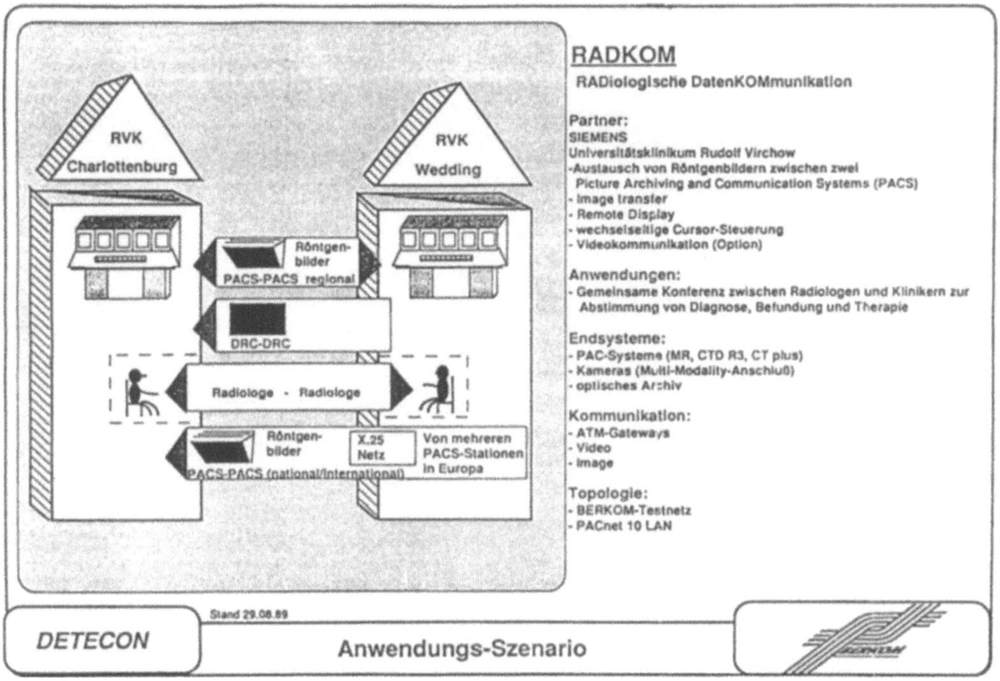

Fig. 7 The Radiological Data Communication System RADKOM

References

1. Astinet F, Langer M, Hunger J, Pack G, Kuritke D, Scharnberg B: BERKOM/RADKOM - A Pilot Project in Berlin to Improve Fast Image Communication between Two or More Hospitals. S/CAR'90, Computer Applications to Assist Radiology, Anaheim, California. pp.84-88. Symposia Foundation, Carlsbad, California, 1990.
2. DETECON (Technisches Zentrum Berlin): BERKOM - Telecommunication within the Optical Fiber Network, 1990.
3. Deutsche Bundespost TELEKOM: ATM - a universal transfer mode, 1989.
4. Mollenauer JF: Metropolitan area network update: The global LAN is getting closer. Data Communications International, Dec. 1989.
5. Siemens Zentrallaboratorium: The intelligent integrated broadband network - Telecommunications in the 1990s. Munich 1989.

PACS and Related Research in Italy

Paolo Giribona[1], Diego Bravar [1], Claudio Giuricin[1], Iole Pinto[1], Fulvio Stacul[2] and Walter Ukovich[3]

[1] Centre for Evaluation of Biomedical Instrumentation, Research Area of Trieste, Padriciano, 34012 Trieste, Italy
[2] Department of Radiology, Cattinara Hospital, 34100 Trieste
[3] D.E.E.I., University of Trieste, 34100 Trieste

Abstract: A survey of PACS and related research in Italy has been carried out during the first half of 1990 in order to quantify the diffusion of commercial PACS and the state-of-the-art of non-commercial PACS experimentations in health care institutions.

The current status of PACS installations in Italy is reported, as well as a description of PACS-related research activities carried out by several institutions in different fields: economic evaluation, metropolitan PACS, operational analysis, RIS/PACS integration, teleradiology.

Keywords: PACS evaluation, technology assessment, RIS/PACS integration

1.PACS installations in Italy

The development of Picture Archiving and Communication Systems in Italy was initiated during 1987 with some delay with respect to other similar experiences in Europe.

Nevertheless, during the last two years we recorded a fast introduction of commercial PACS systems, mainly in public hospitals, due to a very active strategy of sales carried out by industry. As a consequence, at present, Italy is the country in Europe with the greatest number of sold systems.

A particular difficulty in summarising the actual PACS situation in Italy, as well as in other countries [4], is represented by the uncertainty in making a clear distinction between actual level of use, aspirations of users teams, industry market policies, and actual available funds in each site.

From a technical point of view the first requirement for making a survey of PACS projects is a definition of what PACS is.

Greinacher [5] has defined a minimum PACS installation as one which comprises all four of the following components, namely, 1) a storage system, 2) a workstation, 3) a shared database image management unit and 4) a digital communication network system. Using this definition, nine PACS installations have been located during this survey. Eight of them are commercial installations (operational or in progress) from Philips/AT&T and one is a non- vendor specific PACS developed in the Department of Digital Radiography of the "L'Aquila" University.

Table 1 reports a summary of PACS installations in Italy.

NATO ASI Series, Vol. F 74
Picture Archiving and Communication Systems (PACS) in Medicine. Edited by H. K. Huang et al.

TABLE 1

PACS INSTALLATIONS IN ITALY

CITY	SITE	SUPPLIER	STATUS
Bari	Mater Dei Clinic	Philips/AT&T	being installed
Bologna	Maggiore Hospital	"	operational
Castelfranco	Castelfranco Hospital	"	being installed
Ferrara	Sant'Anna Hospital	"	operational
L'Aquila	Collemaggio Hospital	non-commercial	operational
Messina	Messina Hospital	Philips/AT&T	partially installed
Monza	S.Gerardo Hospital	"	being installed
Trieste	Cattinara Hospital	"	operational
	Maggiore Hospital	"	operational
Milano	S.Paolo Hospital	"	being installed

The installations were mainly funded by local government Regional Health Authorities and, in some cases, this created some problems of acceptance by users not directly involved in the mechanism of acquisition and selection of the equipment to be installed.

The most of information concerning the experimentation of such systems is available from the four sites in which a fully operational PACS is already installed (i.e.: Bologna, Ferrara, L'Aquila, and Trieste). A detailed description of the systems and of the related activities is given in the following pages.

TRIESTE PACS INSTALLATION

Project description: In order to assess the impact of PACS technology on Health structures, a project of experimentation of a PAC System has been planned in 1987 at the Local Health Unit of Trieste which administrates the four public hospitals of Trieste.

The PACS facility is presently installed in the Radiology Department of the Hospital of Cattinara with a remote connection to the Maggiore Hospital (the second largest hospital of the city).

Five different diagnostic modalities (CT, MRI, DSA, US1 and US2), two Display Workstations, and an optical Jukebox (178Gbyte) are connected to a central Data Management System (DMS). Two Remote Viewing Stations (RVS) are connected via an Ethernet gateway to the DMS. In particular, the Ethernet gateway allows a distribution of images in the whole hospital by means of a fiber-optic Ethernet LAN which connects all the hospital departments. A Consulting Workstation (CW) is installed in the Radiology Department of the Maggiore Hospital (connection via optical fiber link), enabling the access to the DMS archive and the acquisition of images from three modalities (CT, ultrasound, laser film digitizer).

Goals: Experiment PACS technology and define guidelines for a national policy in this field. A contract of research with the Ministry of Health (Dept. of Central Health Planning) has been stipulated for this pourpose and a technology assessment report will be developed by the end of 1991. The long term goal is the full integration of PACS with the Hospital Information System and the implementation of a Metropolitan Area PACS connecting the four hospitals of Trieste for image transmission to/from a common archive.

Planning and delays: The presently working installation has been splitted onto two phases: a) Entry-level (planned Jan 1988, installed Sept. 1988), b) Present system (planned Mar 1989, installed Feb 1990). The delays were mainly due to late delivering of equipment from manufacturer.

Critical points: Low speed of workstations, no hard-copy connection available from manufacturer, data safety of the short-term magnetic archive, only analog connections with imaging modalities, poor technical support from manufacturer.

Research activities: Research activities are related both to the evaluation of PACS technology and to the development of the PACS system already installed [1], [3], [7], [8]. Some of these research activities have been developed in the framework of national and European projects: A.I.M.: Project A 1030; COMAC-BME: ISCAMI Concerted Action; Italian Ministry of Health: PACS Technology Assessment Project.

L'AQUILA PACS INSTALLATION

Project description: At the Department of Radiology of L'Aquila University a fully integrated non-commercial PACS/RIS has been installed. The adopted LAN is Ethernet with DECNET protocol: early experiences shows a data transmission rate of about 800 kbit/s. The RIS system is implemented in SQL+ language (Oracle), building a relational database. Each workstation has its own filter to the DR-DB with local processing capabilities and different features such as the access to specific tables and software modules. At present time, five workstations are connected to the LAN: 1 for secretary work, 1 for film digitizing, 1 for diagnostic image reporting, and 2 for image consultation.Three PIU (PACS Interface Unit) allow the connection to three imaging modalities: a digital fluoroscopy unit (1024x1024), a D.S.A. (1024x1024), and a digital radiography prototype (4096x4096) based on photostimulable plates.

The adopted image file format is ACR-NEMA compatible. The file server (MicroVAX II) is connected to a film laser printer and to a jukebox with 16 optical disks 2Gbyte each. A gateway connection with a Token Ring network supported by an IBM 6150 is in progress in order to experiment image restoration algorithms [6] .

Goals: To interface all digital imaging modalities located in the Digital Radiography Center. Medium and long term goals are: To evaluate and optimize PACS performances in image transmission, storage and retrieval. Filmless radiology and archive.

Planning and delays: Two years elapsed from the planning of the project (1987) to the development of the system (1989) due to the prototype nature of the equipment used. Delays mostly depended on: agreement with partnership, funding research, solution of technical problems and adaptation of non-medical devices (e.g.: workstations) to radiologists specifications.

Critical points: Connection of imaging modalities to the digital network, large differences in digital image format (512x512 to 4096x4096), training of medical staff, low data transfer rate to/from optical disk drive

BOLOGNA AND FERRARA PACS INSTALLATIONS

Project description: The installation in Bologna and that at Ferrara Hospitals, 30 km away, is funded by the Emilia-Romagna Regional Authority, and is associated with a multihospital site incorporating 4500 beds. The initial aim of this very enterprising project was to create a common image archive and patient data-base, and to make the images available at any one of the hospitals within the group.

At Bologna Hospital a re-usable medium processor, two CT scanners and three ultrasound units are linked to the data management unit. The system has recently been linked to an RIS system which utilizes a relational data base. Medium speed links connect the central system to two remote hospitals (Bellaria and S. Orsola), in one of which there is a CT scanner and an MRI scanner, and in the other, two CT scanners. Both remote sites have diagnostic viewing stations and are also capable of transmitting radiographic images to the central site utilizing a video film scanner. At the central site there is a laser film scanner, and an optical jukebox (76 GBytes), as well as four diagnostic viewing consoles.

At Ferrara Hospital the system is identical in scope and configuration details to the Bologna project described above, except that there are no hospitals links and there are five on-site diagnostic consoles. On this site there are interfaces to two gamma camera systems, a DVI system and a re-usable medium processor.

Goals: Implementation of an on-line emergency service available at the hospitals connected via PACS; development of computerized radiography.

Planning and Delays: The system became operational in first 1990; delays were mainly due to late delivering of transmission lines between hospitals. The purchase of the equipment was managed by the Regional Health Authority with the technical support of an external consultant from C.N.R. (the National Research Council). The system is now managed by the medical staff of the Radiology Departments.

Critical points: Some problems arised in managing this very large project: the massive financial involvement of the Regional Government, the unsatisfactory link between the phase of purchase of the equipment and the phase of utilization by users, the lack of a system administrator, some technical limitations of the system making difficult its acceptance by users (no image hard-copy, low speed of workstations, not friendly user interfaces).

2. PACS Related Research in Italy

Several institutions in Italy are presently involved in PACS related research. Some research activities are carried out in clinical sites that are planning to install a PACS in the near future. In some other cases, research activities are carried out by groups interested in analyzing the economic impact of PACS, or in developing industrial products.

A common characteristic of such research activities is that a large part of them is carried out in the framework of European research projects supported by the E.E.C.

A short overview of PACS related activities is presented in the following pages.

CeRGAS - Bocconi Economic University, Milan

In the framework of AIM- A1030 Project, CeRGAS has developed a study on the economic impact of PACS on health care [1]. The problem of PACS cost/benefit analysis has been studied from a methodological point of view defining a matrix of cost/saving sources.

The research is carried out in co-operation with the Trieste PACS Project where a quantitative determination of costs related to PACS introduction has been carried out during the first half of 1990. Present activities of CeRGAS are related to the development of models for the semi-quantitative evaluation of risk-interest factor in PACS projects, and to the realization, in co-operation with the Trieste group, of a technology assessment report. this latter activity is supported directly by the Ministry of Health.

S. Raffaele Hospital - Polytechnic University, Milan

A group of researchers from the Department of Radiology and the Department of Electronics is involved in image processing and multimodality image matching studies. For this pourpose a digital link is used connecting a PET scanner, a MRI scanner and a CT scanner to a common image processing system.

Dept. of Physiopathology - Dept. of Electronics, University of Florence

This group is involved in the RACE-Telemed research project, supported by E.E.C. Main activities concern the experimentation of an interactive link, connecting 8 sites in Europe by broadband connections, for the transfer of digital medical images and the experimentation of remote expert consultation. In this framework, a relational model for a radiological scientific data base has been developed for teaching and research applications.

Dept. of Radiology - Collemaggio Hospital, L'Aquila

In co-operation with an Italian software house (CITEC), this group is involved in the development of radiological workstations. Up to now, three workstations were developed: a reporting w.s., a consulting w.s., a digitized film w.s. This research and development activity is directly related to the existing PACS project and is a good example of interesting results coming from a close co-operation between radiologists and computer specialists.

Dept. of Radiology, Universty of Brescia

The activity of this group is mainly focussed on the analysis of the behaviour of a Radiology Dept. in order to quantify the impact resulting from the introduction of PACS.

Research activities concern: operational aspects, work volumes analysis, flow analysis, economic justification of partial PACS installations [2].

I.B.M. Scientific Center, Rome

The R&D activity at IBM in Rome is one of the few industrial activities on PACS carried out in Italy. A prototype of a general pourpose system for image management, named IDB, has been developed and is available at IBM Rome Scientific Center.

The system, not developed for specific medical applications, was developed on the basis of three main requirements: management of large volumes of images, integration of specialized and general pourpose equipment, multi user environment.

Up to now no experimentation of the system in a clinical environment took place.

Philips Medical Systems - Milan

Philips is the main contractor of A.I.M. Project A-1030 (Operational Evaluation and Basic Requirements for Prospective Evaluation of PACS Technology) promoted by E.E.C. during 1989. Italian partners involved in the project are the Research Area of Trieste, the Universities of Florence and L'Aquila, and Italsiel (a public software company).

The aim of this project is to evaluate the results of the introduction of PACS in the hospital environment and to guide its evolution towards an IMACS (Image Management and Communication System) integrating all image sources of a whole hospital.

3. CONCLUSIONS

In surveying PACS activities in Italy it became apparent that such systems had a large diffusion, sometime not justified by a real need or by an experimentation project aimed to the assessment of this technology. From the point of view of policy makers this can be viewed as a lack of strategy in the assessment of new and expensive health technology.

Another interesting issue to consider is the lack of a national project on PACS. Up to now, nor the Government nor the Italian Industry took seriously in consideration the strategic importance of a national project in concentrating and supporting efforts in this field. Neverthless, several research activities started during the last years carried out spontaneously by different research groups. This is a sign of the scientific interest in the research related to PACS and the demonstration of a common feeling between researchers, radiologists, and hospital administrators that PACS represents the future of Radiology and that its introduction in the following years will be unavoidable.

Such experiences, carefully analyzed with a reasonable amount of criticism, can give useful information in order to assess this technology and to forecast future trends.

REFERENCES

1. D. Boullier, J. de Certaines, M. del Vecchio, P. Giribona: "PACS Technology Assessment on Pilot Sites". Commission of the European Communities - A.I.M. Project A-1030, W.P. 1040 Final Report

2. A. Chiesa, G. Battaglia, G. Moscatelli, R Maroldi: "Economic Giustification of a Partial PACS". Abstracts of EuroPACS 90, 8th International Meeting, Trieste, 1990

3. P. Giribona, D. Bravar, C. Giuricin, N. Carbi, R. Cuttin, L. Dalla Palma, F. Stacul, W. Ukovich: "A Global Project for the Assessment of a PACS System". Proc. of the Annual International Conference of the IEEE Engin. in Med. and Biol. Society, Seattle WA, 1989, pp.789-790

4. H.I. Glass: "PACS - European Perspective". Proc. IMAC 89, IEEE Computer Society Press, Los Alamitos CA, 1990, pp.14-24

5. C. Greinacher: "Siemens PACS Installations - Practical Experience". Proc. Int. Symp. CAR'89, Springer Verlag, 1989

6. R. Passariello, and Col.: "State of Progress of PACS Pilot Project at L'Aquila University". Abstracts EuroPACS 90, 8th International Meeting, Trieste, 1990

7. W. Ukovich, D. Bravar, N. Carbi, R. Cuttin Zernich, P. Giribona, F. Stacul, L. dalla Palma: "Trieste PACS Project: a functional assessment". Med. Inform. n.15, 1990, pp. 31-37

8. W. Ukovich, D. Bravar, R. Cuttin Zernich, P Giribona, M. Maffessanti, F. Stacul, L. Dalla Palma: "An experimental evaluation of the impact of a PAC System on the operation of Radiology Department". Computer Assisted Radiology. Springer Verlag, Berlin, 1989, pp. 537-540

PACS and Related Research in Japan

Tetsuo Okabe

R & D Center, Hitachi Medical Corp.
2-1, Shin-toyofuta, Kashiwa-shi, Chiba-ken. 277 Japan

Abstract: Various PACS activities by medical and industrial sites are introduced first. The Japanese standard 'MIPS' is presented in comparison to ACR-NEMA standard. The authors also commented the present status of 'ISAC' and 'PHD' activities. Some hospitals and manufacturers have developed and evaluated prototype PACS. On these achievements, about 100 hospitals have already introduced various sized PACS so far.

Keywords: PACS, Japan-PACS, MIPS, ISAC, PHD

Introduction

It was early 1974 when Dr.T.Iinuma introduced first his idea on "digital archiving and communication of medical images" on the Japanese journal. Many persons of medical and industrial sites thought that, sooner or later, owing to the development of high technologies, "computerization and/or systematization"of some management and execution would inevitably come into hospitals too. To his unhappiness, a good large memories for medical images were not available at that time. The advent of optical disk really made us possible to develop and implement PACS.

Other reasons why PACS has been welcome in Japan seem to be such that the number of radiology examinations for a patient has been relatively large, which bring many difficulties to image management for better image diagnosis. So radiologists feel it important to introduce PACS in the hospitals. The high level industrial technologies of Japan could naturally support their efforts.

PACS Activities in Japan

Japan Association of PACS(J-PACS) was established by manufacturers in 1982,which has been aiming at promotion of standardization for interface and data format on the network as well as Public Relations activity for PACS in the hospitals.

In 1984, the committee for Medical Image Processing System(MIPS) standardization was established by Japan Industries Association of Radiation Apparatus (JIRA) and J-PACS on the instructions of Japanese Ministry of International Trade and Industry (MITI). The committee had made MIPS standard with three WGs and was dissolved to transfer the execution of the standard to JIRA in 1990. Since then, JIRA has organized the working group, RC-109, for the execution of the standard.

NATO ASI Series, Vol. F 74
Picture Archiving and Communication Systems (PACS)
in Medicine. Edited by H. K. Huang et al.

In medical laboratories,physicians and scientists have usually used Floppy Disks and Magnetic Tapes as transfering media of X-CT and nuclear medicine images for research and conference. They have thought of using erasable Magneto-Optic(MO) disk and wanted to replace MT and FD by MO disk. MO disk having standard data format and hardware for this purpose is called Image Save And Carry (ISAC) system. J-PACS has organized the committee for ISAC since 1989.

Dr.S.Ikeda, National Cancer Institute of Japan, proposed his idea that every citizens should have his record which files every data concerning his health such as disease history, examination data, typical diagnostic images and etc., and organized the committee for Personal Health Data (PHD) recording system in 1982. PHD committee had promoted to develop large capacity memory and high precision display for PHD.

Committees and Working Groups concerned with PACS and related technologies are shown in Table 1.

Table 1 : Various Activities on PACS in Japan	coordinator
1. J. Association of PACS (J-PACS)	(Dr.Tsuji-uchi)
2. Image Save And Carry (ISAC) committee	(Dr.Inamura)
3. J. Radiology Soc. (JRS) : PACS committee	(Dr.Akisada)
4. J. Radiology Technologists Soc. : PACS WG	(Dr.Takizawa)
5. J. Medical Imaging & Technology (JAMIT) : PACS WG	(Dr.Akisada)
6. Medical Imaging System Develop. Center (MEDIS-DC) : PACS WG	(Dr.Iinuma)
7. J. Industries Assoc. Rad. Apparatus (JIRA): RC-109 WG	(Mr.Shigemura)
8. Committee for Medical Image Communication System (MICS)	(Dr.Kuwabara)
< dissolved WGs or Committees on finishing projects >	
* MIPS committee	(Dr.Takahashi)
* J. Ministry of Education : PACS project of Univ. Hospital	(Dr.Irie)
* Committee for PHD system	(Dr.Ikeda)

At present, we have some conferences and symposia concerning PACS and related technologies regularly and/or temporarily as listed on Table 2. Besides, big PACS sessions have often been held at many meetings on medical images and related items.

Table 2 : Conferences and Symposia on PACS and Related Technologies

1. JAMIT Annual Conference	(JAMIT)	Annual, Summer
2. JAMIT PACS Special Program	(JAMIT)	2 times/year
3. JRS PACS Symposium	(J. Radiology Soc.)	Annual, Autumn
4. Symp. on Medical Image Filing Sys.	(J. Soc. Appl. Phys.)	Annual, Spring
5. IMAC '91	(J. Radiology Soc.)	April '91. Kyoto

Outline of MIPS Standard

While WG-2 was discussing how to make interface standard on the base of Open System Interconnection, the proposal of ACR-NEMA standard was transfered to Japan, and so, one of members of MIPS committee sometimes attended at ACR-NEMA meetings as an observer. Considering the situations that the proposal by ACR-NEMA committee was advanced and export/import of imaging modalities were going increased each other, WG-2 thought that MIPS standard had better be constructed along with ACR-NEMA standard, especially at least its hardware level, unless otherwise problems happened. So MIPS standard may be thought to be a Japanese version of ACR-NEMA standard.

The committee had made draft '87 and its revised draft '89 of MIPS standard. The statements of it is consisted of 10 chapters and appendix. The differences between MIPS and ACR-NEMA are not so large as shown in Table 3. Items with mark* are revised mainly along with Japanese ways of applications.

Table 3 : Comparison of MIPS Standard with ACR-NEMA Standard

Layer	MIPS –'89	ACR-NEMA v 2.0
Presentation	* Command(send,get,find, echo,cancel) * Image(just original image) * Shorten message * To add position code * To adopt JIS code	* Command(send,get,find, move,echo,cancel,dialog) * Image(original,modified, composite)
Session	not-revised	
Transport & Network	* CCR,CCI,OCR,OCI,CMD, data,pause,resume, * Just channel 0	* RCR,RCI,CCR,CCI,OCR,OCI,CMD data,pause,resume, * Channel 0–225
Data link	not revised	
Physical	not revised	

* Clinical experiences on MIPS standard

Dr.Kimura,Tokyo University Hospital, has installed the system such that GE work-station is coupled with Toshiba workstation using MIPS standard on Ethernet. GE work station is operated for three GE X-CTs using ACR-NEMA standard. Dr. Irie, Hokkaido Univ. Hospital, has also evaluated MIPS standard with his system which coupled GE X-CT with NEC data input terminal. Several difficulties and problems have been pointed out for both MIPS and ACR-NEMA standards. Although RC-109 WG is now improving these items, they can not be inevitable to some extent. How to acquire and how to use images seem different from radiologist to radiologist and from hospital to hospital.

Activities for ISAC Standard

ISAC committee has organized four WRs for making standard as follows;

WG-1 for Disk format
WG-2 for Image data file format
WG-3 for Software development
WG-4 for Technology assessment

International Standard Organization has developed the standard hardware for 5.25" and 3.5" MO disks. WG-1 has promoted to use this ISO standard, aothough performances of 5.25" disk itself are still improving to be probably those of 600 MB/side memory capacity, 2 MB/s transfer rate and 40 ms access time on '92 at latest. Meanwhile, WG-2 has proceeded that image data file format on MO disk in ISAC standard should have been promoted by following the situation that ISO has been proceeding the so-called three layered package media. WG-3 and WG-4 will begin their activities this year.

Activity for PHD Recording System

Dr.S.Ikeda and his colleagues had proceeded the development and evaluation of PHD system between '82-'87 under the supports of MITI and Japanese Ministry of Welfare. Three WGs were organized for performing this purpose.

WG-1 for Large capacity memory media for medical image archiving
WG-2 for High precision CRT display and input/output of images to the system
WG-3 for introducing PHD system into existing medical regimen

With many accumulated data, they had concluded such that digital chest X-ray image with 2k x 2.4k pixels at 10 bit density should be displayed on the CRT with 2k x 2k pixels at 8 bit density in order to be used for image diagnosis. 4.5 GB Digital Tape Recorder was developed for image archiving of PHD system.

The work done by this committee has been followed by ISAC committee on the view point of image data handling.

Installations of the systems in Japan

96 hospitals have installed various sized PACS as shown in Table 4. 84 hospitals are aiming at digital image filing for research and conference. This system is stand alone located in medical laboratory or Radiology Department and is composed of one workstation with single optical disk unit and film image digitizer. Although some persons insist that this is not PACS but only image processing system, this kind of system is thought to be the first step for PACS.

12 hospitals have installed the systems which coupled various imaging modalities with image database and workstations through optical fiber cables. Radiologists are trying to use them for routine diagnosis and conference. Only Hokkaido Univ. Hospital has installed a hospital-wide PACS for routine diagnosis and conference. Dr. Irie will introduce this system later in this program.

Recently, teleradiology has become major concerns for hospital administrators as well as radiologists, because small hospital located in a rural country could not have fullflegded radiologists in their radiology department as the number of radiologists in Japan is too small to respond such facilities. Whereas, MRI, X-Ct and NM need the experts for being diagnosed.

Table 4 : Statistics of Installed PACS and Related Data

* Type and Scale of PACS		96
1. Stand-alone system		84
2. Network system	1. inside Radiology Dept.	11
	2. hospital-wide	1
* Teleradiology		(5)
* Manufacturers		installations
1. Fuji Electric Corp.		6
2. Fujitsu Corp.		3
3. Hitachi Medical Corp.		7
4. NEC Corp.		22
5. Toshiba Corp.		40
6. Shimadzu Corp.		2
7. Siemens Asahi Meditec.		3
8. Yokogawa Medical System.		13

(courtesy of Shin-Iryo(J), July, '90)

Govermental Policy for PACS

Hirota M.D.,Ministry of Welfare, says "We know PACS must be future trend in the hospitals. We expect that diagnostic images on PACS should be standardized to be universaly accessed, keep good image quality enough to be diagnosed and have long life on archiving to be used as an evidence". He has often attended at PACS meeting to show positive but severe attitude with digital archiving. We believe Goverment will pay if we could clear their regulations for technology.

Summary

In Japan,so many discussions about PACS performance and implementation have been made on various occasions so far. PACS installed in the hospitals could not always satisfy both radiologists and administrators because of mainly poor technologies--throughput and image quality---. However, "Communication and/or Systematization" of image management in the hospitals will be inevitable and which will bring us better medical environment. Many Japanese radiologists, physicists and engineers are coninually trying to overcome various difficulties to meet the age of PACS.

Acknowledgement

The author would like to express his appreciation to Dr.H.K.Huang for making this opportunity for me. He also thanks to Drs. Akisada, Inamura, Kaneko and Mr. Kohsaka for refering their articles and slides for this presentation.

Contact to:

Tetsuo Okabe FAX: 81-471-33-2543, TEL: 81-471-31-4159 (Japan)

PACS and RELATED RESEARCH in the NETHERLANDS

Albert R. Bakker

BAZIS, Schipholweg 97, 2316 XA Leiden, The Netherlands

Abstract: This paper gives a brief overview of PACS research activities in The Netherlands. Apart from recent projects also major problems met in this field of research are indicated.

Keywords: PACS, The Netherlands, modelling/simulation, costs, technology assesment

INTRODUCTION

PACS research and PACS experiments have attracted a lot of attention in The Netherlands from the early days of the PACS on. Several groups are involved in this field, their level of involvement is changing over time, mainly because the effort is depending on grants. In this paper first the major PACS activities are described one after the other. Of course the description can not be detailed, fortunately descriptions of the projects and their results are being reported regularly at the Medical Imaging conferences and the Europacs conferences. Researchers from The Netherlands have plaid a very active role in Europacs from the start on.

UUH, BAZIS and PHILIPS, the DUTCH PACS PROJECT

The research and development activities of Philips as to digital optical recording lead in 1982 to the formation of a working group to study the applicability of such a device as the kernel of a system for storage and retrieval of digital images. In this group participated apart from Philips itself the Utrecht University Hospital and BAZIS. BAZIS is a non-profit organization developing and supporting an integrated HIS that is at present in use in 7 out of the 8 university hospitals and some 30 general hospitals in The Netherlands. Although this working group started rather early and was mainly triggered by new technical opportunities it stimulated a master's thesis project at the Technical University Twente on image compression algorithms(1983) and lead to a grant of the Dutch Ministry of Health Care (WVC) to BAZIS for research on modelling and performance evaluation (simulation) of PACS (1984,1985).

The cooperation between the partners (UUH, BAZIS, Philips) resulted in a large joint project sponsored by the ministry of health care. The Delft University of Technology (DUT) participated as a subcontractor of BAZIS. This project, known under the name 'The Dutch PACS project' [1, 2] was carried out during the period 1986-1989 and aimed on one hand at the clinical evaluation of the Philips PACS prototype named MARCOM in the UUH and on the other hand at research on the following topics: modelling and simulation, coupling of PACS and HIS, evaluation of

NATO ASI Series, Vol. F 74
Picture Archiving and Communication Systems (PACS) in Medicine. Edited by H. K. Huang et al.

diagnostic quality and technology assessment. The ministry contributed over 2.3 M $ for personnel costs at UUH and BAZIS; Philips made the prototype equipment available free of charge. The research topic technology assessment was added to the project on special request of the Ministry. Initially it was intended to spend attention also to teleradiology however because of the lack of interest for this subject (The Netherlands are a small country, heavily populated) this subject was dropped.

The set-up of this project and its results have been widely published; some highlights are mentioned here:

- the clinical evaluation was based on digitization of all images (both new and historical ones) of one of the wards of the department of Internal Medicine (15 beds). These images would be made available to the radiologist at a multiscreen workstation and to the referring clinician at a simpler workstation in the ward.
- the prototype MARCOM equipment showed all the characteristics of early technology; the reliability was a source of concern and the resolution supplied turned out to be insufficient for primary diagnosis.
- the referring clinicians however appeared to be very enthousiastic about the better accessibilty of images that was realized. A 1024*1024 resolution 8 bits deep helped them a lot. The delay in supply of the images was drastically reduced.
- a PC-based software package was developed for supporting image quality evaluation (based on ROC analysis); this package called FEASIBLE was applied for a number of quality assessment studies [3]. The package is available on request.
- even for the limited number of images to be handled performance was a point of concern; by adaptations of the configuration and organizational measures the most serious problems could be solved. Modelling and simulation revealed some of the problems and was also applied for analysis of proposed next versions of MARCOM.
- a one-way coupling between PACS and HIS was realized and demonstrated during the 1988 Europacs conference. This coupling fed the PACS with patient identification data, reports on radiology examinations and data on patient activities and movement (both ADT and appointment data). The identification data were used by the PACS; the reports could be viewed on the PACS workstation. The other data were meant to be used in prefetching algorithms to be used in the image management software of PACS. Due to the fact that during the project Philips decided to stop its own PACS development and market as their PACS product the AT&T COMVIEW system this prefetching algorithm was not developed.
- analysis of the image flow revealed that PACS will have a very benificial effect on availability of images and timeliness. These effects were confirmed in the experiment. It is expected that because of earlier availability of diagnostic images the clinical diagnosis and treatment will be speeded up leading to a reduction of the average length of stay.
- technology assessment further dealt with considerations on costs and savings when a full-scale PACS would be introduced. A PACS cost model was developed and supporting software for application of this model is available: the package CAPACITY [4]. In this package expected trends in cost components can be taken into account. It was concluded that at this point in time a full-scale PACS for the UUH would be 4 times as expensive as the present film based system. Around the year 2000 costs of PACS can be expected to become equal to the film based system. Effects on quality of care and length of stay were not taken into account in these cost considerations.

At the end of the project the partners proposed a second phase, covering a much larger percentage of the im ages, based on the new Philips product line (COMVIEW). It turned out to be impossible till now to find the necessary finances.

THE DUTCH SCIENCE FOUNDATION

Within the scope of its program 'the national facility for informatics' the Dutch Science Foundation supports PACS research by financing study tours, working visits and congress participation. From this support (1985-1990) a study tour was organized to the USA and two study tours to Japan. Participants were affiliated with UUH, BAZIS, DUT and the Dutch Society of Radiologists.

Recently the foundation decided to support a 4-year project for developing a new method for modelling, simulation and design to be applied for PACS [5].

FOLLOW-UP/SPIN OFF OF THE DUTCH PACS PROJECT

Already during the first phase of the project the UUH developed an interest in 3D research. This lead to a large project together with the DUT and the Utrecht University. The project is financed by the Dutch government with an amount of several M $ over the period 1989 -1992.

Apart from a development project within the department of nuclear medicine, the direct interest in PACS at the UUH is dormant till finances can be found.

The research at BAZIS on modelling and simulation is being continued, partly with support of the Dutch Science Foundation. BAZIS is interested in international cooperation in this field and spends part of its own research budget (1 fte).

The technology assessment activity of the Dutch PACS project is being continued by BAZIS with financial support of the Ministry of Health Care. As part of this research an international workshop on PACS TA will be organized in the spring of 1991. Apart from this grant of the Ministry BAZIS is expanding its activity in the TA field on its own research budget but applying for additional external funding.

At DUT the interest in PACS research is expanding. The research deals on one side with the construction of a high-speed hardware ACR/NEMA interface [6] and on the other side with modelling and simulation (in cooperation with BAZIS). A distributed PACS image database is the subject of a PHD project.

PACS AT THE ST.RADBOUD UNIVERSITY HOSPITAL NIJMEGEN

A so-called bottom-up approach of a PACS development was started in 1985. Based on existing clinical questions to improve the quality of the digital images from DSA, CT and MRI, conversion programs were developed to convert images from all modalities to a central general purpose minicomputer with a general matrix manipulation software package [7,8]. The satellite console of the CT-scanner was used as a diagnostic display station for processed images.

From then on the system was gradually expanded step by step. At first connecting one SomatomDR3 CT-scanner via a 500 meter Ethernet to a Diagnostic Reporting Console (DRC-20). Now a PACS network of 1000 meters connects the central department of diagnostic radiology with the peripherally located department of neuroradiology. Besides the clinical network a research environment has been realized to digitize conventional images from 64 * 64, 8 bits deep to 4096 * 4096, 12 bits deep to enable quantitative clinical measurements and advanced image processing. The thin Ethernet research network enables direct access to the image database of the DRC. In the clinical environment image communication is initiated by the integrated network commands within the software on the 4 connected modalities (2 CT, 2 DSA). Digitized conventional images are sent to the central image database by special software that adds ACR-NEMA header information.

The central image database consists of a 500Mbyte Winchester disk and two 1.2 Gbyte removable optical disk units. The SQL based DBMS keeps track of the location of images, both on-line and off-line.

Actually the diagnostic value of the DRC involves mammography, images from dual energy quantitative CT and MRI studies of carcinoma of the bladder. For ROC analysis the FEASIBLE software package is used [3]. Results will be presented at the end of 1990.

Current projects are the installation and connection of a Magnetic Resonance System with four workstations for imaging and spectroscopy, investigations to connect the departments of radiotherapy and orthopaedics, nuclear medicine and the connection to the HIS (a two-way coupling).

The project is part of the own research program of the hospital and not severely dependent on direct government financing.

PACS AT OTHER UNIVERSITY HOSPITALS

At several other university hospitals it is being considered now how they can prepare themselves for the PACS era without investing too large a percentage of their budget. The interest in the general hospitals is still limited.

The experiences with the Dutch Pacs project and the conclusion of that project that cost/effective full-scale PACS will not be available within 5 years determines the policy: partial PACS with at least partial external funding. The definition of a project requires careful discussion with the vendor involved. At least one new project can be expected within a year from now.

THE HIPACS PROJECT

Within the exploratory phase of the AIM (Advanced Informatics in Medicine) program of the European Community one of the projects is HIPACS, a study on a hospital integrated PACS. This project is carried out by a consortium with partners from 5 member countries, prime contractor is the PRIMIS institute of the Free University of Brussels. BAZIS is the Dutch partner in this project and mainly involved in two subprojects:

a. the specification of a generic HIS-PACS interface
b. modelling and simulation; focussed on network performance.

In this exploratory phase the budget was limited (about 2 manyears for BAZIS).

Although there is a proposal for an AIM main phase the future financing of AIM is uncertain.

OBSTACLES FOR PACS (IN THE NETHERLANDS) AND OPPORTUNITIES

Like most other groups working on PACS we in The Netherlands have been too optimistic about the moment a full-scale PACS would be both realizable and cost justified. The findings from the Dutch PACS project indicate that such a situation will occur around the year 2000. This has lead to some hesitation of radiologists to invest in PACS research. Partial PACS now seems the right direction.

The withdrawal of Philips from the field of active PACS research and development has reduced the possibilities to find financial support of the government. Especially the Ministry of Health Care feels PACS is too much a technology push, its contribution to the quality of care still to be proven.

For fruitful R&D on PACS a direct link with a vendor is important, the departure from Europe of major PACS developments puts PACS research in a weak position, not only in The Netherlands but in all European countries. Perhaps we might turn this into a strong point by sharing the efforts of European PACS groups to design and develop an open PACS architecture. This conference might be a unique opportunity to draft a plan; HIPACS might be a stepping stone. The Braunlage conference brought us Europacs will Evian bring us even more, a EURopean Open PACS ?

References

[1] BAZIS, AZU, Philips: CONCLUDING REPORT Dutch PACS project. Picture Archiving and Communication System, Phase I (1986-1989), ISBN: 90-70690-07-1.

[2] Kouwenberg J.M.L., Ottes F.P., Bakker A.R., Haar Romeny ter B.M., The Dutch PACS project, results of phase I (1986-1989). In: Proceedin Medical Imaging IV Conference, February 1990, Newport Beach, in print.

[3] Ottes F.P., Valk de J.P.J., Kroon H.M., Barneveld Binkhuysen F.H., Seeley G.W., Winter L.H.L., Schultze Kool L.J., Diagnostic image quality evaluation chain: applications of the FEASIBLE package in practice. In: Proceeding Medical Imaging III Conference, January 1989, Newport Beach, SPIE Vol. 1093.

[4] Van Poppel B.M., van Gennip E.M.S.J., Bakker A.R., Wilmink J.B.M., First results of the software package CAPACITY for cost modelling of PACS. In: Proceedings Medical Imaging IV Conference, January 1990, Newport Beach, SPIE Vol. 1234, in print.

[5] Stut jr. W.J.J., Bakker A.R., Steen van M.R., Groenewegen L.P.J., A semantic modelling method to guide simulation-based system development. In: Proceedings of the 3rd European Simulation Congress, D. Murray-Smith et al (eds), September 1989, Edingburgh, pp.101-106.

[6] Reijns G.L., Santilli D., Schellingerhout G., Jochem A.J., High Speed ACR/NEMA interface. Medical Imaging IV Conference, February 1990, Newport Beach, SPIE Vol. 1234, in print.

[7] Van Erning L.J.Th.O., Thijssen M.A.O., Karssemeijer N., Guijt W., PACS in practice: the status of the PACS project at het Radboud University Hospital in Nijmegen. Part A: Introduction and the picture system. Medical Informatics 13, pp. 255-264.

[8] Van Erning J.J.Th.O., Knots M.J.J. Guijt W., Ruijs J.H.J., Pacs in practice: From Software link to On-line Communications. In Proceeding Computer Assisted Radiology CAR 1989.

PACS AND RELATED DEVELOPMENTS IN SWEDEN

Silas Olsson

Spri, The Swedish Planning and Rationalization Institute
for Health and Social Services,
P.O. Box 70487, 107 26 Stockholm, Sweden

Abstract

Large PACS (Picture Archiving and Communication System) installations do not yet exist in Sweden, but some hospitals have had experience with limited PACS activities. At present there are 4 mini PACS installations in radiology departments and about 12 teleradiology systems in use in Sweden. A couple of small Swedish enterprises work in the market segment of digital imaging including PACS and teleradiology, although the radiology market is dominated by the large international companies.

Interest in PACS and teleradiology in Sweden has increased during the last few years, along with advancements in technology and international experience. However, radiology is organized very differently in the USA, Japan, Southern Europe, and Scandinavia. Because of this, PACS will be introduced in different ways, and experience with PACS gained in one health care system may differ from that gained from other health care systems.

This paper reviews the status of PACS and related developments in Sweden.

Keywords: Radiology, PACS, Teleradiology, Telemedicine, Medical Informatics, Diffusion, Evaluation, Quality Assurance.

Introduction

About 4.7 million radiological examinations are performed each year in Sweden (population 8.4 million). This does not include dental X-ray examinations or X-rays by the occupational health services. Most radiological examinations in Sweden, approximately 82%, are carried out at the 105 acute care hospitals. Another 7% are done within primary healthcare settings, 6% in private healthcare, and approximately 5% are mammography screenings, some of which are performed by the private healthcare sector [11,13].

Apart from conventional radiological examinations, a full spectrum of imaging modalities is used, such as ultrasound (US), computerized tomography (CT), digital subtraction angiography (DSA), single photon emission computerized tomography (SPECT), positron emission tomography (PET), magnetic resonance imaging (MRI) and computed radiography (CR) with phosphorous image plate technology.

If one excludes the digital imaging workstations in nuclear medicine units, digital imaging was first introduced into radiology in Sweden in 1973, when a CT scanner was put into operation. Approximately 90 CT scanners are now installed. Diffusion of modern dynamic diagnostic ultrasound equipment started to increase in the early 1980s and is now used in practically every X-ray department [4].

NATO ASI Series, Vol. F 74
Picture Archiving and Communication Systems (PACS) in Medicine. Edited by H. K. Huang et al.

DSA was introduced in 1983, and approximately 25 DSA units are presently in operation [5]. The first MRI unit was put in use in Sweden in 1984, and now 14 such units are installed [7]. At present only one CR system has been installed in Sweden.

In Sweden digital imaging modalities are used in approximately 15% of all radiology examinations performed in hospitals. This ratio is the same in other Nordic countries, i.e. Denmark, Finland, Iceland and Norway [3,13]. On the other hand, digital images account for approximately 40% of total images. The portion of examinations using digital imaging modalities is expected to increase.

With this situation, Swedish radiologists, hospital administrators and central authorities, are concerned about the efficient handling of images, as well as image distribution, archiving, etc. A solution to this problem may be the use of PACS (Picture Archiving and Communication System) which may provide new image management, better information logistics, and improved diagnostic possibilities.

Swedish Companies in the Imaging Sector

In Sweden the radiology market is dominated by large international companies such as Siemens, Philips, General Electric CGR and Toshiba. However, some smaller Swedish companies with 5-35 employees are developing equipment for digital imaging in the medical sector, for example the following:

Bildsystem (Image Systems), based in Malmö, has since its start in 1978, specialized in developing video-based products such as video character generators, mixers for the display of ECG curves and X-ray fluorographic imaging simultaneously on the same CRT-monitor, and X-ray film markers. One widespread product is a TV demonstration unit used to display radiological images for medical conferences (rounds). Other Bildsystem products are TV surveillance systems for the industrial and horse racing (trotting) sectors.

Imtec (Image Technology) now based in Uppsala, was formed in 1982. An interactive image workstation was developed on the basis of the development of a tight-coupled image memory, image processor and image display processor. This workstation and the fiber optic image network (Imnet) are key components in the company's approach to PACS. In the last couple of years Imtec, has been busy in the teleradiology sector, and some such systems have been delivered to Swedish hospitals. Imtec is also active in the field of computerized cell image analysis.

Sectra (Secure Transmission) based in Linköping has been developing systems of image transmission for the press media since the beginning of the 80s. The break-through for this company came during the 1984 Olympics in Los Angeles, when Sectra products were used to transmit digital images to different newspaper editing departments via ordinary telecommunication networks. Their special areas are include data encoding and image data compression, data security, cryptography and information integrity. Sectra has recently completed a several years project in teleradiology and the first system is now being tested in Swedish hospitals.

Bildbehandling, (Image Processing), based in Linköping develops systems for image processing within thermography for both medical and industrial application.

Helax is an Uppsala based company developing and marketing image processing products primarily for oncology departments.

Scanex is a Helsingborg based company developing and marketing for example interface and network products primarily for nuclear medicine departments.

Sarastro is a Stockholm-based company, developing and marketing confocal microscopy systems.

Some Leading Image Research Centers in Sweden

University of Linköping

One group in Linköping has centered research activities on the development of PICAP (Picture Array Processor), regarding hardware, software and application. One of the earlier areas of application was in fingerprint recognition.

Another Linköping group has developed a special computer architecture in the GOP (General Operator Processor) concept. This signifies, among other things, the automatic detection of local structures, relating to magnitude and direction, and can be described as a complex image.

A third group in Linköping has focused on image communication and specializes in encoding and the compression of data. This group has also developed a CCD chip with built-in parallel processor for image analysis, used for example in industrial production.

University of Stockholm, Institute of Technology

There are different groups in Stockholm working with image processing. For example, one group at the Institute of Technology is working with basic research on computer vision and another group at the same institute is working within the confocal microscopy area.

University of Uppsala

The Centre for Image Analysis was started in 1988, but a research team had already been established in the early 70s. The present concentration of research is on the use of image technology in medicine and on the environment in a broad sense. Two current projects are the development of image processing methods that follow the healing process of wounds and the improvement of segmenting methods in medical imaging (magnetic resonance).

University of Gothenburg

A group in Gothenburg has been working for some years on medical applications in image processing. One research project in progress concerns the analysis of ultrasonic images in cardiology.

University of Lund

A group in Lund is working on the development of a high resolution ink jet paper printer for possible application as a hardcopy unit in medical imaging (radiology).

Training Availability in Image Processing

Basic courses in image processing are helt at the universities in Linköping, Stockholm and Uppsala, among other locations.

Training at the pregraduate level can also be found at Linköping, Stockholm, Uppsala, Gothenburg and Halmstad, and elsewhere.

There are also about 5-7 other training associations in this field operating on a commercial basis.

The radiology organization in Swedish hospitals

The general input-output flow in a radiology department in Sweden as well as in other countries consists of a request from the referring physician. If the request is accepted by the radiologist the radiology procedure is performed. All images taken are read by the radiologist who reports the findings in writing to the referring physician.

A key element of Swedish radiology is the round- procedure (conferences). In these daily rounds, mostly held in the mornings, radiologists meet with referring physicians to explain and discuss findings of the examinations done the day before. This is also a forum to propose further examinations if needed. These conferences provide an important opportunity for radiologists to meet clinicians to get feedback about their findings after treatment is complete and the final answer is known. Such conferences also allow referring physicians to ask for the most efficient and cost-effective radiology procedure and to avoid unnessesary examinations. The interaction in the round-procedure is a very important quality assurance activity in radiology. Such cooperation in rounds is essential to establishing high diagnostic quality and to continuously improving quality.

All the images taken in the radiology departments and copies of the written requests from and reports to the referring physicians are stored in the radiology image archive or in the central hospital archive. Although loss of films occur, this is not generally regarded as a major problem compared with countries where the films belong to the patients or to the referring physicians. Also, the fact that the health care system in Sweden up to now has been organized in such a way that ones residence very much direct patients to a particular hospital has facilitate to keep most of the X-ray films from one patient at one hospital. However, this organization is about to be changed to allow more freedom to patients to choose their preferred hospital.

The way the health care system and radiology are organized will significantly influence the need for PACS and the way PACS will be introduced. Consequently, the advantages and disadvantages of PACS in one system may differ from experience obtained in other countries. The organization of radiology in Sweden and increased international experience with PACS may explain why the general interest in PACS in Sweden is now growing for some applications.

PACS in Swedish Hospitals

Larger PACS installations do not exist in Sweden, but some hospitals have experience with limited PACS activities. At present 4 mini PACS are in use in radiology departments, and other hospitals plan to start PACS projects. However, it is difficult to raise funds for investment in such projects.

Karolinska University Hospital, Stockholm

Planning for a PACS project at the Karolinska Hospital in Stockholm started in 1984 and the project commenced in 1986. The aim of the project was to investigate the possibilities of fast transmission and access of images, and to gain experience about the PACS-environment as a base for further developments. The project referred to the integration of four image sources into a network (Imnet) together with a work station

(Imtec). It was planned to extend the system with an image archive and a link to a Radiology Information System (RIS).

It was found to be difficult, however, to develop special interface with the image sources. Difficulties in obtaining necessary documentation from some of the suppliers made this work either impossible or very difficult. Two image sources could however be connected to PACS, a CT and an MR (both Siemens).

Mainly due to financial problems, it was not possible for the project to be as extensive as planned. For example, the image archive and RIS could not be connected. The general experience gained from the project, was that an individual development of interface of image sources without standards could be complicated and costly, that standards should be used as far as possible, and that handling procedure (human interface) must be very simple [2]. This project has now ended.

Another image network development at the Karolinska Hospital Neuroradiology Department, started in the mid 80s. The long term goal was to integrate existing and new digital imaging sources as the medical need became pronounced, and as technology and financial resources became available. At present an Ethernet network (with DECnet or TCP/IP protocol) links together 1 MR, 2 CT, 2 PET, and 1 DSA. Images from those sources can be retrieved and processed on an image workstation (HP) and can be transferred to a treatment planning system and be stored on an Exabyte magnetic tape cassette system (about 2 Gbyte per tape). As yet, no optical discs, jukebox or RIS are connected to this installation [6].

University Hospital in Lund

Since 1987 there has been a CR system in operation at the Lund University Hospital. This was supplemented in 1989 with an image workstation (Siemens). The system also contains an optical disc drive unit for image archiving for research purposes. There are plans to link the CR system to an image display station at the intensive care unit.

Currently, a digital CR chest stand (Fuji) is being installed and connected to the same laser film printer as the previously mentioned CR system.

Another project in Lund is centered around an image workstation (Imtec). On line with this workstation is, an MR (Fonar) via an Imnet network (16Mbit/sec) and a teleradiology system via ISDN (64Kbit/sec) connected to a CT unit (Siemens) in a hospital in Ystad, a town less than 100 km from Lund. A CT (Toshiba) is also interfaced to the image workstation via floppy discs. There are also plans to connect DSA to the same workstation. As yet no image archive or RIS is connected to the network. Separately, a RIS is in use in Lund.

University Hospital in Uppsala

The University Hospital in Uppsala has recently established a network (Imnet) between a CT and an MR unit (both Siemens) and to an image work station (Imtec). The same work station is also used for teleradiology with Gävle and Östhammar about 100 and 75 km respectively away from Uppsala. A RIS system is being installed (Rados).

County Hospital in Jönköping

Since 1989 there has been an Ethernet link between an MR and a CT unit (both Siemens) in the new county hospital in Jönköping. A 4 CRT-monitor PACS (Siemens) consol is connected to this link. The PACS consol is placed in the MR building and is used to send CT images to the MR site. The consol is also used for

PACS in Nuclear Medicine Departments

About 35 hospitals in Sweden offer nuclear medicine services and are equipped with gamma cameras. Together these hospitals have about 100 cameras.

Small PACS are installed in about 7 hospitals in nuclear medicine departments. The first installation was in 1989. At present, a typical such PACS consists of a network (Direct Linc) connecting 2-3 gamma cameras (General Electric) to a PC-based image work station which also serve an optical 5-inche disc drive (no jukeboxes installed).

The first organizational consequence is to eliminate the floppy discs used mainly by large nuclear medicine departments for storing images for potential image processing and research. However, film is still used as the main image archive media.

Teleradiology in Sweden

The development and use of teleradiology in Sweden started in the early 80s, with a small Swedish company developing equipment based on input of analoge images from a TV-camera, digitizing the images into a 256x256, 6 bit deep matrix. The equipment and method was tested and found useful in certain applications of image transmission between hospitals [8]. Image quality was, however, limited and the public network allowed only a relatively slow image transmission. These conditions prevented the diffusion of the method in Sweden.

However, the interest in teleradiology has increased sharply during the last few years. This is mainly due to the fact that technology now permits much better image quality than previously, and that tele networks today allow much faster image transmission [9].

There are now about 12 teleradiology systems in Sweden, provided by three suppliers (two local and one international company). As we have no established standard for image transmission these three systems work with disparate image transmission data formats [9, 10]. This naturally impedes further expansion and restricts the utilization of a teleradiology network. Faster standardization of this area would, therefore, appear to be urgent.

Program for Developing PACS and Related Technologies

The Swedish Government has stimulated the development and use of information technology (IT) in different ways during the 80s. One such action is the so called IT-4 national program with a three-year budget (1988-1990) quivalent to about 75 million dollars [12]. This program focuse industrial developments and is not specificly medical oriented.

However, the Swedish National Board for Technical Development established a support program in 1989 with the intention of stimulating the development of digital imaging products for the healthcare sector. The program will extend over 5 years, and has a budget equivalent to about 4 million dollars. After about one year of operation the program has distributed funds to about 15 small companies and research teams.

Sweden is participating in some common European projects. One such project is Telemedicine within the RACE program (Research and Development in Advanced Communications Technologies in Europe).

During the spring of 1990, the Government financed 7 pilot studies with the aim of obtaining proposals for projects of importance for development and the use of information technology in the public sector. The pilot study addressing health care, conducted by Spri resulted in proposals for about 15 projects regarding PACS, telemedicine, and teleradiology [1].

Spri, The Swedish Planning and Rationalization Institute for Health and Social Services, is a fully-public health-care research and development institute which advises and supports the Swedish county councils in their work to change and improve health care delivery. Spri has followed the PACS development during most of the 80s. The first Spri report on PACS was published in 1987 [3]. Currently, Spri is conducting a new PACS project with the goal of reporting in 1991 on the status of PACS in an international perspective.

Conclusions

Due to technical achievements and increasing international experience, PACS and teleradiology has received more interest in Sweden over the last few years.

However, radiology is organized very differently in the USA, Japan, Southern Europe and Scandinavia. The way radiology is utilizied as well as the way that radiology service is organized influences significantly the need for PACS and the way PACS will be introduced in each health care system. Consequently, the advantages and disadvantages of PACS in one system may differ from other health care systems.

Although some very limited tests have been carried out with PACS in Sweden, we lack experience in using somewhat larger PACS in clinical routine applications. It is necessary to gain experience from daily routine work in order to evaluate the suitability of PACS, telemedicine, and teleradiology, and to determine whether they can contribute to improving in the Swedish health care environment.

References

1. Bildteknologi i offentlig verksamhet : telemedicine/ teleradiologi [Telemedicine and teleradiology]. Stockholm: Agency for Administrative Development, 1990. In Swedish

2. Dahl, M. - PACS skall byggas på standard [Design PACS with standards]. Medicinsk teknik 1989;6(6/7):56-62. In Swedish

3. Den nya radiologin (The new radiology (PACS). Stockholm: Spri, 1987. (Spri report 234)

4. Diagnostiskt ultraljud [Diagnostic ultrasound]. Stockholm Spri, 1987. (Spri tryck 147)

5. Digital angiografi (Digital angiography). Stockholm Spri, 1985. (Spri report 189:1-2)

6. Hägerdal, E., Nilsson, J. - PACS in neuroradiology department at Karolinska University Hospital. Linköping: the University, 1989.

7. Magnetisk resonanstomografi (Application and diffusion of magnetic resonance imaging in the Nordic countries). Stockholm Spri, 1990. (Spri report 279)

8. Nilsson, U., Nyman, U., Nilsson, M. - Teletransmission of radiographic images. Acta radiologica diagnosis 1986;27(3):357-60.

9. Olsson, S. - Ny vår för teleradiologin [Teleradiology]. Medicinsk teknik 1989;6(1):22-25. In Swedish

10. Olsson, S., Svahn, G. - Radiologin knyts ihop med PACS [PACS links radiology together]. Medicinsk teknik 1989;6(4):28-34. In Swedish

11. Olsson, S. - Hvordan har andre land håndtert spørsmålet om mammografiscreening? [How have other countries handled the matter of mammographic screening?]. In: Konsensus- konferansen om mammografiscreening 8-10 februar 1989 / B Backe (red.). Trondheim : Norwegian Institute for Hospital Research, 1989. (NIS-report 1989:2). In Swedish

12. Plan för industriella utvecklingsprojekt inom informationsteknologi (IT4) [Planning industrial development projects in information technology]. Stockholm : IT Delegation, 1988. In Swedish

13. Radiologin i Norden (Radiology in the Nordic countries). Stockholm: NEMT, Spri, 1989. (Spri report 275)

Acknowledgements

The author wishes to thank the hospitals, companies, and others mentioned above who supplied information about their developments and products, and to Prof. Ewert Bengtsson, University of Uppsala, for information about the research centers in Sweden.

PACS and Related Research in Switzerland

Osman Ratib, Denis Hochstrasser, Jean-Raoul Scherrer

Digital Imaging Unit, Center of Medical Informatics, University Hospital of Geneva
1211 Geneva-4, Switzerland

Abstract

A brief review of current activity in Switzerland is presented in this paper. A further description of a hospital-wide PACS project integrated with a Hospital Information System in Geneva is presented.

Keywords: PACS, Hospital Information System, Digital Imaging

PACS in Switzerland

A recent survey among the major hospitals in Switzerland showed that there is still very little development in picture archiving on a departmental or hospital scale. There is however a significant effort toward the development of communication and image transmission networks. Several institutions have undertaken serious development in this direction (University Hospital (UH) of Zurich, UH of Bern and UH of Geneva). Major University Hospitals have a real need for the transmission of radiological images to affiliated clinics and hospitals geographically dispersed. This effort toward the development of metropolitan and international image communication systems for teleradiology and remote consultation is strongly supported by the Swiss government and national telecommunication companies (Swiss PTT). Several European project supported by the European Economic Community (EEC) have also promoted research and development work in this direction on a European level. The UH of Geneva and the UH of Zurich are currently involved in image telecommunication projects with other European countries. The UH of Zurich is involved in a medical image communication project with the University of Freiburg in Germany. The UH of Geneva is taking an active part in a European project called TELEMED for transmission of medical images and development of remote expert consultation systems involving eight other European countries.

Several major hospitals have some plans for picture archiving and communication systems in the next few years (UH of Zurich, UH of Lausanne). Most of them, however, plan to implement commercially available systems provided by the major manufacturers. These systems are usually dedicated to a limited number of imaging modalities (mostly CT and MRI scanners). Several Nuclear Medicine departments (UH of Bern, UH of Zurich) have already started the development of departmental image archiving and communication systems.

The UH of Geneva is currently the only institution that has a hospital-wide PACS under development. The project is supported by the hospital who is committed to develop an integrated Image Management System for radiological as well as non-radiological medical images. It is also the only PACS project that is directly part of a large scale Hospital Information System (HIS).

NATO ASI Series, Vol. F 74
Picture Archiving and Communication Systems (PACS)
in Medicine. Edited by H. K. Huang et al.

Since 1975, the Medical Informatics division of the hospital has developed one of the widest hospital information systems in Switzerland, called the Diogene system [1]. The evolution of the Diogene HIS from a centralized system to a distributed architecture has promoted the concept of the integration of the PACS directly as a part of the HIS. In this architecture the images are viewed as another type of information to be handled by the HIS. It leads to a new concept of Ward Information System (WIS) where physicians and nurses could review and update their patient's medical record directly in digital form.

The Geneva PACS

The PACS under development at the University Hospital of Geneva is based on an open architecture with heterogeneous systems and multi-vendor equipment. It is built around a set of widely available industry standards, namely: Unix as the main operating system of all computers, TCP-IP as network protocol currently implemented on Ethernet networks but soon supported transparently by high speed networks as well, and an SQL-based distributed database (INGRES) that handles both the PACS and the HIS. The PACS is based on a distributed architecture of large number of servers. Two main types of servers are designed: the archive servers connected to the sources of images and equipped with large storage capacity using optical disk libraries (Juke Boxes), and display servers distributed over the hospital for temporary storage of images of current patients (see figure 1).

	Size (pixels)	Images/year	Gbytes/year
X-ray films	2048^2 x 8	400 000	1600
CT	512^2 x 16	400 000	200
MRI	256^2 x 12	100 000	6
US	512^2 x 8	100 000	25
Nucl Med.	variable	170 000	2
Angio*	1024^2 x 8	30 000	30
Cardio*	variable	500 000	240

Table 1: Summary of the volume of images generated per year at the UH of Geneva. The data for Cardio and Angio are a subset of the total amount generated based on an estimate of the volume of relevant images that should be selected for archive.

The Archive Servers

Several archive servers will allow for the acquisition and storage of image from different imaging modalities. Each of these servers is designed with a hierarchical cache system that consist of sufficient RAM memory space (32 to 128 MBytes), large capacity magnetic storage disks (2 to 4 GBytes) and very large optical disk Juke Boxes (aprox. 1 TeraBytes). This cache system allows a better performance and more rapid access to recent images that are kept in RAM and magnetic storage space.

A special file transfer management software is developed for the distribution of the images to predefined display servers where images can be accessed by the users. Images can thereby be distributed to several Display Servers simultaneously to be available at different locations in the hospital such as the radiology reading room and the medical ward for example. Images from previous examinations are extracted from the image database and sent through the same path either together with the current images or in advance prior to the requested examination (using a "prefetch" algorithm).

A unified image storage format was developed based on the ACR-NEMA standard of image communication. This file format called the PAPYRUS format, allows to store sets of images as a sequence of standard ACR-NEMA messages in an "encapsulated" file format [2]. A "folder" structure allows to keep a directory of related images in a single file or in different files. All images obtained from different imaging modalities are converted to the PAPYRUS format prior to archiving.

The Display Servers

Display servers are equipped with magnetic or opto-magnetic disks for temporary storage of images. One server will support a cluster of several workstations. The user of a given workstation can only access the images available on the corresponding Display Server. If the user requires an image that is not available on the local Display Server, provided he has the proper privilege to access these images, a special request can be posted from the workstation to the archive server through a query to the PACS database. The corresponding images will be transferred from the Archive to the Display server in batch mode according to priority schemes and depending on the network load. The user will then be notified when the requested images are available on his local server. A queueing system allows to control the network load at all time and avoid excessive traffic due to large number of simultaneous requests that may be submitted to the Archive Servers.

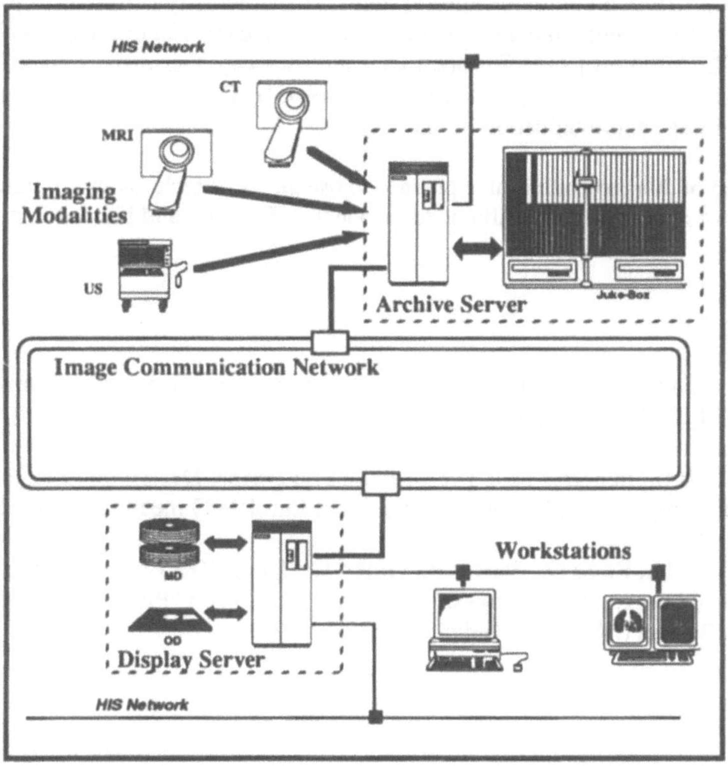

Fig. 1: Schematic diagram of the general architecture of the PACS in Geneva. Two types of servers are implemented: Archive Servers for acquisition and long term storage of images, and Display Servers for local storage of images to be accessed by several workstations.

The Imaging Workstations

Different types of workstation are required for different usage in a hospital-wide PACS. High performance workstations equipped with multiple high-resolution displays are needed for review and primary diagnosis by the radiologists. Multipurpose workstations for accessing images and other clinical information are to be designed to support the WIS. Finally there is an increasing demand for more quantitative analysis workstations, for image processing and computer-assisted image interpretation.

In order to provide a more uniform user interface on a variety of different workstations we elected to develop a common platform for image display and manipulation that can be ported on different workstations. A software called OSIRIS is currently being developed based on X-11 windowing system and OSF/Motif extension [3]. Such a platform is designed to be portable to any computer running Unix and equipped with a graphic display system running X-11. Also because we develop this software in object oriented C++, it is easily expandable and easily adaptable to different needs and requirements.

Teleradiology System

A prototype project for teleradiology and remote consultation is currently under evaluation as part of the European TELEMED project. This projects includes developments of videoconferences and remote expert consultation systems, establishment of a radiology reference image database and PACS to PACS communication systems. The University Hospital of Geneva will be able to test high speed communication lines (2 Mbits) for the transfer of images with other partners of the TELEMED project in Europe (Sweeden, Norway, Germany, France, Spain, Italy and Greece). The PAPRYUS image file format proposed by Geneva was adopted as a standard file format for image communication in this project.

As an extension of this project it is also planned to use the same technology for transferring images to and from remote clinics affiliated to the Geneva Hospital and located in different parts of the city.

References

1. Scherrer JR, Baud R., Hochstrasser D., Ratib O.: An Integrated Hospital Information System in Geneva. M.D. Computing, vol. 7, No. 2: 81-89, 1990.

2. Appel R., Ratib O., Scherrer J.R.: The Use of a File Format Description Language for easy Handling of Images from Multiple Sources. Proc. EuroPACS meeting, Triest, 1990.

3. Ligier Y., Funk M., Ratib O., Perrier R., Girard C.: The OSIRIS user interface for manipulating medical images. Proceedings of NATO ASI meeting on "Picture Archiving and Communication System (PACS) in Medicine". Evian, 1990.

Medical PACS Research in Turkey

Kemal Oflazer[1], A. Enis Çetin[2], Hayrettin Köymen[2], and Münci Oran[3]

[1]Department of Computer Engineering, Bilkent University, Ankara, Turkey
[2]Department of Electric and Electronic Engineering, Bilkent University, Ankara, Turkey
[3]Department of Radiology, Faculty of Medicine, Hacettepe University, Ankara, Turkey

Abstract: We present an overview of RAGARİS, a radiological image archiving and communication system being designed with the cooperation of two universities in Turkey. The system is being designed and will be implemented by Bilkent University and will be be deployed at the Radiology Department of Faculty of Medicine at Hacettepe University. We describe briefly the system requirements and present the system functions and organization.

Keywords: Medical Image Databases, PACS, Distributed Systems, Optical Disk Storage.

Introduction

The advent of low cost and powerful personal workstations and very high-density storage media like optical disks have profoundly influenced many fields and applications. The enormous capacity of optical disks have made feasible the construction of large electronic archives for paper documents in the fields of banking, law, engineering where documents on paper or similar media scanned and stored into such archives are accessed, and viewed or modified through workstations [6]. In the field of medicine, diagnostic radiology has seen the impact of such technologies, and many institutions have started applying these technologies to the problem of medical image archiving, access and communication (e.g., [5, 4] — see [3] for a comprehensive survey.)

In this paper, we describe RAGARİS[1], a medical image management system that is being designed with the cooperation of two universities in Turkey. Our research group at Bilkent University is designing RAGARİS with input and requirements specifications coming from the members of the faculty at the Department of Radiology of Hacettepe University. After a brief overview of the Hacettepe University Hospitals environment in terms of its facilities and workload, we present our considerations and requirements for the system. Subsequently, we present a functional description of basic system components and an overview of the system organization.

---

[1]RAGARİS stands for "**RA**dyolojik Görüntü **AR**şivleme ve İletişim Sistemi" (Radiological Image Archiving and Communication System).

NATO ASI Series, Vol. F 74
Picture Archiving and Communication Systems (PACS) in Medicine. Edited by H. K. Huang et al.

Image Source	Number of Patients	Images per Patient[2]	Total Images	Image Size (KBytes)	Comp. Size (KBytes)[3]	Total Stg. Req. (MBytes)
X–Ray	222,934		501,000	1000	34.13	17121.33
USG[4]	24,143	4	96,572	256	8.53	824.08
CT	8,855	24	212,520	256	8.53	1813.50
VA	1,655	8	13,240	256	8.53	112.98
ESD	2,835	17	48,195	256	8.53	411.26
IVP	2,856	4	11,424	256	8.53	97.48
BE	1,788	10	17,880	256	8.53	152.58
Total	265,066		900,831			20,533.22

Table 1: Radiodiagnostic workload at Hacettepe University Hospitals in 1989

The Application Environment

The application environment for RAGARİS is the Radiology Department of Faculty of Medicine of Hacettepe University, which provides diagnostic radiology service to Hacettepe University Hospitals. Although patients are mostly admitted through referral by other hospitals, for various reasons the workload at these hospitals is substantial as indicated by the 1989 statistics given in Table 1.

The Radiology Department at Hacettepe University Hospitals has the following imaging units:

- 13 Bucky Tables
- 2 Computerized Tomography Unit
- 2 Skull X–Ray Units
- 1 Digital Angio Unit
- 1 Urology Table
- 2 Chest X-Ray Units
- 1 Mammography Unit
- 7 Flouroscopy Tables with Video Chains

RAGARİS Requirements and Benefits

The main functions of RAGARİS are the *capture*, *archival* of and *access* to medical images. The system will be able to interface with a variety of medical imaging devices, and provide rapid access to any or all radiological information associated with a patient, while concurrently providing sufficient capture bandwidth for the workload input shown in Table 1.[5] As seen in Table 1, with an aggressive lossy compression ratio, the daily input to the system is estimated to be 85 MBytes and will be higher with smaller compression

[2]These numbers are the average number of archivable images from respective imaging devices.

[3]Assuming a lossy compression ratio of 30.

[4]USG – Ultrasonogram, CT – Computerized Tomography, VA – Vascular Angiography, ESD – Eusphagogram Stomach Duodonography, IVP – Intravenous Pyelography, BE – Barium Enema.

[5]The statistics in Table 1 indicate that just the X-Ray image capture rate is about 4.5 exposures per minute assuming 240 workdays per year and 8 hours per day. The total rate is about 8 images per minute.

ratios. The expected amount of (compressed) input to the system may easily reach *several hundred MBytes/day* in the next 5 years.

The system will also provide a number of image processing functions like scaling, enhancement, estimation of tumor sizes, 3–D image reconstructions from 2–D images etc. The system is expected to be used from about 20 radiologist offices, and from a number of reading rooms with expansions being planned to some of the operating rooms. The system has to have enough capacity to store information about up to 1 million patients, with around 10 or more images per patient on the average, for a total of about 10 million images. The system should be tolerant to most faults at least on the capture side so that device failures should not adversely affect patient imaging service.

The system is expected to eliminate most of the film usage in the hospital environment in the long term. Considering Hacettepe University Hospitals spent about US$1 Million for X-Ray films alone, it is obvious that the RAGARİS will recover its cost within 2 to 3 years with such savings. Furthermore, elimination of unnecessary exposures due exposure or positioning errors will imply further savings both in time and material. If necessary, alternative hardcopy outputs (e.g., on silver paper) will be provided at a cost of about $1/10^{th}$ the cost of X-Ray film.

RAGARİS Functions and System Organization

RAGARİS Functions

Figure 1 shows the basic functional block diagram of RAGARİS along with the system organization. The system has the following functional units:

• *Image Database Server:* This is the main information database that lets the radiologists *query* the system, *identify* patient(s) and then *access* the image information associated with the selected patient(s). The information accessed may either be the image data or any other relevant information like any previous diagnosis information etc. The main component of this unit is a *relational database* that stores information about patients (like name, admission numbers, social security numbers, admission dates, diagnoses etc.) that can be queried by the radiologists for identifying a patient based on various criteria.

• *Storage Servers:* These servers provide transparent access to all the image (and possibly other) files stored on various storage media. The *Optical Disk Server* manages the optical disk drives and the associated jukebox system. RAGARİS also requires substantial amount magnetic disk storage, mainly as a fast cache and buffer for captured images. The diagnostic images taken during one day are frequently used during the following few days for various reasons. Hence, from a performance point of view it is advantageous to keep these image files on magnetic disk storage and then commit to archival optical storage after some time. The *File Server* manages such magnetic storage.

• *Image Processing Server:* This unit provides a number of functions for various image manipulation tasks - the most crucial being fast compression and decompression of images. We will employ both irreversible and error-free coding for medical images. For *irreversible coding*, we have developed a subband image coding method for images of natural scenes

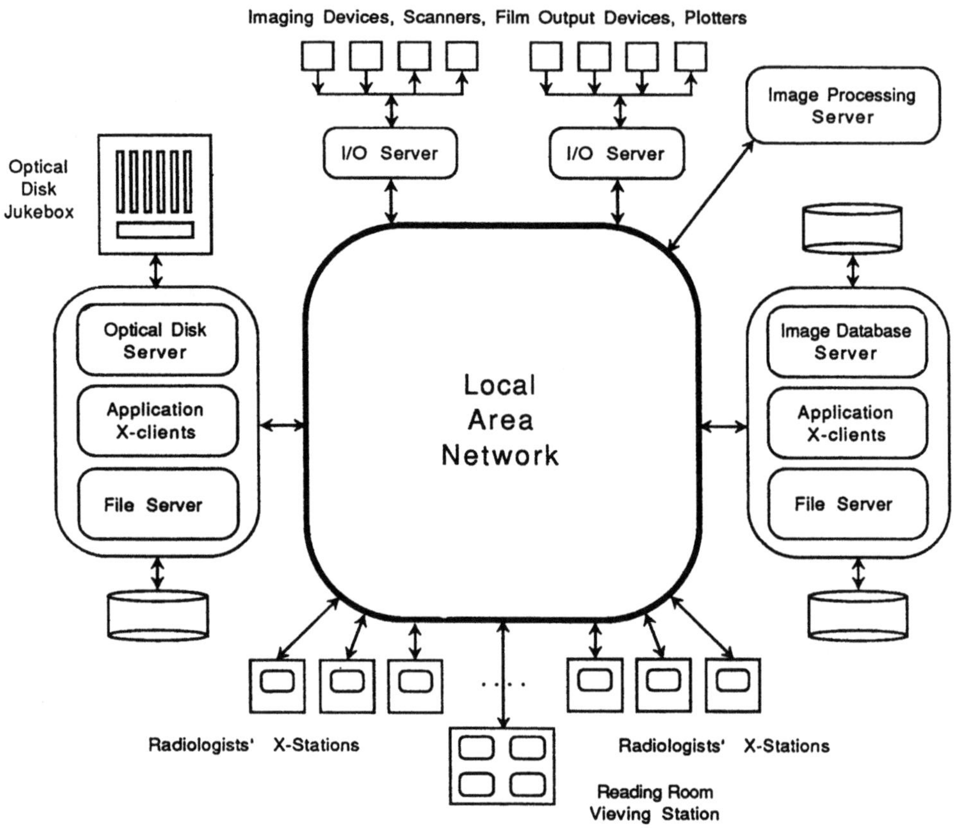

Figure 1: RAGARİS system organization and functional block diagram

[1]. A modified version of this method will be used for medical image compression. In this method, a medical image is filtered by a digital filter bank and a set of sub-images are obtained. These sub-images are coded by using a Discrete Cosine Transform based algorithm. We expect to get a compression ratio of around 30 without introducing any visual degradation. The method will be tested by using the medical images which will be determined by the ISCAMI-COMAC-BME research program of the European Community [2]. For *error free coding*, we will use a multiresolution image decomposition technique called pyramid representation. The sub-images resulting from pyramid decomposition will be coded using an arithmetic coding method.

• *Input/Output Server:* This unit interfaces all imaging input units (scanners, CT, etc., and the image output units (film output devices, plotters) to all the other units in the system. Any conversions between a standard image format and device dependent formats will performed by this unit.

• *Radiologists' Image Access Terminal Units*: Radiologists query the Image Database Server through these units and when the requested images are identified, the corresponding image file are retrieved from the Optical Disk Server or the File Server and displayed on high-resolution screens.

System Organization

RAGARİS is a distributed system with a number of server and user workstations connected on a local area network. The radiologists terminals are X-Stations that provide high-resolution 256-level gray raster graphics. The service load for such X-Stations is distributed (via the application X – clients to the two main high-performance server class machines (e.g., SUN Sparcserver 490 or IBM 6000/530). The Input/Output server will be implemented on one or more lower performance workstations (e.g., SUN or IBM workstations). The reading room display station is a high-performance system with 4 high-resolution, 12-bit gray level display screens, (e.g., MegaScan MS211-8P).

Conclusions

We have described the basic motivation, requirements and system level architecture for RAGARİS – a medical PACS system being designed with the cooperation of two universities in Turkey. The system is to be employed in a university hospital environment with a substantial workload. The system provides the functionality for image capture, image identification and access through querying a database, image processing and manipulation.

References

[1] R. Ansari, A. E. Çetin, and S. H. Lee. Subband coding of images using nonrectangular filter banks. In *Proceedings of 32nd Annual International Technical Symposium.* SPIE-The International Society for Optical Engineering, 1988.

[2] J. Demongeot. Integrated system for computer assisted management and manipulation of medical images —ISCAMI. Annual Report, 1989.

[3] H. K. Huang et.al. Picture archiving and communication systems (PACS) for radiological images: State of the art. *Critical Reviews in Diagnostic Imaging,* 28(4):383 — 427, 1988.

[4] P. S. Cho et.al. A PACS for chest radiology. In *Proceedings of Conference on Medical Imaging - IV.* The International Society for Optical Engineering, 1990.

[5] S.-L. Lou et.al. PACS in neuroradiology. *Hospimedica,* 7(7):25 – 39, 1989.

[6] K. Oflazer. Document management — issues and experiences. Invited Paper in *Proceedings of International Conference on Computer and Information Sciences – IV,* pages 157 – 171, 1989.

PACS AND RELATED RESEARCH IN THE UNITED KINGDOM

H.I. GLASS[1] AND N.A. SLARK[2]

[1] **North West Thames Regional Health Authority, 40 Eastbourne Terrace LONDON W2 3QR UK**

[2] **Medical Devices Directorate, Department of Health, 14 Russell Square, LONDON WC1B 5EP UK**

ABSTRACT

The Government supported PACS projects are at Leeds General Infirmary and at Hammersmith Hospital. The Leeds group are carrying out an assessment of computed radiography units, and developing image processing algorithms.

The Hammersmith group is installing a whole hospital PACS. The economic basis for this project is presented. The special problems arising from the installation of PACS in a postgraduate teaching hospital are discussed.

Keywords: PACS, WHOLE HOSPITAL, ECONOMICS

INTRODUCTION

There are two major Government supported PACS projects in the United Kingdom. One is at the Leeds General Infirmary and the other at Hammersmith Hospital, London. The Leeds General Infirmary Project is concerned with the objective assessment and development of imaging devices and in the development of image processing algorithms. The Hammersmith Hospital project is concerned with the implementation of a whole hospital PACS, to be operational in December 1992.

The requirement to provide a new x-ray department at Hammersmith Hospital, presented an opportunity to instal a whole hospital PACS. The expenditure enables a full assessment to be made of the effect of PACS on the operational efficiency of the whole hospital. It is considered likely that cost savings arising from PACS will be due to increases in the efficiency in the hospital.

PACS RELATED ACTIVITIES AT THE LEEDS GENERAL INFIRMARY (FAXIL)

THE "FAXIL" GROUP

The Department of Health's Facility for Assessment of X-ray Imaging, Leeds (FAXIL) was set up in the Department of Medical Physics of the University of Leeds in 1973.

NATO ASI Series, Vol. F 74
Picture Archiving and Communication Systems (PACS) in Medicine. Edited by H. K. Huang et al.

The work undertaken includes research into equipment and methods for it's assessment, image processing, image intensifier development, and computed radiography assessment. An essential element of the success of the Philips DSI digital image intensifier system has been the development at FAXIL of the image processing algorithms necessary to obtain the highest quality images. Another project relates to the development of image processing algorithms for mammography.

In co-operation with Philips Medical Systems, the digital spot imaging system (DSI) was developed which obviates the use of x-ray film during fluoroscopic procedures.

FAXIL is carrying out a technical and clinical evaluation of two Computed Radiology Units, namely: The Siemens Digiscan Unit and the Philips Graphic IIE Unit. The quality of the images are being assessed and evaluated objectively.

PACS ACTIVITIES AT HAMMERSMITH HOSPITAL

THE HAMMERSMITH HOSPITAL CAMPUS

Hammersmith Hospital shares the campus with The Royal Postgraduate Medical School, The Medical Research Council, and The Institute of Obstetrics and Gynaecology.

Hammersmith Hospital serves the local population of 100,000 approximately, and provides regional, national and international specialist referral services. It combines the functions of a District General Hospital with its role as a centre of research over a wide range of specialties at postgraduate level. For 1989/90 the total revenue expenditure of the was £59 million. In 1989/90, the number of patients treated were:-

General and acute inpatients admitted:	18,266
Outpatients attendances:	118,751
Accident and Emergency attendances:	31,399

The number of acute specialty beds is 534. The average length of stay is 8.3 days.

DIAGNOSTIC RADIOLOGY DEPARTMENT PROFILE

The Diagnostic Radiology Department contains all the modern radiology imaging modalities. The annual attendances are indicated in Table 1.

About 100 folders per day require to be extracted from the main film store for out-patients clinics. Joint radiological/clinical conferences require approximately 35 patient folders to be made available each day.

The proposed new department is expected to comprise twenty one imaging rooms. Cine angiographic cardiac images will not be stored or distributed by the PACS system, nor will obstetric ultrasound images.

TABLE 1 - RADIOLOGICAL ATTENDANCES (1990)

Inpatients (excluding Mobile examinations)	19,516
Outpatients	20,656
Casualty (excluding Mobile examinations)	13,166
Mobile Examinations	5,856
General Practitioners	6,634
Angiography	1,320
Ultrasound (Outpatients 43%)	4,580
CT (Outpatient 41%)	2,572
MRI	2,000
Radioisotopes Imaging (Outpatients 55%)	2,102
TOTAL:	78,402

THE ECONOMIC BACKGROUND

INTRODUCTION

The model used in this study has been developed from the work of Saarinen et al [1], but some simplifications have been introduced.

The model was used to estimate the economic effects of the installation of PACS on the operational pattern of Hammersmith Hospital. The changes required in the operational pattern of the hospital to provide indirect cost benefits are vital in the economic justification of installing a PACS.

GENERAL DESCRIPTION

There are six discrete direct cost/saving components in the model, namely archive material and film use, labour, maintenance, operation and supplies, space, and capital (equipment and building).

There are also seven indirect saving components which have been included in the model which represent opportunities for savings which would need to be developed, and transformed into real savings by the introduction of administrative and operational changes within the hospital. These indirect benefits relate to saving of junior medical staff time, reduction in unnecessary investigations, saving in radiologist, radiographer and clinician time, re-designation and change of use of a number of acute beds, and reduction in the length of stay.

No account has been taken of the potential increase in income generation capacity created by the improved efficiency on the part of the radiologists, radiographers and clinicians. The potential savings have been estimated from the the salaries elements only, corresponding to the estimated improvement in their efficiency.

ASSUMPTIONS

An annual increase in workload of 2% is assumed. This is reflected in an increase in film costs and storage space requirements and is superimposed on the film costs. Labour costs are assumed to be in line with inflation, except in the case of professional groups (clinicians and radiographers) whose salaries are assumed to be 1% above inflation. It is assumed the cost of equipment maintenance will increase in line with the rate of inflation. The cost of building and space is assumed to be 5% above the inflation rate. This factor is applied to the opportunity space created by the virtual abolition of the requirement of space for a film store.

The major additional expense incurred with the introduction of PACS is the hardware maintenance cost. The equipment is currently very reliable and an overall maintenance cost of 5% of the equipment cost has been assumed. The economic viability of the project is very sensitive to this factor. Due to the high degree of availability required using mirrored CPU's and discs, and the high level of replication of the workstations, expensive maintenance contracts normally associated with isolated items of complex equipment will not be entirely relevant to PACS.

It is assumed that unnecessary investigations or repeated examinations would be avoided by the immediate availability of the results of other imaging investigations on the same patient. The estimate of the saving due to this element has been put at approximately 8.5% of the total departmental costs.

It is estimated that the whole time equivalent of five junior medical staff are employed in searching for and preparing films for the large number of joint x-ray conferences.

The substantial improvement in department efficiency will provide additional time for both radiologists and radiographers to enhance the utilisation of the department's resources. This saving has been expressed as an efficiency improvement and is conservatively estimated as a 15% saving of the labour costs of these two groups.

It is estimated that there will be a significant improvement in the efficiency of the medical staff, a estimated at 2% increase, which has been converted to a comparable "saving" in the labour component of the medical staff cost. This is likely to be a very conservative estimate of the real cost benefit.

It is estimated that there will be an improvement in efficiency of the clerical staff serving the wards and out-patient clinics of the order of 25%.

A significant number of acute beds are occupied by patients undergoing radiological investigations and requiring minimal nursing care. These could be accommodated in five-day beds, under the organisational control of the radiology department at a significantly lower cost.

If a minimal reduction of length of stay were achieved by the re-examination of operational procedures, arising from the rapid availability of all images, then this is equivalent to a significant reduction in cost of hospital operation which has been conservatively estimated at 0.5%.

RESULTS OF APPLICATION OF THE MODEL

The total direct and indirect costs/savings for each of the main headings discussed above are summarised in Table 2. The system is assumed to have a lifetime of ten years. The NPV goes negative after nine years from the start of the project, indicating a NPV of -$0.34M after ten years. It can be seen that the contribution of the indirect savings is an essential factor in achieving a justifiable investment decision.

The economic analysis of the hospital PACS installation is based on a conservative but realistic model. The results derived from it depend on the ability of the management to implement the operational changes to ensure the achievement of the indirect benefits. The maintenance cost expressed as a proposed percentage of capital cost is considered achievable.

What emerges from the study is that it is unlikely that PACS could be economically justified purely on the basis of direct cost savings.

TABLE 2
ACCUMULATED COSTS/SAVINGS OVER TEN YEARS ($M)

	Whole Hospital System
Direct Costs/Savings	
Archive / Film Materials	-2.00
Labour	-0.80
Operations	-0.25
Space Saving	-1.35
Capital Expenditure	+10.80
Maintenance	+4.60
Sub-Total	**+11.00**
Indirect Savings	
Junior Medical Staff Time	-1.55
Radiologists' Time	-1.10
Radiographers' Time	-1.25
Medical Staff Time	-2.70
Hospital Clerks	-1.25
Unnecessary Investigations	-2.30
Reduced Length of Stay	-3.20
Redesignation of Acute Beds	-1.90
Sub-Total	**-15.25**
Net Total	-4.25
NET PRESENT VALUE	-0.34
EQUIVALENT ANNUAL COST	-0.037

ORGANISATION OF THE PROJECT

The project is under the direction of a Project Team which is Chaired by the Hospital Director of Finance and Information . The project is under the day-to-day direction of a Project Manager.

The first patients are expected to be accepted within the new department and the PACS system brought into operation in December 1992. This will be followed by a period of amendment and adjustment based on operational experience, which will be completed in August 1993.

A pre-installation evaluation survey will be carried out between July and October 1992 and a final evaluation will be initiated in January 1994.

The main activities of the Project Team are carried out by a series of sub-groups which are co-ordinated by the Project Manager.

Close contact is being maintained with the United States Armed Services MDIS project, and the specification being developed for Hammersmith relies heavily on the MDIS specification for the technical content. The proposed utilization of the system is however likely to be significantly different because of the characteristics of Hammersmith Hospital. The heavy commitment to teaching, joint conference activity, and the dispersal of specialist radiological expertise throughout campus at Hammersmith introduces significant operational demands.

CONCLUSION

There is an increasing awareness of the potential advantages of PACS. Several groups are currently investigating the possibility of introducing PACS when new x-ray departments are commissioned in the future. The project at Leeds will yield invaluable results on the limitations of currently available CR systems. The whole hospital PACS project at Hammersmith will help to identify the extent to which the indirect savings, which provide the economic justification of PACS can be achieved or will exceed the predictions. This project will provide invaluable experience in the organisation and implementation of a whole hospital PACS project in a large tertiary referral teaching hospital.

[1] Saarinen, A.O., Haynor,D.R., Loop, J.W., Johnston, L., Russell, n., Mitchell, K., Nemereer, M. : Modelling the Economics of Pacs: What is Important? Proc. SPIE Medical Imaging II, Vol. 1093; 1989.

HIPACS: Hospital Integrated Picture Archiving and Communication System. An R & D Effort of the European Communities

R. Mattheus [1], F.P. Ottes [2], A.R. Bakker [2], M. Osteaux [1]

[1] PRIMIS/Department of Radiology, University Hospital Brussels (VUB), Laarbeeklaan 101, B1090 Brussels, Belgium
[2] BAZIS Central Development and Support Group Hospital Information System, Schipholweg 97, 2316 XA Leiden, Netherlands

Abstract: A PACS research project under the name "Foundations for a Hospital Integrated Picture and Communications System" (HIPACS) is currently being conducted. The overall objective is the modular conception of a experimental knowledge-based medical information management and distribution system. Image/text document access and display will be based on an intelligent user interface for diagnostic image workstations. The concept will comprise an interface with the Hospital Information System. It will be based upon standards and be realized by an interactive collaboration between 8 European research groups from universities, hospitals and industries. The pluridisciplinary groups consist of engineers, computer scientists, medical doctors and physicists.

Introduction

In the hospital of the future, diagnostic imaging procedures based on picture archiving and communication systems (PACS) are expected to replace the film-based procedures of today. This will require a tight integration of PACS with the other hospital structures. The current state of technology only allows the realization of partial, small-scale solutions in the clinical environment.

The aim of this HIPACS project is therefore the identification and -whenever possible- elimination of the bottle-necks that hamper the implementation of a full scale hospital integrated PACS. In order to achieve this goal, coordinated research efforts in the fields of digital network design, data management software and diagnostic imaging workstations are needed.

The project has been set up as a part of the European Community's Advanced Informatics in Medicine (AIM) framework initiative. The Pluridisciplinary Research Institute for Medical Imaging Sciences (PRIMIS) is the prime contractor, with M. Osteaux as Project leader and R. Mattheus as Project manager.

Eight partners from five countries have joined forces. They bring together complementary skills in medicine and information technology (IT), drawn from industry (Philips Hamburg (D) - BAZIS Leiden (NL)), Hospitals (Klinik fuer Radiologische Diagnostik Aachen (D)- University hospital VUB Brussels (B) and Universities (University of Ulster (GB)- Institute for computer Science Crete (G) - Lehrstuhl fuer Messtechnik RWTH Aachen (D)- PRIMIS Vrije Universiteit Brussels (B)).

This paper first explains the backgrounds of the AIM program, Next, the objectives of the HIPACS project are reviewed. The cooperating partners and their

NATO ASI Series, Vol. F 74
Picture Archiving and Communication Systems (PACS) in Medicine. Edited by H.K. Huang et al.

(scheduled) research activities are listed. Finally, the continuation of HIPACS within the scope of the main phase of AIM is discussed and some preliminary conclusions are given.

Europe Community R&D: "AIM program" evaluation phase

The main objectives of AIM are:
- Improvement of the efficiency of the health care,
- Enhancement of the position of the European communities in the field of medical, biological and health care informatics.
- Creating a favorable environment for the rapid introduction and appropriate application of informatics in health care.
- Stimulating the collaboration between European companies, R&D institutes and Hospitals

Several of the 43 tasks as defined in the AIM workplan relate to PACS systems:
- Clinical data and process modelling
- Medical image databases; functional requirements
- HIS integration and management
- Development of medical information protocol
- Medical image and signal interpretation, pattern recognition
- Development of a user interface generator system

The deadline for submission of the proposals was February 14, 1989. More than 200 project applications were received. These proposals were evaluated by 60 experts and 42 consortia were selected to participate in the AIM exploratory actio. The majority of the projects was started in June 1989 and have a duration of one or one-and-a-half year. The budget allocated for this first phase of the AIM program amounts to 20 million ECU (at this moment 1 ECU roughly equals $ 0.9)

The major achievements of the exploratory phase of the AIM program will be presented of the EuroFORUM Conference, that will be held in Sevilla, Spain on December 13-15, 1990. The HIPACS project has a budget of ECU 1.3 million. of which 50% is provided by the Commission of European Communities (CEC).

Each AIM project produces quarterly reports. In addition to these, a review report should be realized at the end of each year on the activities in that year, and on the workplan for the coming year. At the beginning of each year a technical review will be organized to decide whether the concerned project should be continued, be modified or be stopped.

HIPACS technical summary

In order to define a second generation PACS, coordinated research efforts in the field of digital network design, system integration, data management, and diagnostic imaging workstations are needed.
Five main tasks efforts are defined:

The modeling of the PACS/HIS and the related coupling of PACS and HIS.
When introducing an image management world, an important point fur using the PACS in clinical routine is the integration with the patient management procedures of the hospital, the Hospital Information System (HIS) and Radiology Information Systems (RIS). Patient information must be shared between these four worlds.

For transporting this enormous amount of image information, classic networking is not sufficient. We are working in two directions.

A prototype of a high-speed network for image transfer will be implemented and evaluated.
A fast image communication technique is being developed using image management software based on optical fiber.

A model of a heterogeneous network structure will be developed and used to study routing optimization of image and other data in existing and future networks.
Knowledge based network management techniques are being developed for routing these images to places where normally the images are needed, based on the organization and medical information in the hospital.

Multimedia data-base software with facilities for intelligent information retrieval will be developed.
Images (digital radiographs), Alphanumerical data (reports) graphical data (eg.heart diagrams...., all these multimedia data must be taken into account for storage and retrieval.

An adaptive user-interface for increased diagnostic efficiency of existing image workstations will be designed.
Medical knowledge is not only important for network management, but it is also useful information for defining the behavior of the user-interface. Suggestions are made as to which images of the examination must be displayed at a certain place of the reporting stations.

The HIPACS project has produced four reports. These reports are called:
- "State-of-the-art"
- "Specifications of interactions (analysis and functional models)"
- "Requirements and definitions"
- "Concept for HIPACS architecture"

The activities and results will be laid down in a book ("A Second Generation Picture Archiving and Communication Systems", mid-1991, Springer-Verlag, Berlin).

HIPACS topics, a first results

Each of the six main tasks is, although placed under the responsibility of one partner, carried out by a group of partners in collaboration. In this way tasks will interact and the results can be integrated in the final report. The tasks and their targets are:

PACS-HIS-RIS flow composition
A well-defined, complete, consistent and valid set of information requirements and functional specifications will be defined. An architecture of a RIS integrated with both the HIS and PACS, will be created. The PACS, HIS and RIS have to be divided into manageable subsystems for reasons of maintenance and operation. The demarcation of what information (database) belongs in a subsystem, which functions belong there and which computer peripherals are most convenient to achieve integration is an important aim.

This primarily analytical activity will employ some modelling and prototype techniques.
The results will comprise:
- A model for representing the data architecture (ER-approach of P.P. Clen)
- A data activity/flow diagram representing the application architecture (2)

A data flow model has been set up for various tasks in the hospital, eg. request formulation, visit and patient registration, retrieve patient information, scheduling of an examination, execution of an examination and reporting For each flow model all actors, the input objects and the output objects are listed. A schematic diagram of the relation is given.

Integration PACS-HIS

Successful implementation of PACS in a hospital can only be realized when it is coupled to the HIS. The main reasons for this coupling are:

- Identification of PACS images needs demographic HIS data
- Images in the PACS and alphanumerical related patient information from the HIS should be displayed coherently on the same display station. The correlation and interpretation of data will offer new opportunities to support the medical decision making process.
- Uncoordinated storage of data of different types, eg. alphanumerical and image, in dedicated uncoordinated databases should be avoided.
- HIS functions, such as report editing, should be available on the PACS workstation
- Acceptable response times will require intelligent image management strategies (prefetching) based on data from the HIS, eg. patient location and appointment (3) Images should be transferred from the central long-term and slow archive to the fast but small memory of the workstation, hours before the user actually evokes the command to display the images.

Deliverables on the following issues have been realized:

- Literature survey on PACS sits with an actually used PACS-HIS coupling. Six carefully sites in the USA, Europe and Japan are described in detail (Georgetown-USA, Seattle-USA, Brussels-Belgium, Graz-Austria, Osaka-Japan).
- Data that should be exchanged between HIS and PACS
- Requirements of a generic HIS-PACS coupling
- Design of a generic coupling
- Previous images needed for diagnosis of a new case
- Pilot study on the design and the expected efficiency of an image management strategy, namely prefetching
- Network model and simulation results.

Medical image communication

Activities within this task deal with:

- Requirements for the communication of medical images in a hospital
- Evaluation of existing communication protocol standards, and the proposition of appropriate modifications or extensions
- Developments of a concept for a prototype PACS network
- Modelling and simulation of the network concept
- Performing a prototype implementation
- Within the scope of the future AIM main phase, to test the network under real clinical routine work load.

Achievements until this moment include:

- The network requirements as a consequence of image data acquisition, storage and retrieval at the Aachen University Hospital have been analyzed. The total average network load has been estimated.
- An overview of the existing network management techniques has been given, also mentioning where they are used, and for what purpose.

These techniques are evaluated for their use in managing a PACS network.

- Evaluation of the organization of various radiology departments with respect to image storage and retrieval.
- An information flow model for images and image related alphanumerical data has been set up and the critical paths in is this model have been investigated.
- A knowledge based network control technique has been developed and partly implemented at the PACS in the University of Brussels (VUB((4)

Adaptive user interfaces
The knowledge needed to design user models and task models has been and will further be acquired. These models will support the development of adaptive user interface for medical image workstations.

A user interface scenario for adaptive cooperative work- stations, and strategies for the technical realization of such workstations have been described (5). Based on the present situation in radiology, a projection to the future situation in a fully digital systems has been made. Workstation user interface and functionality has been defined for various user groups and tasks. The behavior of the radiology staff has been studied by means of observation, interviews, simulated reading sessions and discussions.

The user behavior with respect to data (image and text) handling has been recorded. The organization of the radiology départments of the Aachen and the Brussels hospital have been investigated.

Requirements for multimedia databases
The state-of-the-art in the field of multi-media databases has been described (6). The material was subdivided in: Multimedia databases, Image databases, Distributed databases, Object-oriented databases and User requirements.

Existing architectures have been evaluated on their applicability in a PACS. Preliminary prototyping on text and image handling has been conducted based on federate databases. The basic parameters of data communication in a typical hospital have been analyzed. Research has been performed on the construction of a digital image archive for hospitals.

Image indexing by content
An image data base system that can be accessed not only by key words, but also by using an index represents the content of the image, will be designed. Image processing and analysis techniques must be applied to the images to extract the parameters representing the content.

The database will be interfaced with an image processing and interpretation system for coding images prior to storage and the extraction of appropriate image descriptions. The document database will consist of physical database and logical database. A query concerning a document may specify the semantic content, the structural content and various image features and primitives. Exploiting the prototype, so that it can be implemented on a variety of computers accessible by different groups of users, will be interesting.

Continuation of the AIM/HIPACS activities

The main phase of AIM has been prepared by the AIM Requirements Board. Its task was to address future requirements and options in health care information in infrastructure and services. For the main phase of AIM some 100 tasks in the medical informatics field have been defined (7). Three of these tasks are very closely related to PACS. The three tasks and their backgrounds are:

- Development of a modular IMAC architecture:

The management of the vast amount medical images and information generated by today's hospital department units is a growing problem. The solution to the problem will increasingly require the use of advanced computer-based technologies, heterogeneous and fast networks, large volume storage and high-resolution display techniques and methodologies for integration.

- Functional classes and development of workstations:

In a clinical environment different levels for image and data manipulation, viewing and processing can be defined from a medical. organizational and economical point of view. These aspects have a direct impact on the specification, functional integration and requirements, developments of image- data workstations and their level of integration and interaction in the departmental systems.

- Integration of IMAC with departmental systems:

Methods of electronic data processing have spread considerably in the hospitals. The main activities of EDP in the hospital at present comprise the fields of administration and organization as well as the supply management in a technical sense. The PACS technology has an impact on the departmental systems, because of the demand for functional, technical and conceptual integration.

Conclusion

Technology driven approach

Since 1982, when the "PACS" acronym was defined, research has been conducted in different domains, but most of the installations have experienced a lot of technical problems for clinical use which need to be solved before the real introduction in hospitals can take place.

Large archiving possibilities are needed. The required storage capacity became available with the introduction of the Optical Disk technology. Recently the erasable magnetic Optical disk became available. The bottle-neck was moved to the networking environment, where faster communication is emerging. Until now PACS was principally driven by the evolution of technology.

Medical driven approach

New hardware can fulfill some of the clinical requirements for PACS. To overcome the remaining problems on our way to realizing a PACS, we should first of all focus on the further development and application of software that is based on medical knowledge.

Image management network traffic regulation and workstation user-interface software should be based on information concerning the patient, such as its medical status, its location in the hospital, and information concerning the actual image flows. A prototype for a partial small-scale PACS should be created to demonstrate the benefits of such software for the performance of that PACS. This work could be one of the targets for the main phase of the HIPACS project.

REFERENCES

1. AIM workplan, Advanced Informatics in Medicine, (Directorate General XIII/F) October 1988.
2. Gare C. and Sarcon T., "Structured systems analysis: tool and techniques" Prentice-Hall Inc. Englewood Cliffs 1974.
3. A.R. Bakker, H. Lodder, J.L.M. Kouwenberg, "Traffic between PACS and HIS/RIS; data or information? "Medical Imaging IV Conference, February 4-9 1990, Newport Beach, SPIE vol 1234, pp. 495-500.
4. R. Mattheus, F. Moyson, Y. Temmerman, M. Osteaux: Hospital Integrated Picture Archiving and Communication Systems; A. European Project; pp. 396-404; SCAR'90 Computer Applications to Assist Radiology, R. Arenson, R. Friedenberg, Symposia Foundation 1990.
5. Wendler Th., Cooperative human-machine interfaces for medical image workstations; a scenario Proceedings CAR'89, 3rd International Symposium on Computer Assisted Radiology pp. 775-779, 25-28 June 1989.
6. Bell D.A. and Abul-Huda B.A.H., Multi-media databases - Present and Future, Database technology journal 1, 1987.
7. Operation 1992, Health Care and Medical Informatics, Specification of Priority Research and Development Tasks (AIM Central Office) September 1989.

The HELIOS European project on software engineering

P. Degoulet [1], J. Coignard [2], F-C. Jean [1], M-C. Jaulent [1], L. Lucas [2], M. Ben Said [1], H- P. Meinzer [3], U. Engelmann [3], A. Springub [3], R. Baud [4], J-R. Scherrer [4]

[1] Broussais University Hospital,
[2] CAP-SESA Tertiaire, Paris, France
[3] German Center for Cancer Research, Heidelberg, Germany
[4] Geneva University State Hospital, Geneve, Switzerland

Abstract: The objective of the HELIOS European project is to create a Software Engineering Environment (SEE) to alleviate the development of medical applications. The resulting system can be viewed as a layer above an operating system providing services to the application designer, and finally to the application end-user. HELIOS is constituted of a set of Software Elements communicating through an abstract channel called Software Bus. The SEE kernel is made of three main software elements, the Information System, the Documentation Facilities and a component which provides the application user with a convenient user support interface, the Interface Manager. The Services consist of a collection of tools providing the necessary services to the medical application designers including an image processing toolbox.

The object oriented paradigm is used both as a methodology for the modeling of the entities and procedures found into an application and as the basic structure for building the Information System and the Interface Manager. Development standards include UNIX as the operating system and X-Window and OSF-MOTIF as windowing environments. The target application for the SEE is the development of a hospital ward information system with its ward image manipulation subsystem.

NATO ASI Series, Vol. F 74
Picture Archiving and Communication Systems (PACS) in Medicine. Edited by H. K. Huang et al.

TELEMED project

O. Ratib

Hospital Information Center, Geneva University State Hospital, 1211 Geneva 4, Switzerland

Brief summary: The TELEMED project is an application Pilot (R 1086) of the RACE (Research and development on Advanced Communications) program of the European Community (CEC) and is partially funded by the Commission. It consists of 15 partners in 8 European Community countries and 4 partners from 3 EFTA countries.

The overall objective of this project is to examine the problems which arise when participating physicians communicate over IBCN. The features and retirements specified above shall be analyzed in particular, so that an overall Cooperative Diagnosis and Therapy can be realized as application pilot. In addition, the consequences of the corresponding organization and operation of these services on the clinic work must be examined.

In this connection the following problems are dealt with as individual objectives:

- For the storage of images (x-ray photographs etc.) appropriate methods and data compression techniques are being developed, allowing the transfer of bulk data to physical storage media.
- For the storage of records uniform data structures are being selected or rather defined in order to permit multi-media applications.
- For the access to the records suitable retrieval techniques and a query language for medical applications are being examined.
- High-level protocols are being analyzed in order to allow access in an "open system", i.e. the retrieval of data records stored in different data bases by means of access methods and various manufactured display-consoles will become possible, (independently from the manufacturer)
- Transmission protocols are being selected and developed
- Data base architectures are being selected allowing access times necessary for an application as e.g. image transfers, (frequency of picture change 3 seconds)
- High-speed transmission of real time video in real-life applications are being defined and examined.

The work undertaken in this project will comprise also the problems concerning the special needs for security, accuracy and integrity of the record data. Here the objective "appropriate techniques and management methods" has to be proposed, which may be the first step towards an integrity of patient data.

The broadband communication (BC) support for medical record transmission, management and control developed by this pilot research project enables clinical and research staff to cooperatively work together on diagnosis and therapy methods and techniques, using multimedia workstations.

TELEMED brings together medical staff, telecommunications professionals and computer scientists in a multi-disciplinary team. This team has identified the user requirements and fixed the details of four application scenarios. The three scenarios "Remote Expert Consultation", "Radiological Image Reference Database", and

NATO ASI Series, Vol. F 74
Picture Archiving and Communication Systems (PACS) in Medicine. Edited by H. K. Huang et al.

"Transfer of patient Data and Images" deal with digital image communications, image databases, medical image workstations including their human interfaces, and a common medical image file format based on ACRNEMA for all TELEMED applications.

Picture Archiving and Communication Systems (PACS), workstations, and video conferencing equipment located in remote sites (Berlin, Heidelberg, Tromsoe, Lund, Geneva, Montpellier, Florence and Barcelona) will be connected by broadband connections (2 and 140 Mbit/s). Physicians will use the interactive link for the transmission of new medical techniques, for remote expert consultation, and for the transfer of digital medical images. The first trials with video conferencing and with PACS-to-PACS communication have been carried out 1989 in Berlin as well as between Tromsoe and Lund.

A radiological Image Reference Database (IRDB) for use in teaching and in care applications is under development in Montpellier and in Florence. Medical images from modalities and from PACS systems will be stored in the IRDB, and physicians in Barcelona, Montpellier, Florence, Geneva, Heidelberg and Berlin will have access to it by 2 Mbits links using a retrieval language close to the current medical natural language. An off-line model of the graphical interface and of the access language has been supplied to physicians to assess its functionality.

Digital cardiac images (stills and films), biochemistry results and patient data will be transferred between Leeds, Rotterdam, Berlin, and Patras by broadband links for remote image processing and expert consultation. Research on the unification of images generated by different modalities has been completed, and implementation has begun in order to display images from different medical modalities on one workstation.

The pilots extend across national boundaries, initially using the available communication facilities and developing towards broadband communications in cooperation with related PTT's and EBIT. The completed analysis of local area networks in the application sites, of trunk networks in the involved countries and of available satellite links support TELEMED in the planning of the required network and in the specification of the network access units.

At the current stage several work groups have already achieved several of the proposed objectives. The complete specifications for remote expert consultation and video conference have been reported. The need for digital transmission of the images between different partners has lead to the development of standards in image format and image communication. Other developments in the standardization of a dictionary of descriptive and anatomical nomenclature for image description are also underway.

A standard image file format (called the PAPYRUS format) [1] based on an extension of the ACR-NEMA image communication standard [2], has been developed. These development was undertaken at the Geneva University Hospital with the collaboration of several academic and industrial partners of the TELEMED project.

Specifications of the needs and requirements of the image manipulation and management software are also underway. These specifications should define a standard user interface with a set of image manipulation tools that are required for remote image consultation.

References:

[1] Ratib O,: PAPYRUS Image File Format, version 2.2. Int. Rep. HCUG-CIH, Reg. 900016, 1990.
[2] Digital Imaging and Communications. ACR/NEMA standard publication No. 300-1988.

ISCAMI: a European Collaboration on PACS

A. Todd-Pokropek

University College London, Gower Street, London WC1E 6BT, UK

Summary: Under the auspices of COMAC-BME, a joint European collaboration on PACS has been initiated in 1990 under the title Integrated Systems for Computer Assisted Management and Manipulation of Medical Images or ISCAMI for short. This collaboration has participants from most of the (extended) EC countries, including the COST countries, and is due to run for two years. The main purpose of the project is to enable participants to communicate and exchange ideas, rather than funding research as such. However, four specific aims have been defined.

The first working group is intended to define the clinical requirements of a PACS system: what is it for, what response time is required in different clinical areas, what are the resolution and contrast requirements etc. It is NOT intended to define technical requirements (matrix size etc.) but rather to specify the clinical needs from which these technical requirements can be derived.

Working group 2 has the aim of trying to define and generate a series of test data sets, for use in evaluating PACS systems, and eventually in defining a sets of specifications for quality assurance of such systems. Clinical data, physical phantoms, and simulated phantoms are envisaged. These data sets would be circulated in digital form, and provide a test bed for the data exchange methods and protocols envisaged in PACS.

The third working group is intended to provide a set of recommendations about the installation of PACS systems and guidance for purchasers of such systems throughout their planned life (and replacement). This group is currently undertaking a survey of PACS users with a view to preparing a set of documents covering all practical aspects of purchasing, installing, running, maintaining, and replacing a PACS system.

Finally the fourth group is concerned with communications and multi-modality image matching. For demonstrating the communications facilities required in PACS, the Berkom system will be used. Data format and protocol specifications are being assessed, in particular with respect to acquiring data in 3-D form different image modalities such as CT, MR, PET etc. Using this data, various image matching methods are to be assessed.

The project as a whole has organized two scientific meetings and is planning two more in 1991 all of which are to be published. Links between ISCAMI and other related organizations (EuroPACS etc.) are being encouraged. A regular newsletter is also published.

NATO ASI Series, Vol. F 74
Picture Archiving and Communication Systems (PACS)
in Medicine. Edited by H. K. Huang et al.

PACS-IMACS: Operation evaluation and basic requirements for prospective evolution of PACS technology

R. Passariello[1], C. Politi[2], D. Caramella[3], P. Giribona[4], D. Baleriaux[5], P. Hernando[6], J. Carreno[7], L. Micheli[8], J. de Certaines[9], C Florin[10]

[1]L'Aquila University, [2]Philips Sestemi Medicali, [3]Florence University, [4]Area Ricerca Trieste, [5]ULB Brussels, [6]Telefonica Sistemas, [7]Barcelona University, [8]Italsiel, [9]Rennes University, [10]Hewlett-Packard.

Abstract: The aim of the PACS-IMACS project was to evaluate the results of the introduction of PACS (Picture Archiving and Communication System) in the hospital environment and to guide its evolution towards an IMACS concept (Image Management and Communication Systems).

Nowadays PACS technology is available on the market and our first objective was to evaluate and assess the currently available installations in the relevant pilot sites.

A characterization study has been carried out to clearly identify users and needs. The main aspects related to the integration of PACS with the HIS (Hospital Information System) have been considered to allow a bidirectional data exchange between PACS and HIS.

Particular emphasis has been devoted to diagnostic image presentation in order to verify its clinical validity and to define a common technical baseline. This experience has taken place in the European PACS scenario defined by the partners.

Experimental activity on sophisticated image presentation, processing and communication platform has been performed to explore new diagnostic possibilities.

Results and experiences gathered during the project will be used to set up a knowledge base to enhance and to rationally manage the technology required by the implementation of IMAC systems possibly interconnected via metropolitan area networks.

Introduction

The current Radiology system is undergoing a fast evolution phase primarily concerned with information exchange and diffusion between the vast range of clinical Divisions referring to Radiology and the Radiological Department itself.

Therefore the state of the art of PACS technology cannot be considered as definitive; the advantages coming from the adoption of PACS technology would be probably better assessed and exploited with the development of an integrated IMACS. The concept of an integrated IMACS should be not dependent on any peculiar operating field to ensure validity and continuity. The basis to obtain such a result is to define a model after having deeply assessed all the benefits and critical issues in a number of relevant clinical setting as depicted in the deliverables of this project.

NATO ASI Series, Vol. F 74
Picture Archiving and Communication Systems (PACS) in Medicine. Edited by H. K. Huang et al.

1.0 Organization of the project

The main idea behind the PACS-IMACS project was to exploit the experience of a number of PACS European working groups in order to evaluate PACS IMACS evolution harmonizing it to the medical needs and requirements.

Three action lines have been followed:
- to define a composite European Scenario
- to establish relationship with other AIM Consortia
- to participate to the initiatives concerning this topic with particular attention to the standards.

The following criteria have been declared for the identification of the European Scenario:
- To ensure sufficient commitment from the sites to participate in a survey: at least a contact person was appointed per each site.
- to work on a technical infrastructure adequate for the evaluation of different suppliers and modalities.
- to provide a representative sample in Europe covering several countries.
- to build on existing research programs at each site, covering subjects as those addressed by our work packages.

The outcome of the project and the activities above mentioned will contribute to the formulation of a strategic planning for the new PACS generation and IMACS developments.

1.1 Composition of the Consortium

The complexity of the Radiological environment and of the PACS-IMACS concept itself determined the choice of heterogeneous partners. The PACS-IMACS Consortium composition offers backgrounds and competencies and therefore "VIEWS" from the following branches:
- Clinical Radiology Experts
- Professional Evaluators of Information systems
- Bio-medical Experts
- Sociological Experts
- Economists

This pluridisciplinary approach has often brought to discrepancies during the development of the activities; nevertheless this "Horizontal" approach should be maintained in the European concerted action that should take place to steer and coordinate appropriately the evolution of such a complex scenario. The presence of professional evaluators has clearly contributed to ensure a realistic and non-passionnel approach to the evaluation.

1.2 External Contacts

The PACS-IMACS Consortium has given significant importance to establish contacts with other consortia in order to facilitate exchange of information, results and to set up the most appropriate bases for preparing a deeper activity in the next phase of the AIM program.

Relationship with AIM HIPACS:
The cooperation with HIPACS in approaching a crucial item such as the coupling concerning PACS with an HIS/RIS has been outlined in a common deliverable titled: A REVIEW OF STANDARDS FOR PACS, HIS AND RIS COMMUNICATIONS.

Relationship with AIM TELEMEDICINE:
The main goal was to assess the feasibility of transmitting radiological and medical diagnostic images on the communication infrastructure generally suitable for TELEMEDICINE.

Relationship with RACE TELEMED
The investigation areas of the RACE project that appear to be particularly interesting in the scope of IMACS are:
- database of a radiological reference images
- use of video conference in medicine
- remote expert consultation

Relationship with TELEMED (Italian local Consortium)
The TELEMED Consortium is deeply involved in promoting TELEMEDICINE in South of Italy where the nature of the Country requires the implementation of a reliable infrastructure to facilitate patient remote monitoring, tele consultation in emergency cases etc.

2.0 OBJECTIVES

The main objective of the PACS-IMACS project is to help PACS evolution process by evaluating PACS technology on the basis of operational prototypes and by defining an effective strategy for PACS management and strategic planning.

The project's objectives can be summarized as follows:
* Identification of a European PACS Users Scenario (Fig. 1)
* Assessment of the impact of PACS Technology in a medical imaging Department:
- clinical aspects
- technical aspects
- organizational aspects
- economical aspects
- psychological aspects
* Identification and evaluation of new techniques for image processing
* Definition of the requirements for integration with RIS/HIS
* Identification of indicators and evaluation criteria for PACS management and strategic planning

3.0 ACHIEVED RESULTS

The results of the project are exhaustively reported in the deliverables (public domain) here listed.

The deliverables' numbers 1020, ...1100 are reported in order to facilitate their request through the AIM Office.

1020: An Overview of Picture Archiving and Communication systems Pilot Sites specification for a European PACS IMACS Survey

1030: Taxonomy of Users, Equipments and Services

1040: PACS Technology Assessment in the Pilot Sites
- Institutional conditions for PACS development
- Work organization and image distribution
- Technical aspects
- Cost-effectiveness analysis

1050: Clinical Evaluation of current Technology
- Image quality evaluation
- Data compression techniques

1060: Evaluation of Advanced Technology effectiveness
- Assessment of advanced image processing techniques
- Development of a noise reduction algorithm

1070: Basic requirements for PACS and RIS/HIS integration
- High level RIS and HIS description
- Identification of Macrofunctions and ViewPoints relevant for the PACS environment
- Detailed models of the selected Macrofunctions of RIS/HIS and of their PACS interface
- Detailed specifications of Software functions
- Data definition for HIS/RIS and PACS interface
- External interface specifications

1100: PACS Technology management and strategic planning
- Approaches for a PACS evaluation model
- Key issues for PACS evolution in Europe

4.0 CONCLUSIONS: the CONCEPT of PACS 'revisited

The complexity of the subject and the difficulty encountered in finding more than few PAC systems routinely operating for the achievement of a filmless radiological department has not allowed to examine in depth some of the topics targeted by the project, nevertheless the opinions of the partners of PACS-IMACS can be considered as a feedback from the field in order to contribute to the evolution of the new generation of PACS.

4.1 PACS: Three axes model

"Volume" (amount of data), "Space" (networking domain) and "Research" (i.e. 3D imaging development) are the three functionalities used as component indicators to determine the position on the three axes of any PACS project.

This model shows that the complete functionality along the three axes at the same time will not be possible on a large scale before a few years. Therefore the cooperation of clinical users, manufacturers, researchers and other actors in the Health Care Environment (HCE) will drive the PACS evolution towards the most appropriate point defined by the three-dimensional model.

This representation is useful to develop PACS model, to analyze the evolution of the concept of PACS or the variations of a project during its phases and to determine the strategy of any PACS evolution.

4.2 Organization in the HCE

Looking at the informative, organizational, economical and technological challenges that we are facing it is relevant to point out that a model of management and organization for selection, archiving and communications of images has to be defined. After the definition of the model, an IMACS will be designed and eventually implemented in order to resemble the model itself.

4.3. Open PAC Systems

The openness of PAC systems allow to design and implement systems offering the best solution for the individual users exploiting the contributions from any scientific, technical and commercial source.

At present every PAC system examined is capable of running on standard networks (i.e. ETHERNET network or segments) or to provide gateways to standard networks.

A second step towards the achievement of an open system is the capability provided to an external component (or subsystem) to be embedded in a 'HOST system environment' via EMULATION of the hosting environment.

Both of the above mentioned approaches do not guarantee the achievement of an open system. Indeed the OPEN system will result by the design of an OPEN environment; in other words rules to which the 'subsystems' should comply with have to be established and accepted.

These 'rules' will not only include communication protocols, record layouts, image formats etc., but mainly the definition of the FUNCTIONS of the subsystems which are supposed to build up an IMACS.

4.4 PACS + RIS

The experience carried out outlined the necessity of connecting and integrating a PACS with a RIS as a milestone towards the achievement of a PACS fully operating in routine.

To emphasize this concept it is worthwhile mentioning the experience in Trieste where a self-made RIS has been connected to the PACS.

4.5 Clustered PACS

The proliferation of imaging systems and the diffusion of the PACS concept have affected also medical environments outside Radiology like Neurosurgery (RENNES experience), Cardiology, Radiotherapy, Nuclear Medicine.

This will contribute to work for the design of an IMACS.

4.6 User Interface

The difficulties encountered in accepting the PACS 'as it is' arose from the recent introduction of PACS and from the user interface which is at present not so friendly as the users require.

Therefore emphasis has been stressed on window based interface and on sophisticated voice recognition systems capable of understanding either the commands given during the reporting session for manipulating the images either the words dictated concerning the report itself.

To address this topic is fundamental for the satisfaction of the PACS users and for the efficiency brought in the routine operations.

AIM: PACS - IMACS

FIG. 1

SUMMARY OF PILOT SITES

Pilot Sites	PACS	Modalities	Date	Storage (Gb)	Research
L'Aquila	Prototype	MR/CT/US/DSA/DF	1988	132	AIM PACS,AEMI
Barcelona	Philips	CR	1989	2.4	RACE Telemed,AIM PACS
Bologna	Philips	MR/CT/US/CR	1989	10.4	Regional
Bruxelles/VUB	Prototype	MR/CT/US/DSA	1987	7.4	AIM HIPACS
Bruxelles/ULB	Philips	MR	1989	2.9	AIM PACS
Florence	Teleradiology	MR	1990	na	AIM PACS,RACE Telemed
Montpellier	Siemens	CT/DSA	1988	2.4	RACE Telemed
Trieste	Philips	MR/CT/US/DSA	1988	178	AIM PACS, Regional
Utrecht	Prototype	Scanner	1987	na	National

FIG. 2

THREE AXES MODEL

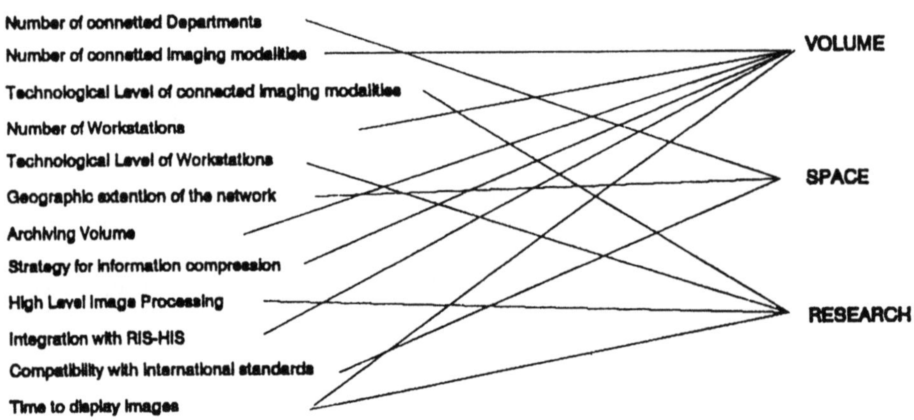

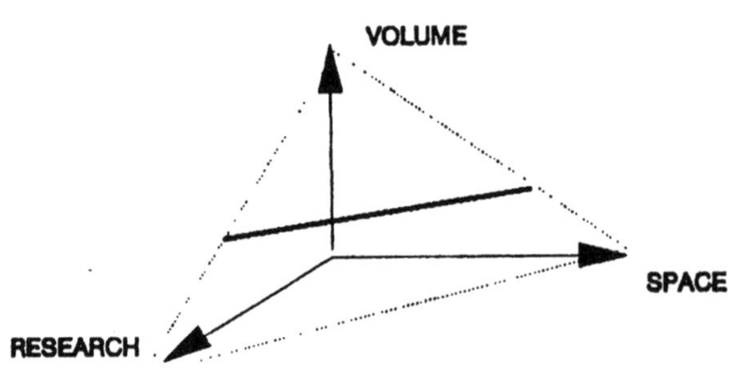

Present Position of most PACS Projects on the 3-Axes Model

COST-B2: Quality assurance of nuclear medicine software

A. Todd-Pokropek

University College London, Gower Street, London WC1E 6BT, UK

Summary: Under the COST umbrella a European collaboration on the quality assurance of nuclear medicine software has been undertaken by a group of about 18 European countries, being most of the EC countries themselves, COST countries such as Austria, Scandinavia, Switzerland, and is now being extended toward the Eastern European democracies.

The aim of this project, known for short as COST B2, is to provide tools to evaluate software provided in a number of clinical areas for the analysis of nuclear medicine data. There are a number of working parties, some of them related to specific types of clinical studies such as brain, heart, bone and kidney, the remainder being more general, for example, on data formats, using requirements, and communications.

A central idea has been the exchange of data in the form of software phantoms between different groups. Thus, for example, a series of validated cardiac studies would be collected and then redistributed to various participants with known (expected) results, to validate the software on that participants sytems. For this reason a common interchange file format was required. Since it was judged that ACR/NEMA was inappropriate, an alternative definition based in AAPM report 10 was adopted and has become the COST B2 Interfile format.

Working group 1 is responsible for maintaining this format, which has been published [1]. An attempt is now being made to harmonize this format with that of ACR/NEMA. Translators have already been written between Interfile and most of the commercially available systems, and data has been distributed in this form in a number of clinical areas. In particular, as part of the SPECT group's effort, a program for simulating a Jaszczak phantom was written and distributed which generated the test images in Interfile format. The definitions in Interfile were sorted out in a very short time primarily since this was carried out by dialogue throughout the world on electronic mail over EARN/Bitnet and Internet. This is perceived to be one of the strengths of the COST B2 project.

With respect to the clinical areas, the main task has been to define the requirement for the test data sets themselves, what kind of studies acquired in what conditions, and how they can be validated. It must be stressed that the aim of the project has been to develop tools for testing software, but not as such to carry out the testing in general.

A further project has been in the development of documents to describe the user requirements and specification for system in use in nuclear medicine with a view to eventual standardization.

1. Cradduck TD, Bailey DL, Hutton BF, Deconinck F, Busemann-Sokole E, Bergmann H and Noelpp U, A standard protocol for the exhange of nuclear medicine image files. Nucl. Med. Commun. **10**: 703-713 (1989).

NATO ASI Series, Vol. F 74
Picture Archiving and Communication Systems (PACS) in Medicine. Edited by H. K. Huang et al.

CHAMPOLION: an Image File Format Translator

Ron D. Appel, Osman Ratib, Jean-Raoul Scherrer

Digital Imaging Unit, Center of Medical Informatics, University Hospital of Geneva
1211 Geneva-4, Switzerland

Abstract

The University Hospital of Geneva is developing a PACS which will be able to handle medical images from different sources. All images are stored in one unique file format, the Papyrus image file format. We are designing CHAMPOLION, an image file format description language to describe the formats used by the various scanning devices employed in the Hospital. A compiler will read the description files and produce read/write routines in the C programming language for storage and retrieval of the medical images.

Keywords: PACS, Image File Format, Digital Imaging, Compiler

Introduction

The University Hospital of Geneva is developing a hospital-wide Picture Archiving and Communication System (PACS) which will be part of the existing Hospital Information System (HIS). The PACS's architecture is built using widely accepted industry and academic standards. All computers will run under the Unix™ operating system; the programs will be written in the C and C++ programming languages; the user interfaces will be built according to the XWindows standard, using OSF/Motif™ look and feel [1].

The Geneva PACS will be built around a network of Unix™ graphic workstations and of heterogenous multi-vendor image acquisition devices (CT, MRI, x-ray,...). Each of these devices produces image files in a different format. In order to standardise the PACS at the storage level, and to simplify subsequent retrieval of the images on the graphic workstations, we have decided to convert all images to one unique file format before storage. For this purpose we designed the Papyrus image file format [2], based on the ACR-NEMA specifications on medical image communication [3]. It is therefore a tagged file format, containing a header with variable length fields preceded by a label. One Papyrus file may contain multiple images using the folder concept introduced by Siemens and Philips in the SPI format. The Papyrus format will evolve in the future, when more precise recommendations will be published by the ACR-NEMA for medical image file formats.

Problems associated with a Unique File Format

For each scanning device a different set of read/write routines is needed to convert images from the device's format to a Papyrus file. Many such routines will have to be implemented to handle the images from all scanning sources. A large library of conversion programs will have to be written, maintained and documented. For each update of a source file format, the corresponding conversion routine will have to be modified. More importantly, each time the Papyrus format is updated, all the programs will have to be corrected.

NATO ASI Series, Vol. F 74
Picture Archiving and Communication Systems (PACS)
in Medicine. Edited by H. K. Huang et al.

One possible solution is to implement a unique conversion program for all image sources. Nevertheless this would result in a very large program and the maintenance problems would quickly become uncontrollable.

CHAMPOLION: an Image File Format Translator

We have chosen to design a description language for medical image file formats [4]. The language will allow us to easily describe any foreign image file formats. A compiler for the Champolion language will then be implemented, which will produce automatically read/write routines to convert foreign image files to and from the Papyrus format (Figure 1). An ASCII description file will be written with the Champolion language for each image file format, including the Papyrus format. A modification of one of the file formats will easily be carried to its description file. The Champolion compiler will then produce a new version of the read/write routines for that format. In case of an update of the Papyrus format, only its description file will have to be modified. The rest of the work will be performed by the compiler.

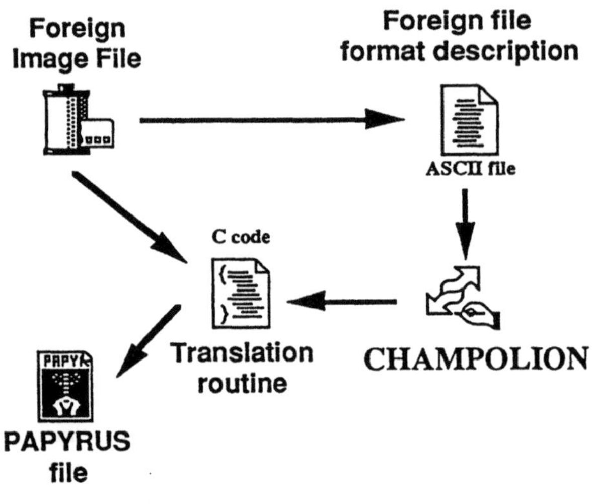

Fig. 1: The Champolion foreign file format description language and compiler

The issues in designing the language are bound to its specificity. All existing image sources handled by the PACS have to be taken into consideration. Future image formats will have to be described with the Champolion language, as well as format updates. Also, the compiler developed for this language has to produce efficient read/write C routines.

The main challenge of the creation of the Champolion language is the balance between its specificity and its ease of use. The formats used in the Hospital are very different from one another. Some are very simple fixed field formats, while some more complex image formats have variable fields, with or without labels. The Papyrus image file format is one of the most complex ones, because of its definition of folders to allow storage of more than one image in a unique file. Furthermore, the folders may be stored in one or several files. In some other formats, a field may sometimes contain different types of values, depending on the content of some other fields. In the TIFF format, for example, a field contains either a pointer to its value, or the value itself, if it is small enough to fit in the field. All these variations must be describable by Champolion. Also, exceptions like the one of the TIFF format have to be described easily. Despite all these peculiarities, the description of an image file format with the Champolion language has to remain simple. If describing a format is more difficult than writing read/write routines in the C (or any other) programming language, then Champolion misses its goal. This is the main and most challenging problem of this project.

We are currently writing the specifications for the Champolion language, taking into consideration as many foreign image file formats as are currently available. A first image file format description language (which is a subset of Champolion) has already been designed and its compiler implemented. It has been successfully tested with images stored in simple formats.

Conclusion

We are developing a PACS which will be able to handle medical images from different scanning sources. The images are stored in one unique file format, the Papyrus format. We are designing CHAMPOLION, an image file format description language to describe the formats used by the various scanning devices employed in the Hospital. A compiler will read the description files and produce read/write routines in the C programming language for storage and retrieval of the medical images. The main difficulty in creating the language is the requirement to design a language which is both precise and easy to use. A prototype compiler has already been implemented for a subset of the Champolion language. It has been successfully used for simple image file formats.

References

1. Ratib O., Hochstrasser D., Scherrer J.R.: PACS and Related Research in Switzerland. Proceedings of NATO ASI meeting on "Picture Archiving and Communication System (PACS) in Medicine". Evian, 1990.

2. Ratib O.: PAPYRUS, Image File Format, version 2.2. Int. Rep. HCUG-CIH Reg. 900016, 1990.

3. ACR-NEMA standard publication #300-1988.

4. Appel R., Ratib O., Scherrer J.R.: The Use of a File Format Description Language for easy Handling of Images from Multiple Sources. Proc. EuroPACS meeting, Triest, 1990.

A NUCLEAR MEDICINE PACS AND RIS

F. Bitter, B.A. Bitter, R. Weller, L. Almasi*, E. Mate* and W.E. Adam

University of Ulm, Nuclear Medicine, Ulm, FRG
*Dept. of Nuclear Medicine, University of Szeged, Hungary.

Keywords: PACS, RIS, NUCLEAR MEDICINE

Introduction

Currently the images of nuclear medicine institutions are stored on magnetic media. The documentation is done with hardcopies. Officially magnetic media are not accepted for documentation purpose and hardcopies do not allow further processing. The access to the original data is time consuming and personal intensive. There is a demand to store at least the original information cheaply, safely and direct accessibly. The appearence of the optical disks together with a jukebox facilitates digital archives at least for nuclear medicine. As a prerequisit it is necessary to connect all imaging devices through its data systems within a nuclear medicine institution to a local area net (LAN) forming a PACS (Picture Archiving and Communication System) Data compatibility must be fulfilled as well as a positive patient identification. Quality controls must provide for data security and reliability. An effective retrieval system must serve for fast access to the archives. PACS and RIS (Radiological Information System) must built a unified system. Such a local net in addition should be connected to the HIS (Hospital Information System) in order to reduce the administration work in nuclear medicine. A prototype of a nuclear medicine system is in use at the University of Ulm since 1988, wich will be hooked to the HIS in 1990. Since PACS currently is developed mainly for the other modalities (CT, MR, DSA, DR) through the industry, nuclear medicine should participate in this development with its particular needs.

Current State

The manifold of nuclear medicine instruments nowaday is reduced essentially to the versatile gamma camera. The spectrum of invstigations spans static scintigramms as well as the most complicated methods as the gated SPECT [3] studies. Indispensible hereby is

NATO ASI Series, Vol. F 74
Picture Archiving and Communication Systems (PACS)
in Medicine. Edited by H. K. Huang et al.

the nuclear medicine data system, which still is utilized almost only for image acquisition, processing and documentation. Usually these are single user systems eventually with foreground/background processing option, designed for real time image acquisition and processing. If several gamma cameras are installed within a department each one usually is served by one system leading very often to a heterogeneous installation.

Other tasks are managed either offline as e.g. image archiving with floppy disks or magnetic tape, or are manually fulfilled as the entire administrative work which is necessary partly because of organisational reaseons, partly beause of regulations as quality assurance and radiation protection and last not least for financing. All of these duties are time consuming and labor intensive.

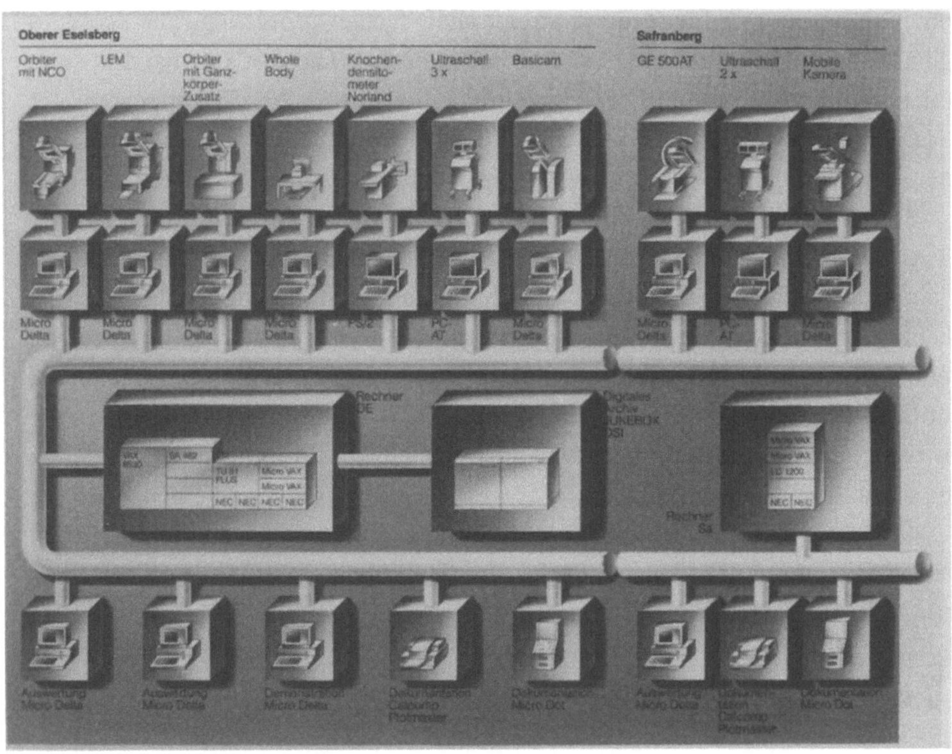

Fig 1. University of Ulm, Nuclear Medicine Departmental System.

Very often demographic data have to be written severel times. For radiological departments special Radiological Information Systems (RIS) have been developped as RADOS [5] or SIMEDOS [6] , but not for nuclear medicine.

Therefore, for our new clinic, which was set in operation in 1988 we have implemented a concept covering all the tasks mentioned above. Before our solution is introduced, we want to point out some problems.

Nuclear medicine systems which are used today are designed through the particular manufacturer who sometimes performes market analysis, but most often asks consultants for assistance due to missing standards. Hence very different clinical protocols are originated with heterogenous data formats. The data system from different manufacturers are not compatible and cannot be networked. Still the storage capacity is not sufficient and redundancy does not exist. The images are archived if at all offline onto tape for floppy, which is very time consuming and leads relatively often to data losses. Database systems usually are not provided for administrative data and results. If demographic patient data are provided through a Hospital Information System (HIS) but printed on paper, the data have to be input again into the nuclear medicine acquisition programs and errors of around 15% result. Therefore we must claim a consistent patient identification at least within the nuclear medicine department based on a DP-concept. Up to now the industry does not offer these systems.

Realization

Under such conditions we have realized our own system which is described here. It is based on the draftwork [1] performed in 1979 for the BMFT (Federal Ministry of Research and Technology FRG). A small solution has been installed in 1983 [2]. We had the chance to make our experiences and to build up knowhow.

Figure 1 now shows the complete configuration. Computers from DEC are used, a VAX 8530 and Maxdelta-systems from Siemens consisting of Micro VAX II and several Microdelta slave terminals serving the gamma cameras, for image acquisition and processing, for reporting and programming. Ultrasound images are digitezed with PC's of the AT-class. All the original images are archived by means of optical WORM-disks (Write Once Read Many) and a jukebox with 16 disks each one of 2 times 1 GB capacity, 32 GB in total. The computer are networked and connected in a local cluster. Ethernet provides sufficient speed for nuclear medicine. Our department is subdivided into two parts (6 km distance). The communication is performed through a 9600 Baud line, which will be replaced in 1990 by a 2 MBaud line and further on by 100 MBaud or 144 MBaud providing also fast image communication. Within this configuration data are administered by means of the relational data base management systems ORACLE. Office communication and word processing is provided through ALL-IN-1 from DEC. Statistical calculations are performed with help of SAS. The documentation of nuclear medicine studies is centralized with color printers from Calcomp and HP. Color printing is standard in nuclear medicine for image documentation. Each image generating and processing system is able to send the screen content to the printer queue. Matrix and laser printers facilitate text output.

Already 2 years of experience with the entire system demonstrate the advantages. Since 1984 demographic data of more than 50 000 patients are stored together with data from more than 127 000 studies. The digital image archive contains at the moment (July 1990) 28 000 studies including around 11 000 radionuclide ventriculogramms (RNV). With these heart studies we made our experiences concerning the use of the departmental system. From 9000 RNV-studies 250 parameters each are stored in addition.

Shortly one application [4] is described: For a clinical study a physiscian requested patients with left bundular branch block (LBBB). The diagnosis was not stored, but the phase of the base frequency of the left and right ventricle. A simple query retrieved 41 patients within few minutes with 95% hits. Lowering the phase difference from 36 to 18 degrees resulted in 257 patients with 40% LBBB. The patients could be indentified and their studies could be used to answer the clinical question. Hardly one can estimate the expense for searching these patients in conventional file archives.

Structure

Schematicaly figure 2 demonstrates the structure of our system. Using the data flow through the departement the relations are described as follows:

Patient data are input at the adminission of the department using forms provided by alphanumeric terminals. The patient is scheduled, labels are printed and the image data of previous studies may be retrieved from the digital archives. The requested radioactive material is ordered and the delivery is registered in the data base. At the day of investigation lists are generated automatically for the working places, printed for security purpose and transferred to the Microdelta's as menues. The patient identification automatically is copied to the acquisition programs and stored within the images on magnetic disk. At night the new studies are stored in the digital archives. The images are kept on magnetic disk for two weeks and then deleted. Procesing and patient reporting can be performed with each of the data systems and results can be archived automatically.

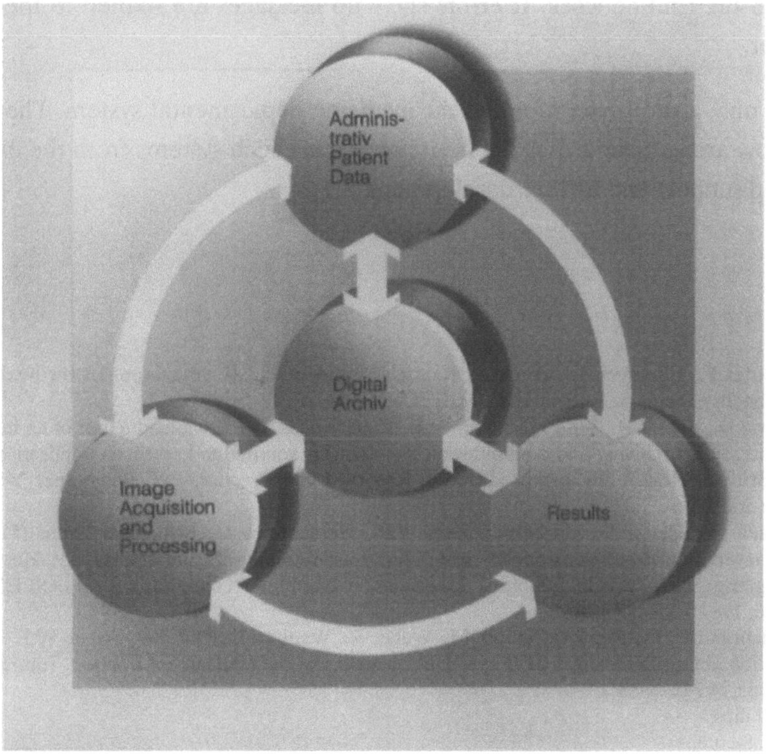

Fig 2. Departmental System: Structure.

Reports are necessary for the acquisition of radioactive material, laboratory statistics and so on and may be generated automatically at the requested date in order to avoid completely manual writing. The task of the staff is directed to control and to correct errors. Patient reports are prepared by means of word processing within the office communication. The administrative data allow a patient oriented access to the studies and the stored results provide access through study parameters as in the example of the LBBB patients. Patient demographic data and in general all administrative data as well as the results are maintained through the data base. The image data in the digital archives are controlled through the data base utilizing the patient identification and the physical adresses within the optical disks and therefore providing a fast access. The database is connected to the image acquisition and processing programs through files, which are used as header of the images. Besides the special image processing language MEDICL we still use FORTRAN and the high language interface to the database. A spreadsheet provides access to ORACLE and to SAS. Periodic tasks as backups, reports, archiving, deleting from magnetic disks and so on are performed automatically at night and do not disturb the routine work. If errors show up messages are mailed to the responsable operator.

Our system is only a prototype of a nuclear medicine departmental system. The experiences up to now are sufficient. For the future we expect such systems from the industry with guaranted support and further development.

References:

[1] Adam W.E., Bitter F., Ellebruch H., Huck A., Kaatsch P. und Weller R.: Pflichtenheft für nuklearmedizinische Datensysteme, im Auftrag des BMFT, DVM 141, 1979

[2] Bitter F., Hellwig D., Weller R., Wild H., Adam W.E.: Experiences with the introduction of an image acquisition, processing, archiving and communication system for routine and research application, in Schmidt, Ell, Britton (Hrsg.), Nuclear Medicine in Research and Practice, F.K. Schattauer Verlag Stuttgart NewYork, 1986

[3] Graf G., Clausen M., Henze E., Bitter F., Adam W.E., Heidenreich P.: Schnellverfahren (Flash-SPECT) zur Gewinnung dreidimensionaler Scans der Fourieramplitude und -phase des Herzens mittels der getriggerten Single Photon Emissions Computer Tomographie (GASPECT), NucCompact 20, 1989

[4] Henze E., Hildebrand P., Hellwig D., Clausen M., Weller R., Wanjura D., Bitter F., Adam W.E.: Regional wall motion abnormalities in LBBB specified by means of the contraction fraction, Journal of Nuclear Medicine 14:241, 1988

[5] RADOS, Fa. Philips

[6] SIMEDOS, Fa. Siemens

Storage and retrieval of medical image data, relationship to image processing and analysis

Yves Bizais[1], Anne-Marie Forte[1], Bernard Gibaud[2] and Florent Aubry[3]

[1] Imagerie Médicale Multimodalité, HGRL, CHRN, 44035 Nantes, France
[2] Unité INSERM U335, CHR, 35000 Rennes, France
[3] Unité INSERM U66, IGR, 92000 Villejuif, France

Abstract : In this paper, we take the standpoint of physicians (radiologists, clinicians, surgeons) to define the required content of medical image data bases, from the needs for retrieving images. We conclude that two main approaches (minimal and generalized) solutions are medically acceptable.

Keywords : PACS, medical image data base, medical multimodality imaging

1. INTRODUCTION

For now ten years, extensive research activity has been devoted to PACS. Significant advances were achieved in standardizing image sources, defining communication protocols and building medical image workstations. However less attention was paid to medical image data bases (MIDBs), although it constitutes the main PACS component. For this reason, we focused our research effort on MIDBs. This paper reports our work and our first conclusions in this field.

We were primarily concerned with *the way information is produced and consumed* in medical imaging : what kind of information is involved ? what are the relationships between data ? who is allowed to use data ? For this purpose, we interviewed physicians resulting in the inventory of data described in section 2. From this inventory we defined two possible approaches to MIDBs presented in section 3. Our main (somewhat provocative) conclusion is that medically useful and cost effective PACS must make use of MIDBs not yet available.

2. INVENTORY OF DATA

The life of an imaging procedure extends from the time it is requested, until the time it is consumed by final users (the requiring physician or a surgeon in most cases). All data produced during this period of time are relevant data characterizing the procedure. The data inventory was aimed at collecting such a data set for several imaging procedures.

2.1 Methodology

In a first step, we defined a questionnaire used to collect data, made up of 6 main sections :

1) *Request* : requiring physician, diagnostic / interventional procedure, delay between request and realization, number of procedures performed yearly. They are mainly RIS items but are useful to understand *why* a procedure is performed, and thus *how* information is used later.
2) *Acquisition* : description of acquired series, useful to understand *why* such images are acquired, i.e. the relationships between acquisition and interpretation.
3) *Processing* : description of image processing, useful to understand *why* such processing is used, i.e. *how* processed data are consumed during interpretation.

NATO ASI Series, Vol. F 74
Picture Archiving and Communication Systems (PACS) in Medicine. Edited by H. K. Huang et al.

4) *Primary interpretation* : description of the data and of the relationships between data, used during primary interpretation by the radiologist. These items characterize information really useful during interpretation.
5) *Distribution* : the fraction of produced data sent to final users, means and urgency of distribution, useful to understand *how* data are used during secondary interpretation.
6) *Storage* : the fraction of produced data actually archived, means of storage and retrieval, useful to understand the medical goals of image retrieval and the rationale for storing data.

In a second step, we selected a set of about 30 representative (i.e. commonly performed) imaging procedures, involving main (CT, MRI, DSA, NM, interventional radiology) imaging modalities and major (brain, lung,heart, abdomen, bone) organs.

Finally we interviewed relevant radiologists for each selected procedure, in three University hospitals.

2.2 Main features characterizing an imaging procedure

There is no place here to describe the actual results of our inventory. In this section we summarize the features considered as important by interviewed physicians to characterize an imaging procedure :

1) From the very beginning of the procedure life (request), the *medical goal of a procedure* is the obtention of a diagnosis or the realization of a therapeutic act. Imaging procedures are defined in such a way by all interviewed physicians.
2) Acquired series are a means to obtain information to achieve this goal. To describe the latter, it is necessary to make relationships between series explicit. For example, it is quite different to know the view angles of two DSA series, and to know *why* these series were acquired with such view angles.
3) Image processing is used to extract information relevant to the procedure goal from the huge amount of data produced. It is therefore necessary to precisely characterize *how* and *why* processing is performed.
4) In almost all cases of digital procedures, image processing is performed. Most often very simple (windowing, zooming, measurements) processing techniques are used. Only in Nuclear Medicine complex processing techniques are commonly performed. However there is no relationship between the algorithmic complexity of a processing technique and the need for storing parameters used during processing (see below).
5) by far, the main medical reason to store images is to compare *sequential procedures* to evaluate the evolution of a pathology. It follows that if quantitative comparison is to be used, the same processing techniques using the same parameters must be run on the two procedures.

2.3 Consequences

To be medically useful, a MIDB must be able to store every piece of information used during the realization of a procedure, and during its consumption. If it is not the case, the medical interest of a PACS is significantly decreased. The above inventory precisely describes what kind of information has to be stored to achieve medical usefulness, for 30 main procedures. By far it exceeds commonly accepted requirements for MIDBs (1). In other words, even though the ACR/NEMA (2) standard is a communication protocol, it defines what kind of information must be transferred and consequently stored. We discuss in the next section to which extent the ACR/NEMA set is enough to fulfill this goal.

3. IMAGE DATABASES

We start this section by stating that a procedure is characterized (identified) by a medical goal. As long as image data are available to achieve this goal, physicians are likely to be satisfied with a PACS. It follows that two solutions can be now defined.

3.1 The minimal MIDB solution

If we agree on that *composite documents* including processed images, ROI, curves, figures and reports are enough to achieve the above defined medical goal, then it is enough to store such composite documents. Their *semantics* is *not included* in the data base, but is obvious to physicians or easily accessible from such documents. This is the way it actually works today with films : only a very limited set of image data associated to the report is used by requiring clinicians and surgeons.

In such a case, composite documents are built by the radiologist, using the image source console or a secondary console. And only the resulting documents are needed to be archived. It is fairly easy to build such an MIDB and the corresponding PACS, because image sources, archives and display workstations are logically disconnected. Such a PACS concept on which most prototypes are based, corresponds to an image extension of HIS.

However there are strong limitations to such PACS. Basically they only allow the replacement of film by digital means, but do not provide any new functionalities. We fully appreciate the advantages of this approach (speed, no loss of data), but it certainly explains the actual lack of interest of most physicians for it.

The ACR/NEMA standard provides the basis for such a solution (3), and even slightly more elaborate ones (including raw data).

3.2 The generalized image database

An opposite way of approaching the MIDB issue is *to store the semantics* of a procedure in the database along with the data. For this purpose, we need to define the goal of the procedure, how acquired and processed data contribute to this goal, and data themselves. In this case relationships between acquired series as well as between acquired and processed data are explicit. Whatever allowed processing is performed, its consequences on procedure interpretation is automatically clear for final users.

Obviously such a solution corresponds to a much more complex data base, and a much more complex PACS, since image sources, archives and workstations are not fully disconnected : archives must know how acquisition and processing are done in image sources and workstations, to make it clear to final users why it was done (in terms of medical goal).

However such a solution is needed for the PACS to support new functionalities such as quantitative comparison of procedures, and multimodality image processing. With such a MIDB and a PACS, physicians would have available a powerful tool to better use image information, and would be more willing to use PACS. For this reason, we started specifying an object-oriented model for medical image data, which is the basis for such a MIDB.

4. CONCLUSION

Our inventory of medical imaging procedure clearly demonstrates that is is not enough to know *how* data are acquired. It is necessary to know *how and why* data are processed, i.e. to specify the *medical goal* of the procedure. It would be useful to check the validity and the generality of our approach by interviewing more radiologists, clinicians and surgeons in other hospitals and for other imaging procedures.

A simple MIDB solution fulfilling basic medical needs consists in storing only *composite documents* resulting from image processing and interpretation (image extension to HIS). But we are strongly in favor of a complex MIDB including the *semantics* of procedures and image data, in order to provide *new functionalities* justifying PACS form a medical and an economical standpoint.

(1) M.P. Zeleznik, G.Q. Maguire, B. Baxter : *Pacs data base design*, SPIE PACS II, 418, pp 287-295 (1983)

(2) *Digital Imaging and Communications* ACR-NEMA publication, No 300-1985, NEMA, Washington D.C.

(3) D.M. Budler, C. Lee Hamilton : *A prototype relational implementation of the NEMA schema*, SPIE Medical Imaging II, 914, pp 1066-1068 (1988)

Subband Coding of DSA Images

A. Enis Çetin

Department of Electrical and Electronics Engineering, Bilkent University, Ankara, 06533, Turkey

Abstract: In this paper we present a non-rectangular subband coding method for Digital Subtraction Angiography (DSA) images. The non-rectangular subband decomposition is a multiresolution image representation scheme. The subimages resulting from this decomposition are coded using a Discrete Cosine Transform (DCT) based image compression algorithm, and pulse code modulation with runlength coding of zeros.

Keywords: Compression of DSA Images, Subband Image Coding, DCT

Introduction

We describe a medical image coding scheme based on the decomposition of the image into components whose frequency domain support ideally consists of nonoverlapping regions. Recently a variety of methods have been proposed based on such a decomposition and are described as subband image coding techniques [1]. In [2] a survey of subband coding based medical image compression techniques is given. We also proposed a a subband image coding method, non-rectangular subband coding [3], for images of natural scenes. In this method the frequency domain is divided by filters whose passbands are diamond shaped regions.

In this paper we present a coding method for DSA images. This image compression method is a variant of the method described in [3].

Description of Procedure and Simulation Examples

We perform a multiresolution decomposition of a given DSA image by using the filter bank structure of [3]. In this way we obtain subimages at different resolutions. The resultant subimages are coded according to their frequency contents.

The DSA image is applied to a filter bank consisting of two filters. One of these filters is a lowpass filter with a diamond shaped passband. The other filter is a high pass filter with a diamond shaped stopband. The outputs of these filters are downsampled by a factor of two on a line quincunx grid. In this way two subimages are obtained. The subimage coming from the lowpass filter is decomposed and downsampled once again by the same filterbank. In effect the procedure splits the image information into three regions in the frequency domain, ideally as shown in Figure 1. The information in the

NATO ASI Series, Vol. F 74
Picture Archiving and Communication Systems (PACS) in Medicine. Edited by H. K. Huang et al.

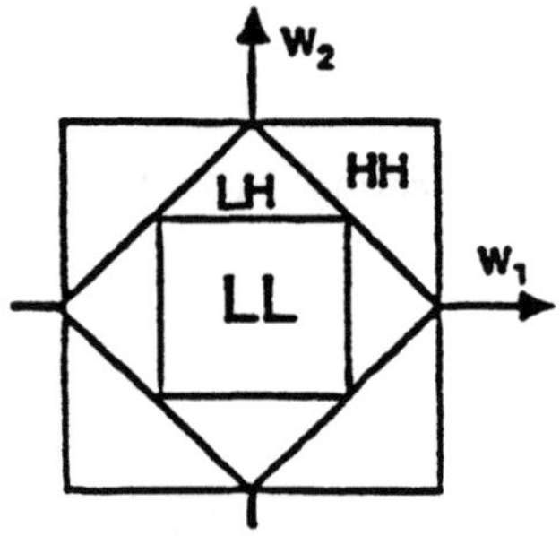

Figure 1: The ideal nonrectangular subband partition used in the coding scheme

high frequency band (HH in Figure 1) is quantized using a deadzone quantizer and the resulting data is coded using runlength coding of zero-valued samples and variable length coding of the nonzero samples. A similar procedure with a different deadzone quantizer is used to code the information in the bandpass region (LH in Figure 1). The lowpass image (LL in Figure 1) is represented by quantizing the coefficients of the Discrete Cosine Transform (DCT) of varying sizes of the data. A variant of the DCT coding scheme of [4] is used. DCT coefficients are uniformly quantized and variable length coded, and zero-valued quantized transform domain coefficients are runlength coded.

In order to perform the multiresolution subband decomposition the following IIR filters described in [3] is used:

$$H_i((\omega_1,\omega_2)) = \frac{1}{2}[1 + e^{j\omega_1}T(\omega_1+\omega_2)T(\omega_1-\omega_2)], \quad i = 0,1 \tag{1}$$

where

$$T(\omega) = \frac{1/3e^{j\omega}+1}{1/3+e^{j\omega}} \tag{2}$$

is an all-pass section. The filter with subscript $i = 0$ ($i = 1$) is the lowpass (highpass) filter. In order to reconstruct the coded image from the subimages a synthesis filter bank which provides exact reconstruction in the absence of quantization errors [3] is used.

The coding procedure was applied to several DSA images. In Table 1 simulation results for an 8 bit/pixel DSA image of a hand of size 512x512 are shown.

Conclusion

In this paper we present a non-rectangular subband coding method for Digital Subtraction Angiography (DSA) images. The subimages resulting from this decomposition is coded using Discrete Cosine Transform (DCT) based image compression algorithm, and pulse code modulation with runlength coding of zeros. The resulting DSA image compression technique is a lossy method. It can be implemented in real-time, if custom designed

DCT Block Size (LL)	Quantization Level(DCT)	LL bit/pel	LH bit/pel	HH bit/pel	Total bit/pel	Compression Ratio	MSE (dB)
8x8	1	2.87	0.786	0.342	1.08	7.4	36.2
8x8	2	2.10	0.786	0.342	0.89	9.0	35.8
16x16	1	2.09	0.786	0.342	0.89	9.0	35.7
16x16	2	1.35	0.786	0.342	0.70	11.4	34.6
32x32	1	1.44	0.786	0.342	0.73	11.0	34.5

Table 1: Coding results for a DSA image. Compressed DSA image is visually indistinguishable from the original one.

VLSI chips are used to realize the digital filter bank which consists of filters with integer coefficients, and DCT.

In DSA screening physicians usually store a sequence of images. A three-dimensional (3-D) subband coding based compression method for DSA images can be developed. This idea will be investigated. By using a 3-D coding method we expect to get better compression ratios.

References

1. J. W. Woods and S. O'Neal, 'Subband coding of images,' IEEE Trans. Acoust. Speech and Signal Processing, vol. ASSP-34, Oct. 1986.

2. Otto Rompelman, 'Medical Image Compression: Possible Applications of Subband Coding,' in *Subband Image Coding* edited by J. W. Woods, Kluwer, 1990.

3. R. Ansari, A. E. Cetin, S. Lee,'Subband coding of images using nonrectangular filter banks,' *Proceedings of SPIE Conference on Applications of Image Processing XI*, San Diego, Aug. 1988.

4. W-H. Chen and W.K. Pratt, 'Scene Adaptive Coder,' *IEEE Trans. Comm.*, vol. 32, pp.225-232, March 1984.

PACS in Britain: A Radiologist's Perspective

Richard M. Dawood

Department of Radiology, St Mary's Hospital, London W2 1NY

Introduction

An important difference in approach between Britain and other countries has been the disproportionate attention devoted to a single issue: should total digital radiology departments be implemented in a single move?

The history of British interest in PACS is inextricably linked with work at St Mary's Hospital, London, which began in 1982 with a feasibility study aimed at achieving a totally digital radiology department by 1987. Tempering ambition with realism, the St Mary's project evolved into a series of smaller scale investigations from which important conclusions can nonetheless be drawn.

Background

St Mary's Hospital is a 1000-bed university hospital in central London, with a radiology department performing examinations on approximately 100,000 patients each year. The original hospital was constructed in 1854. The radiology department was a grim and gloomy place, scattered in a haphazard manner through a basement that pre-dated Röntgen's discovery by half a century. Despite this, it had been a recognised centre of excellence for at least since the post-war years; when the opportunity for redevelopment arrived, these surroundings added impetus to the motivation to make the new department as up-to-date as possible.

Plans were drawn up in 1979, with a new wing expected to open in 1987, housing a new department of diagnostic radiology conceived initially along conventional lines.

NATO ASI Series, Vol. F 74
Picture Archiving and Communication Systems (PACS) in Medicine. Edited by H. K. Huang et al.

Detailed consideration of the equipment to be installed began in 1982, just as awareness and interest in new imaging technology had reached its peak. The PACS concept was entirely new, and the arrival of all the technology necessary to implement what seemed initially to be a very simple concept, was perceived to be just around the corner.

This appeared to be an ideal opportunity to explore the introduction of PACS technology. A formal feasibility study was commissioned by the British Department of Health in 1982, and this reported in 1985 that the objectives were indeed considered just possible. The new St Mary's accordingly came to be planned as a totally filmless hospital, and the timing seemed perfect to implement the PACS ideal in its entirety. An all-digital hospital seemed eminently feasible, and this became the objective. There would of course be a few technical problems, just as in any new technology, but with such rapid progress on every other radiological frontier, it seemed that they would quickly be resolved. Linking all the equipment together, and linking the radiology department up to the rest of the hospital, appeared to be a relatively simple matter.

By 1985 it had become clear that the necessary technology was simply not available. Funding in excess of 9 million pounds had been promised, and had not been the problem; but no manufacturer could supply what was required, with the necessary guarantees that a clinical service could actually be provided in time for the opening date of November 1987.

A limited digital display system was however installed in the new department, and became the focal point of digital research and evaluation in the UK. The objective was to explore one of the weakest links in this emerging technology - the issue of diagnostic image quality in digital displays. From the perspective of the practising radiologist, we set out to evaluate commercially available display systems in terms of their clinical radiological diagnostic performance. The most important question, as far as diagnostic radiologists and their patients are concerned, is simply this: will we be able to make the correct diagnosis?

Equipment

Two different display systems were subjected to examination. Both were supplied, installed and maintained by Philips Medical Systems (UK). The equipment was purchased with funds from the Department of Health.

System 1

The first system, delivered in April 1988, consisted of "1K" and "2K" laser film digitizers (Konica, Japan) - (scanning spot approx. 100 microns, 8 bits per pixel); an optical disk archive with single 12 inch drive (1GByte per side); an image management system; and a display terminal consisting of a 4-monitor viewing station (Raytel Corporation, USA) [1024-line monitors, with band width corresponding to 48MHz pixel clock rate and 8-bit video coding, interlaced, vertical raster]. A proprietary blocked discrete cosine transform (DCT) algorithm was available on the system (approximate ratio, 15:1).

Difficulty adapting this equipment to handle 2K images led to serious problems with performance, and the equipment was replaced with a second system.

System 2

This second system was commissioned in January 1989 and consisted of two laser film digitizers (FD2000, Dupont, USA) (scanning spot size 210 microns, bit depth programmable by user to 8, 10, or 12, and set to 12 throughout duration of our investigation); an optical disk archive with single 12 inch disk drive (1Gbyte per side); 4 x 825 Mbyte magnetic disk buffer; and a 4-monitor display workstation (CommView, Philips Medical Systems/ AT&T) [1280-line monitors with band width corresponding to a 60MHz bandwidth and 12-bit video coding, non-interlaced, horizontal raster]. The system also incorporates a proprietary irreversible compression algorithm incorporating DCPM and Lempel Ziv encoding; this was applied at approximate ratios of 8:1.

Methods

A number of ROC experiments were conducted, comparing diagnostic performance of film and laser-digitized images. Four major series of images were used: hand images from patients with sub-periosteal resorption, chest images from patients with *Pneumocystis carinii* pneumonia, mammograms showing microcalcification, and skull radiographs of patients with vault fractures.

A total of more than 13,000 individual observations was recorded, and many of the experiments have been presented in greater detail elsewhere [1-4].

Results

The results can be broadly summarized as showing significant and substantial differences in diagnostic image quality between the displayed digitized images and original film radiographs. At a false positive rate of 15 per cent, the decrease in diagnostic accuracy ranged from five to twenty-two per cent. Significant differences were also observed when the proprietary compression algorithms were applied .

We also noted a three to four fold increase in the time taken for radiologists to perform identical observation tasks using the display system, compared with film radiographs. Workers in Italy have observed an increase of at least 30 per cent for systems displaying CT images alone, and have stated that an increase even of this size was sufficient to provide a powerful disincentive to daily use [6].

Formal experiments have been conducted in other related aspects of display systems technology, including quality assurance, performance testing, and evaluation methodology.

Discussion

The work at St Mary's includes one of the largest formal assessment of image quality from digital displays. One of the most important conclusions is that 1000- and 1280-line display systems are unsuitable for primary diagnosis of subtle lesions, a conclusion that is in accord with the work of many others (eg Slasky et al [5], who studied digital displays and digital hard copy at higher resolution than feasible on our system).

This leaves many issues unresolved. Neither of the two manufacturers currently aspiring to supply total systems actually yet offers of supports monitors of higher specification as part of an integrated system. There is now a broad assumption that monitors with at least 2000-lines will be necessary, but these in turn will still require extensive evaluation, and we cannot yet assume that they will be satisfactory. No clear indication of their likely cost has yet appeared, and most cost analyses for total systems have been based on the use of 1000-line monitors.

There has also been a widespread readiness to assume that higher resolution monitors are required only by radiologists, and that in the absence of hard copy, referring physicians can make do with display systems of lesser specification. There is in fact no way for radiologists to ensure that such display systems will not be used for diagnosis and as a basis for clinical decisions, and there is a need for extreme caution in permitting a traffic in "dual standard" images.

One situation in which such a system might be tolerable would be if it were possible to guarantee that each image released onto a digital network carried with it a radiologist's report. However, the integration between text and image management software that would be necessary to achieve this remains a serious and unsolved problem; the implications for radiological work practices would also be considerable.

Digital display systems are slow; technological advance in terms of resolution will yield increased volumes of image data, so reducing the practical benefit from technological advance in terms of speed. Our own experience suggests that radiologists quickly become frustrated when their productivity is impaired. There will be an urgent need for "intelligent" work-stations, able to pre-fetch relevant images (with relevant prior examinations), and able to integrate text and image management software. Intelligent work-stations remain in their infancy; though a number of research prototypes have been developed by users themselves.

Setting aside the many other technological problems that have yet to be overcome, it is likely to be several years before display systems can be relied upon to provide the resolution and functionality necessary to satisfy patients and medical staff alike.

Concluding comments

There is an unshakeable conviction among the British health administration that PACS can only be justified when implemented as total systems in a single step. By implication, this also means that such systems are likely to be purchased from, and maintained by, a single supplier.

The British authorities are currently attempting to achieve such an implementation in at least one institution, at considerable cost. Such an exercise has not been attempted in other countries, and in the light of our own experience at St Mary's is ill-advised. On the other hand, many aspects of PACS are now ripe for implementation, and the current pre-occupation with total systems has been at the expense of partial schemes that stand a better chance of yielding useful results, and has deprived many institutions of the opportunity to begin to reap the benefits of applying advanced information technology to radiology.

References

1. Dawood R M, Craig J O M C, Highman J H, Wadsworth J, Glass H I, Todd-Pokropek A, Cunningham D A, Stevens J M, Al-Kutoubi A, Kerslake R W, Choudhri A H, Barber C J, Crofton M E, Porter A W. (1989) Clinical diagnosis from digital displays: preliminary findings of the St Mary's evaluation project. Clinical Radiology, **40**: 369-373

2. Dawood, R M (1990) Digital Radiology. (Leading article) Clinical Radiology **42**, 6-11

3. Dawood R M, (1989) Radiology about to go digital: exciting new technology must be carefully evaluated (leading article) British Medical Journal, **299**: 340-1.

4. Dawood R M, Craig J O M C, Highman J H, Wadsworth J, Glass H I, Todd-Pokropek A, Cunningham D A, Stevens J M, Al-Kutoubi A, R W Kerslake, Barber C J, Crofton M E, A W Porter. Evaluation of VDU displays in radiological diagnosis: a preliminary comparison of two systems. Proceedings of the SPIE: **1091**: 638-371.

5. Slasky B S, Gur D, Good W F, Costa-Greco M A, Harris K M, Cooperstein L A, Rockette H E (1990). Receiver operating characteristic analysis of chest image interpretation with conventional, laser-printed, and high resolution workstation images. Radiology **174**:775-80

6. Ukovich W, Dalla Parma L, Stacul F, Zernich R C, Girivona P (1990) Practical experience with PACS: benefits and pitfalls. Proceedings of the SPIE **1234** (In press)

Beyond PACS: The Electronic Triage System

Gilbert B. Devey[1] and Michael J. Ackerman[2]

[1]Consultant, American College of Radiology, 2801 New Mexico Avenue, Washington, DC, 20007, USA.
[2]Chief, Educational Technology Branch, National Library of Medicine, Bethesda, MD, 20894, USA.

Abstract: PACS is one element of an *electronic triage system* (ETS). Other elements under continuous development are MEDLINE, GRATEFULMED, RIS, HIS, telepathology, etc. There is ongoing research on artificial intelligence (expert systems), outcome probabilities, and practice guidelines as aids to diagnosis and treatment. Each research project and technological development is evolving independently of others, but to be cost effective in the clinical setting they need to be combined in an interactive system. An ETS, perhaps using PC–based computational agents such as *knowbots*, would integrate all data bases and contribute enormously to improving the capabilities of the diagnostician and the therapist. A review is made of the state of development of the numerous technologies, costs, and the time to diffuse ETS into clinical medicine. Gaps in knowledge must be filled, human resources identified, management systems determined, and the roles of the public and private sectors clarified.

Keywords: practice guidelines, HIS, MEDLINE, outcomes, PACS, RIS, data bases, triage.

Introduction

Computers are ubiquitous in health care, but the hoped for major contributions to medical decision–making are yet to be realized.[4,10] Deterrents are the slow and cumbersome procedures for obtaining computer–based medical data together with the specialized and limited scope of the available information. A system is needed to provide fast and easy access to information about <u>all</u> aspects of clinical medicine and health services. All of the necessary technologies now exist to provide state–of–medical knowledge to every clinician in <u>real time</u> through an *electronic triage* computer network *system* serving the entire nation. Every clinician would have immediate access to the entire body of knowledge needed for making accurate diagnosis and planning of optimum treatment protocols, plus evaluating the risks, costs, and likely outcomes from procedures adopted. The quantity of available data has reached gargantuan proportions, and continues to double about every 5 years. The ETS may become the principal tool used in clinical decision–making.

NATO ASI Series, Vol. F 74
Picture Archiving and Communication Systems (PACS) in Medicine. Edited by H. K. Huang et al.

PACS, RIS and HIS are becoming key components of the radiology decision–making process; they could be augmented by interactive 3–D imaging workstations (MIDAS) [7], disease–specific expert systems and devices, data bases for cancer diagnosis and treatment [1], and reference standards like the *Visual Human* [8] and other structural biology projects.[3]

A number of data bases including *MEDLINE, PDQ* (physicians data query) and *Exerta Medica* are available to clinical medicine practitioners, which as a group would gain the most benefit from an electronic triage system. Additional data bases covering all fields of medicine and health care must be prepared, with attention needed in the development of an *outcomes of health services and procedures* data base from projects supported by the Agency for Health Care Policy and Research, as cataloged by the National Library of Medicine (NLM).

Materials and Methods

Today's reference medical data bases operate off–line in a "stand alone" mode, much like a single PC. Clinical data bases such as PACS, RIS and HIS, although on–line in nature, operate essentially in the "stand alone" mode and are akin to a LAN (local area network). What is required for the ETS is a regional, interactive, distributed, integrated services digital network (ISDN), where the region is the entire U.S. health care system. Critical to the establishment of such a system is a national high–speed computer network (1 GB/sec.) similar to that being developed by the National Science Foundation (NSF). [9] Other critical elements for the electronic triage system include: network architecture; affordable and higher performance personal computer workstations; high capacity interactive data bases perhaps using *knowbots* [11]; and, a suitable logistics and management support structure.

Enormous computational power at low cost is required for ETS to be realized in practice. One can predict with high confidence that the required microprocessor chips will be incorporated in PCs and medical data base systems well before the establishment of an ETS. Examples: a single chip with 100 million transistors may be in production by the year 2000; cost/performance gains made during the past twenty years indicate that the same money spent in 1990 should buy 30 times today's performance in 2000 [6]; and, a national high performance computer network should then be in full operation.[9b]

Development of the interactive data bases required for ETS will be a long and demanding endeavor. New and existing data bases must be compatible with current and future network (data) interface standards.[2] Software (knowledge) standards for new data bases should be developed according to requirements still to be determined. Existing bases would be modified as feasible [5]; if infeasible, *knowbots* could be incorporated. [11]

Business methods now employed by the U.S. decentralized phone system could serve as a model for the ETS. A crucial difference between ETS and the phone system is the dominant role of government in the funding of medical science and health care R&D. Still lacking: a conceptual design to provide easy and simultaneous access by multiple users to remote, specialized, self–standing data bases from any point on a network; a financial plan; and, a complete model of a management arrangement.

Results

The availability of an ETS would provide all clinicians with immediate access to an integrated and balanced state-of-medical knowledge covering all pathological conditions. The physician no longer would be limited to recalling a few of his recent cases - he would benefit in real time from the very latest information about proven diagnostic and treatment procedures and outcome probabilities for the particular case at hand however unusual. ETS would contribute significantly to the national goal of providing high quality health care to all persons.

Discussion

ETS might slowly emerge over time through the normal process of technological advances. However, to hasten and assure development, a study group should be convened to determine the technical merits and clinical feasibility of the concept, examine system design parameters, identify specific data base resources now available or under development, determine gaps and how to fill them, analyze computer networks as to the capability to handle the large volume of traffic, estimate costs, outline an organizational structure to manage a phased systems development leading to an operational system, and predict the time to move to clinical application. Leadership in developing ETS could come from a consortia of medical users, government health agencies, and/or the telecommunications industry using a nonprofit entity along lines already proposed to deliver high-speed computer capabilities to every American home. [9c]

References

1. American College of Radiology/National Cancer Institute: Radiology Diagnostic Oncology Group (RDOG) and Radiation Therapy Oncology Group (RTOG).
2. Bell, T.: Technical challenges to a decentralized phone system. IEEE SPECTRUM, September 1990.
3. Brinkley, J. F.: A Framework for the Design of Knowledge-Based Systems in Structural Biology. SCAMC, 1989.
4. Devey, G. B.: Electronic Computers and Medicine for Large Populations. JAAMI, Jan./Feb. 1967.
5. Horii, S.: ACR/NEMA Standards. NATO/ASI Picture Archiving and Communication System in Medicine, Evian, France, October 1990.
6. Magid, J. I.: Directions of Industry. Workshop: 3D Display and Analysis for Cancer Treatment Planning. National Cancer Institute, July 1990.
7. Merical, M.: What's in Store for Imaging Workstations? Diagnostic Imaging, August 1990.
8. National Library of Medicine: Request for Applications (RFA), July 1990.
9. National Science Foundation: a- IBM, AT&T Expected to Join Project. Washington Post, June 9, 1990; b- Upgrading Planned for Data 'Superhighway'. Washington Post, June 13, 1990; c- U.S.Studies Nationwide Computer Link. International Herald Tribune, July 17, 1990.

10 Trani, J.M.: Diagnostic Imaging Trends. Decisions in Imaging Economics, Vol. 3, No. 1, 1990.

11 Wright, K.: The Road to the Global Village. Scientific American, March 1990.

A Multimodality Medical Image Interpretation System as a PACS Component

Anne-Marie Forte, Yves Bizais

IMM, HGRL CHRN, BP 1005, 44035 Nantes Cedex 01, France

Abstract: A PACS provides means for archiving and transmitting Image Data, while Multimodality Medical Image Interpretation Systems (MMIIS) provide tools to analyse and interpret them. To be integrated in PACS, such a system should be a standard tool, compatible with the whole system and able to communicate. In this paper we describe the specification and the implementation of the prototype MMIIS we developed and we demonstrate the versatility of our approach on an actual example.
The MMIIS formally consists of a specific set of Image Processing (IP) objects, a classification of IP procedures, partly describing IP knowledge, a multi-expert system describing the various categories of knowledge.
To ensure portability, we developed an IP language (link between formal and implemented IP objects and procedures) and used standard industrial tools.
The interest of clearly separating application-dependent knowledge and generic imaging tools in Medical Image Analysis and Interpretation is demonstrated on lung SPECT images.

Keywords: Image objects, IP procedures classification, Image analysis, portability, reusability

0. Introduction

A PACS provides means to store and transmit medical images between sources and targets, such as a Multimodality Medical Image Interpretation System (MMIIS). The latter truly makes use of image information as a whole and provides tools to better analyse image data.

To be integrated in a PACS, a MMIIS must make use of :

1. standard image objects in order to communicate with the PACS database (§1),
2. classified Image Processing (IP) procedures, the calls of which are standardized to make applications compatible between them and with the PACS software itself (§1)[1],
3. an expert system to take into account the various domains of knowledge involved in interpretation (§1)[4],
4. portable software tools (operating system, interfaces) to allow its use in various PACS configurations (§2)[5].

Clear specifications of image objects, classification of IP procedures, reusability through well-defined knowledge and *portability* are the required bases to realize a PACS-integrated MMIIS.

This paper summarizes our work in this field which will be examplified using lung SPECT scans in the oral version.

1. Theoritical aspects with actual medical illustration

The definition of a limited and stable set of image objects is extremely important :

- it ensures a simple and globally *consistent* viewpoint of data to the MMIIS required during data exchanges with the database [2] and with the expert system,
- it summarizes many similar image objects found in the literature in a restricted object set, thus avoiding ambiguities.

We designed it to be *complete* and made up of simple objects and structured ones.

Simple objects do exist physically and are defined in the following :

1) *image* : a 2D array associating an intensity to each sampled position,
2) *region* : a set of pixels within an image, verifying a homogeneity condition,
3) *curve* : 1D array (e.g. histogram, time intensity curve),
4) *graphics* : text or unformal drawings annotating images,
5) *attribute* : characterizing IP objects and constituting the basis of image interpretation.

Structured objects describe the organization of simple objects, and are used by IP procedures as actual entities. For *a study data set* belonging to a given *patient folder*, they are :

1) *set* : part of a study data set grouping logically related data,
2) *series* : part of a set grouping simple image objects indexed on time, depth... The "series" object is necessary because most medical images are acquired in such a way,
3) *group* : part of a series in which index is sampled at fixed rate. The "group" object is extensively used by medical IP procedures.

Our classification of IP procedures provides a framework to unambiguously classify *every IP procedure according to its objective* and thus to dynamically build an IP expert ; the latter, during interpretation, runs IP procedures according to their class. It is quite different from previously proposed classifications [3] based on operational aspects.

We define five main classes :	3.[{IP object}] ---> {curve}
1.[{IP object}] ---> {image}	4.[{IP object}] ----> {attribute}
2.[{IP object}] ---> {region}	5.[{IP object}] ----> {graphics}

Each above class is subdivided into two sub-classes : generation of objects by external means (interaction, PACS components, models) and generation of objects by procedural ones.

The expert system used in a study data set interpretation, according to a global objective :

- runs the proper IP procedures on the available image objects (see above),
- makes use of extensive domain knowledge (medical, image source, scene, IP).

In medical imaging, image data strongly depend on acquisition modalities, acquisition devices, on the organ being studied and on used IP processors. It follows that reusability of generic IP procedures and domain knowledge requires a multi-expert system in which application specific knowledge and domain knowledge are truly independent.

The interpretation process control, *both* bottom-up and top-down, is based on the various expert knowledge and also on facts resulting from IP. This is the main originality of the system to run external actions and to use the resulting information for its reasoning.

2. Implementation aspects

Image objects described in §1 correspond to standard data structures easy to represent in conventional programming languages. Moreover these structures are independent of the way of storing objects. IP objects used by MMIIS can thus be stored in a standard PACS database.

A tool based on Lex/Yacc was developed to translate formal IP procedure calls into the corresponding code executable by a target processor using conventional languages (C, Pascal). For this purpose a translation file must be created for each target processor, allowing efficient use of hardware specificities. Our work results in a portable IP library, the scope of which is much more general than MMIIS in a PACS environment.

An expert system generator based on µProlog is being realized. Communication between experts is based on Unix pipes.

3. Discussion

During the design of a Medical Image Interpretation System, we defined formalisms for image objects and IP procedures. They are the main basis for designing a well organized IP knowledge and image objects database to be integrated in a MMIIS. The expert system makes use of various knowledge bases (KB) in an efficient way, and activates application-independent IP procedures. Existing standards were chosen to ensure the portability and efficiency of our system. Other cases of MMI analysis are under development to demonstrate the generality and the validity of the approach. Because it permits a better use of MMI not otherwise possible, it provides a significant added value to PACS.

This work was supported by a grant of the Institut National de la Santé Et de la Recherche Médicale (INSERM).

4. Literature

1. Abmann, K., Venema, R, Höhne, K.H. : the ISQL experience. in NATO ASI Series, Vol F19, Pictorial information systems in Medecine, edited by K.H.Höhnr, Springer-Verlag, 1986, pp333-356.
2. Aubry, F., Kaplan, H., Di Paola, R.: An Image Handling System for Medical IP. SPIE, Paris (Fr), apr.1989.
3. Rosenfeld, A., Kak, A.C.: Digital Picture Processing. Vol1, Second edition, Academic press, 1982.
4. Suetens, P., Oosterlinck, A.: Using Expert Systems for Image Understanding. International Journal of Pattern Recognition and Artificial Intelligence, Vol.1, n°2, 1987, pp237-250.
5. Tamura, H., Sakaue, K.: Design and implementation of SPIDER a transportable image software package. Computer vision, graphics, and image processing 23, 1983, pp273-294.

Interfacing PACS with the DVA Information System

Marcus W. Hedgcock, Jr[1,2,3]., Suzy Smith[4] ; Tod S. Levitt [1,3]

[1] DVA Medical Center, San Francisco, Radiology ; [2] DVA Medical Center, San Francisco, Research; [3] University of California, San Francisco, Radiology Department; [4] Madigan Army Medical Center, Seattle, Washington

Abstract: A robust interface between PACS and the Hospital Information System is critical for making patient information available via PACS and allowing effective distribution of imagery via PACS. The same Hospital Information System, called the DHCP (Decentralized Hospital Computer Program) is used by all the medical centers of the United States DVA (Department of Veterans Affairs) and the Department of Defense. To develop the interface, we first defined the functional goals for our PACS operations and workstations. We determined which information fields from the DHCP were needed for effective PACS operations. Our PACS/DHCP interface design aims to minimize or eliminate redundant data entry and to provide quick access to current demographic information for central system operations and workstations.

Keywords: PACS, HIS, PACS operational requirements, image archives

Introduction

The DVA is the largest multicenter health care system in the world, with 176 medical centers and more than 300 additional affiliated health care facilities. All DVA medical centers use the same HIS/RIS system: the DVA-developed Decentralized Hospital Computer Program (DHCP), a MUMPS-based system. The Department of Defense (20 medical centers) also uses a modified version of DHCP. Our PACS program at San Francisco DVA Medical Center is a pilot and model for the systemwide implementation of PACS within the DVA.

The global goal of PAC systems is to make medical imagery more effective and available. Medical imagery is rendered more effective when workstations are capable of altering medical imagery displays, analyzing imagery using computers, and allowing access to ancillary data from imagery and text records. Availability is increased when imagery is distributed from electronic archives to workstations.

PACS must have appropriate information and instruction available from the Hospital Information System, which governs information flow and scheduling in most medical centers. This access to HIS allows PACS to incorporate ancillary data for image interpretation and augment effective distribution of medical imagery for medical center operations. The HIS does, however, contain a great deal of information, some of which is not germane to imagery use.

The DHCP HIS contains more than 100 fields. Our goal was to determine what information from the HIS 1) is needed for effective distribution of PACS imagery to workstations, 2) is needed for effective use of medical imagery on PACS workstations, 3) is lacking at present in HIS records and should be added to the HIS database, 4) needs to be made available on PACS workstations, and 3) requires a unidirectional or bidirectional interface between the HIS and PACS.

NATO ASI Series, Vol. F 74
Picture Archiving and Communication Systems (PACS) in Medicine. Edited by H. K. Huang et al.

Materials and Methods

To determine the information needed for imagery distribution, we studied imagery distribution patterns from our imagery archives at SFDVAMC (San Francisco Department of Veteran Affairs Medical Center). To determine the information needed for imagery interpretation, we studied imagery interpretation at our own center and three other centers in the San Francisco Bay Area, including a city medical center, a large university medical center, and a large private medical center.

Results

In the following lists, information currently resident on the HIS is underlined; information currently resident on the PACS is in italics. Imagery is distributed from the archive via lists of patients for the following purposes: 1) imagery interpretation, 2) procedures, imaging and nonimaging, and 3) imagery review at a) inpatient rounds, b) outpatient clinics, c) emergency area, c) procedures, d) specialty conferences, e) transfer to other medical centers, f) teaching conferences, and g) ad hoc consultation. All of the imagery uses listed above require 1) *new imagery*, and 2) *old imagery for comparison*. In an earlier study [1], we determined that, except for chest radiographs, which commonly use the three most recent examinations for comparison, the most recent prior examination generally suffices for comparison. Ideally, any CT or MRI studies of the same body part should also be available, and nuclear medicine examinations should be available for bone studies. Imagery interpretation requires additional information from the HIS, including 1) patient demographics, including patient location in the medical center, 2) prior imagery interpretations, 3) current laboratory information, 4) treatment teams or consultants identified with the patient, and 5) current chart notes about the patient's condition. In addition to the above information, imagery interpretation will require the PACS to route the imagery for interpretation to the workstation(s) in the correct reading area. The four medical centers we studied had 11 to 16 separate reading areas.

Discussion

Imagery distribution will require the PACS system to route imagery with appropriate comparisons to the appropriate workstations. Routing according to specific uses (i.e., clinics, rounds, etc.) can be set up once by the system manager and then left in place until modifications are required. The lists of patients for specific uses are in most cases already resident on the HIS 24 hours before the imagery is needed. These lists need only to be sent to the central PACS Data Management System (DMS) the night before to allow imagery to be retrieved from the archive and sent to the appropriate workstation or LAN. For example, appropriate imagery for all patients scheduled for a surgery clinic is retrieved from the archive the night before and sent to the LAN serving the workstations used by the surgery clinic physicians. Since 84% of imagery needed for comparison is known the previous day, the DMS is more effective in handling ad hoc requests, such as those from Emergency. It is fairly easy to develop decision rules about which comparison imagery must be sent with which specific type of imagery and what part of the patient's total imagery file must be sent to each area.

The imagery distribution lists not underlined in the results section above are needed for complete imagery distribution. These are most easily added to the HIS database rather than the PACS because HIS terminals are more widely available; and PACS workstations are best used for imagery interpretation and review, rather than for sending instructions to the HIS. Most physicians felt that it would be useful to have access to the distribution lists on the PACS workstation. We are making these available as we conduct our clinical trial.

The information from the HIS includes both static and transient field information. Determining how often the PACS will need to update its transient fields requires testing in the clinical setting; however, our operating hypothesis holds that updates every twenty-four hours at night may be sufficient.

Conclusion

Effectively operating PACS requires that a large amount of information from the HIS system be available both to the PACS DMS and to users of the PACS workstations. This data transfer appears to be unidirectional from the HIS to PACS. Our operational plan would move the majority of imagery retrieval and database updating to the night hours, thereby reducing demand on the DMS during normal clinical operating hours.

[1] Hedgcock MW, Smith S, Levitt TS: Operational strategies for optical archives. SCAR 90 Proceedings, pp. 362-7.

Tools for Analysis of Requirements for PACS

J.C. Honeyman[1], E.V. Staab[1], M.M.Frost[1], T. Brandt[2], W.D. Bidgood, Jr[1]

[1]Department of Radiology, University of Florida, Gainesville, Fl 32610

[2]Information Services, University of Florida, Gainesville, Fl 32610

Abstract: The Department of Radiology at the University of Florida has set a goal of "filmless" Radiology in an ambulatory care center to be completed in 1993. Networks, archival systems, displays, and workstations are being installed in a sequence that facilitates evaluation of clinical usefulness, acceptability of image presentation, limitations, and optimization of both transfer speed and retrieval. A variety of tools measure the performance of different protocols and media in networks, collect and analyze the volume of images produced for archival, and predict the pattern of retrievals. Observation and questionnaires are used to collect requirements for composition of images on monitors including the number and types of requests for previous images. Various displays and workstations are being evaluated to determine the requirements for different types of images.

Keywords: PACS, Communication Networks, Workstations

Cluster Concept

In order to organize the categories of design and development to be included in a needs assessment, physical, clinical, and technical clusters are defined for the department. Physical clusters describe image acquisition equipment according to placement in the department and are used in network analysis and design for bridge and subnet placement to contain traffic and to investigate the use of alternate network configurations. At the University of Florida, the main department consists of a general radiology operation which includes two CT scanners, several Nuclear Medicine cameras, angio-interventional rooms, and ultrasound devices. Orthopedic radiology is currently located on a different floor, but within the same building as the main department. Two Magnetic Resonance units are located approximately 900 Meters from the main department, and an additional Nuclear Medicine Spect unit is located at the Veteran's

NATO ASI Series, Vol. F 74
Picture Archiving and Communication Systems (PACS)
in Medicine. Edited by H. K. Huang et al.

Administration Hospital, approximately 400 Meters from the main department. In addition, three image research labs are located in remote areas. Locations for the archival systems, workstations for diagnosis by radiologists and display stations for image review by referring physicians are included in physical cluster analysis and design.

Clinical clusters describe radiology divisions and are used to define workstation requirements, placement of reading rooms, and the types of images needed for diagnostic reporting. The Department of Radiology has eight clinical clusters, including general, orthopedic, pediatric, and neuro-radiology, abdominal imaging, mammography, nuclear medicine, and angio-interventional. We define diagnostic workstations as systems with sophisticated displays and high level image processing capabilities. These systems will be used primarily for diagnosis by the Radiologist. Display stations may have lower resolution displays, and limited image processing capabilities and will be used by referring physicians for clinical review. Display stations for clinical review are included in our analysis to provide the best combination of functionality, ease of use, and cost for the referring physician.

Technical clusters are groups of equipment producing similar image types, such as CT Scanners, MR units, digitizers, Nuclear Medicine cameras, digital angiography, computed radiology, and ultrasound equipment. Technical cluster analysis includes investigating sizes, types, and timing of images transmitted and received, and is used in the analysis and design of physical and clinical clusters, as well as for calculating the requirements for archival storage. Workstations capable of creating new images and archival systems are also considered image producing equipment with transmission characteristics.

Physical Cluster Tools

For the physical clusters, tools are required to analyze different network configurations and the anticipated impact of adding additional nodes. In our model, images produced will always be sent to archival before distribution to viewing by physicians.

The network is modeled using the simulation package LANNET from CACI Products Company. LANNET allows the description of a network configuration, properties of each node, and types of communication activities performed by each node. From the reports on the simulation, one can determine the number of nodes a network can adequately support, how the addition of a new node or changes in network configuration will effect traffic, server storage requirements, and network response time.

Clinical Cluster Tools

The goal of clinical cluster analysis is to predict workstation requirements for each division of radiology, including composition of images on monitors, sizes of images, predicted resolution and processing requirements, speed of retrieval, and volumes of images to be stored locally for rapid retrieval. The tools record the current use of alternators and scopes, number of previous studies required, outside film evaluation, types of images viewed, and amount of time required to access all images used. Following implementation of workstations, further tools investigate the display requirements for diagnosis and viewing, the user interface, and image processing features.

To estimate workstation requirements for a clinical cluster, an observer is placed in a diagnostic reading area for several reading sessions. The form in figure 1, is used to record the activities during the reading session. Section A in figure 1, records general reading session information and assigns a study

A
SESSION: ________ DATE: ________ OBSERVER: ______ ST.NO. ____
PERSONS READING FILMS: ____________________

B
START TIME: ______ ☐ SEQUENTIAL ACCESS
TYPE OF PROCEDURE: ______ ☐ RANDOM ACCESS

C

PRELOADED IMAGES			
TYPE	NUMBER OF INDIVIDUAL IMAGES	NUMBER OF SHEETS	INSIDE OR OUTSIDE FILMS
TIME IN SECONDS TO LOCATE AND VIEW ALL:			

D

IMAGES NOT PRELOADED			
TYPE	NUMBER OF INDIVIDUAL IMAGES	NUMBER OF SHEETS	INSIDE OR OUTSIDE FILMS
TIME IN SECONDS TO LOCATE AND VIEW ALL:			

E
WORKSTATION PROCESSING PERFORMED: ☐ YES ☐ NO
COMMENTS: ____________________
____
TIME DICTATION COMPLETED: ________

FIGURE 1: OBSERVER FORM FOR READING SESSIONS

number. Subsequent pages are numbered sequentially and assigned the same study number for bookkeeping purposes. Sections B - E are used for each procedure, where B records the type of procedure and start time for reading. Sequential access is the routine method for a reading session with random access used when a clinician requests information for a specific study. Sections C and D record the information on images preloaded on scopes and additional images used in diagnosis. Section E records comments, end of dictation, and whether or not workstation processing is used. Comments include interruptions, which may refer to another film reading if a clinician requires immediate interpretation, comments overheard by the observer concerning availability of films or quality of images, and the use of a hot light or magnifier. Workstation processing indicates that the radiologist used a remote workstation to further process or adjust the images. Information from this instrument predicts the amount of workstation storage required to store the current and previous images used in diagnosis for a procedure type, the composition of images on monitors, the amount of time a radiologist currently uses to access images, and the type of access needed to complete a typical reading session. We also consider the number of days a study remains on a scope for viewing. Workstation storage must be large enough for studies performed on all patients in recent history, along with the previous and relevant studies for the patient. Any additional cases must be requested from the archival store with a delay for access and transfer. The image memory and display buffers allow rapid paging through images and must be large enough to accommodate all the images used to diagnose a typical study.

Limited Receiver Operating Characteristics (ROC) analysis is used to verify the documented results from other researchers on display requirements for clinical clusters. Our primary interest is comparative evaluation of the use of display stations, workstations and film, including time requirements for the use of each medium, and comments on the user interface. An observer uses the form in figure 2 to record data on the time required for each study read, and the types of image processing used. In addition, the radiologist is interviewed to collect their comments on the user interface, ease of use, and additional features requested for the workstation. The information is collected initially in a controlled environment, then during clinical operation to monitor changes over time, document learning curves for using display and workstations, and fine tune optimal user interfaces for the user.

Technical Cluster Tools

Technical cluster analysis is used to estimate volume and timing of network transfers and total archive size. For image acquisition equipment, we rely on the Radiology Information System (RIS) to report the number of studies per month and log books in the areas of interest to identify patterns of work load. The sizes of images produced depends on the characteristics of the technical cluster. Since all images are sent to the archive prior to workstation viewing, these data will predict network load between image acquisition nodes and the archive. We have determined the number of images to be sent to individual workstations in clinical cluster analysis. From this we can predict network load between the archive and workstations. By observing the volume of new images created on the workstation, we can also predict the impact of workstation archival on network traffic. Total archival size must be sufficient to store all images produced by technical clusters.

FILM OBSERVATION

READER NUMBER: ________________ OBSERVER: ________________
DATE: ____________ STUDY NUMBER: ______________
TIME STARTED: ____________ TIME COMPLETED: __________
CHECK ALL EXTRA ACTIVITIES:
☐ REARRANGEMENT OF FILMS
☐ USE OF MAGNIFIER
☐ USE OF HOT LIGHT

MONITOR OBSERVATION

READER NUMBER: ________________ OBSERVER: ________________
DATE: ____________ STUDY NUMBER: ______________
TIME STARTED: ____________ TIME COMPLETED: __________
MONITOR TYPE: ☐ MEGASCAN ☐ IMLOGIX
CHECK ALL EXTRA ACTIVITIES:
☐ REARRANGEMENT OF IMAGES ☐ ARCHIVAL OF PROCESSED IMAGES
☐ USE OF ZOOM / ROAM
☐ USE OF WINDOW / LEVEL
☐ USE OF IMAGE PROCESSING TYPE OF PROCESSING USED: __________

FIGURE 2: OBSERVER FORM FOR ROC STUDY

Preliminary Results and Conclusions

We present our approach which is modeled on the cluster concept and is verified by information gathering and analysis tools. The tools are developmental and will be expanded and modified. They are a reasonable starting point for departments entering into PACS implementation.

The LANNET simulator allows us to model the effects of the addition of bridges, subnets, and increased server storage space on archival and retrieval times. We can also model the effects of adding increased bandwidth when economically viable.

The initial analysis of the tools to examine the workstation requirements for two clinical clusters (neuro-radiology and nuclear medicine) indicates that we can predict the patterns of retrieval and the number of images used in a typical reading session. Workstations are currently under development and ROC verification and information gathering will begin in the fall, 1990.

Technical analysis of equipment producing images for neuro-radiology and nuclear medicine has predicted the amount of archival space required per year to store images for these two clinical clusters. An administrative decision will indicate how long we will keep images on-line for retrieval and will be used to estimate the total archive space required for all images stored in the future.

The Design and Development of an HIS-PACS Interface: Utilizing the Efforts of ACR-NEMA WGVIII

Betty A. Levine[1] and Karen T. Randall[2]

[1]Georgetown University Hospital, Washington, DC 20007
[2]AT&T Bell Laboratories, West Long Branch, NJ 07764

Abstract: Completion of the efforts by ACR-NEMA working group VIII will increase the usefulness of the ACR-NEMA standard for interfacing HIS/RIS/PACS systems. This paper describes the Georgetown effort to use the ACR-NEMA standard and WG VIII extensions to develop a HIS-PACS interface.

Keywords: PACS, RIS, HIS, Information Systems

Introduction

Georgetown University Hospital installed an AT&T CommView system in 1987. In preparation for acceptance of the system an interface to the existing hospital information system (HIS) was designed and development efforts began. Involved in this development, ASCII messages from the HIS had to be converted to ACR-NEMA formatted messages acceptable by the PACS. Although the creation of the ACR-NEMA logical messages alone was not difficult, the development of such an interface to meet operational requirements was complicated.

The steps required to develop an interface between the HIS and the PACS and the way the ACR-NEMA standard was implemented in our development effort will be discussed.

Methods

The development of the interface required a detailed analysis of both the clinical operations and information movement throughout the radiology department and the effect the PACS has on registration clerks, technologists, physicians, transcriptionists, and referring physicians. Technologists' compliance and the accuracy of image and text acquisition into the PACS are critical to the effectiveness of the PACS. Extra steps are required to capture the images into the PACS, especially duplicate entry of patient registration and order entry information into both the HIS/RIS and the PACS. In a busy Radiology department this is a problem. Therefore it was determined that at a minimum patient registration and order entry information should be exchanged between the HIS /RIS and the PACS.

Our HIS system, maintained solely by an in-hospital group, is a MIIS based system (a subset of MUMPS) and operates on 7 Data General computers. Our PACS is an AT&T CommView system. Selected data elements were chosen to be exported from the HIS and sent to the PACS. However, the information is passed as an ASCII file using a

NATO ASI Series, Vol. F 74
Picture Archiving and Communication Systems (PACS) in Medicine. Edited by H. K. Huang et al.

communications package, developed in-house, that runs over a SYTEK network. The PACS requires data elements to be sent in an ACR-NEMA formatted binary file, over direct RS232 lines, using the KERMIT file transfer protocols. The incompatibility of these data formats and communications protocols led us to develop an interface which uses a personal computer as a protocol converter that receives HIS messages, converts them to ACR-NEMA formatted messages, and transfers them to the PACS.

The creation of the PACS messages required that we determine which data elements from the HIS were required or desired on the PACS, convert these data elements into ACR-NEMA format, and create the ACR-NEMA message. Because the efforts of ACR-NEMA WGVIII have not yet been approved as a standard, the messages required by the CommView system follow the intentions of the working group, but use shadow groups, permissible under the current ACR-NEMA standard, in order to have all pertinent information transferred from the HIS/RIS.

The ACR-NEMA Groups utilized by the AT&T CommView interface specification include command information (0000 & 0001*), identifying information (0008 & 0009*), and patient information (0010 & 0011*). Also, the groups' diagnostic results information (4101), diagnostic report text (4103), and transaction information (4201) were created by AT&T and follow the ACR-NEMA convention as outlined in the standard. The use of non-standard groups in the specification was necessary because the data elements in these groups did not fit into any of the existing groups and the efforts of WGVIII were not completed. Information from our HIS registration and order entry modules were required to create the PACS messages. This includes approximately 6 MUMPS nodes containing all patient demographic and exam information. The nodes are parsed, and the information needed by the PACS is reformatted to ACR-NEMA data formats (e.g., dates as yyyy.mm.dd). Appropriate ACR-NEMA groups are created from the information received and element numbers and lengths are associated with each. All the elements are combined to form an ACR-NEMA binary message which is then sent to the PACS using KERMIT.

One of the problems in creating an ACR-NEMA message is the overhead which accompanies each data element in the message. Every data element is burdened with a group number (1 byte), an element number (1 byte), and element length (2 bytes). Each message must contain group 0000 and every group carries with it an element describing the length of the group. Thus this element cannot be determined until the group has been created. Similarly, the entire message length is required as the second data element in the message. Again, this element cannot be determined until the entire message has been created and thus must be filled in later.

Results

The limitations in the current standard cause the manufacturers of PACS and HIS/RIS to use shadow groups and manufacturer defined groups to a greater extent than originally planned. This is the reason for the existence of WGVIII. The efforts of that group have led to a nearly completed standard which contains many elements found in existing HIS and RIS. The smaller

---

* ACR-NEMA shadow group

the number of shadow and manufacturer defined groups for specific implementations of an interface, the more useful the interface will be.

At Georgetown, technologists' compliance more than doubled with the implementation of our interface. The reduction in work for the technologist was the greatest factor in aiding the increased use of the PACS. Once the technologists put most of the cases on the PACS, the Radiologists were more likely to begin using the system since the cases they required were stored and the PACS became more important to the department. The implementation of the interface also allowed for the automation of our PACS, by allowing for the automatic routing of images and exams to workstations prior to their being needed for review.

The involvement by many PACS and HIS/RIS vendors on the ACR-NEMA WGVIII committee is encouraging. The approval and subsequent acceptance of the standard throughout the PACS/HIS/RIS community will allow for easier integration of a PACS into a Radiology department and the rest of the hospital.

The OSIRIS User Interface for Manipulating Medical Images

Yves Ligier, Matthieu Funk, Osman Ratib, René Perrier, Christian Girard

Digital Imaging Unit, Center of Medical Informatics, University Hospital of Geneva
1211 Geneva-4, Switzerland

Abstract

This paper describes the OSIRIS project developed at the University Hospital of Geneva. Its goal is the design of a software to read medical images files stored in the PAPYRUS format and, more important, to display and analyse them in a window based user interface environment. This software is a main part of the PACS project conducted at the Geneva Hospital. OSIRIS has been designed to work on a wide variety of workstations and window systems (Unix systems with X11 and OSF/Motif and Macintoshs). It provides the traditional modes of visualisation (zoom in, zoom out, invert, etc) and allows more specific manipulations such as a movie mode, filters, measure of distances, modification of the color lookup table. OSIRIS allows to display a second window containing the image interpretation report, a window containing some study information and allows to add annotations to an image.

Keywords: Workstation, PACS, User Interface, Digital Imaging, Object Oriented Programming.

Introduction

The PACS project has been designed as part of a more general project, the DIOGENE project, which is a complete hospital information system, developed at Geneva Hospital. We have decided not to choose a closed PACS system as offered by a few manufacturers, but to develop an open architecture software: OSIRIS. This software will be able to work on a wide variety of existing and future workstations and window systems (mainly Unix based). The present version is available for SUN/Unix with X11 and OSF/Motif and Macintoshs. OSIRIS reads images stored in a standard format: PAPYRUS, which has been developed at Geneva [1]. To insure extensibility and reusability of the software, we elected to develop it using object oriented programming techniques.

User Interface

The different facilities of the user interface [2] are directly accessible either by specific graphic icons or by menus. This allows to provide a user interface adapted to the needs and knowledge of physicians. The figure 1 shows what the OSIRIS main window looks like with its different

NATO ASI Series, Vol. F 74
Picture Archiving and Communication Systems (PACS) in Medicine. Edited by H. K. Huang et al.

icons and menus. OSIRIS allows to display a second window containing the image medical interpretation report, that can be modified by authorized people.

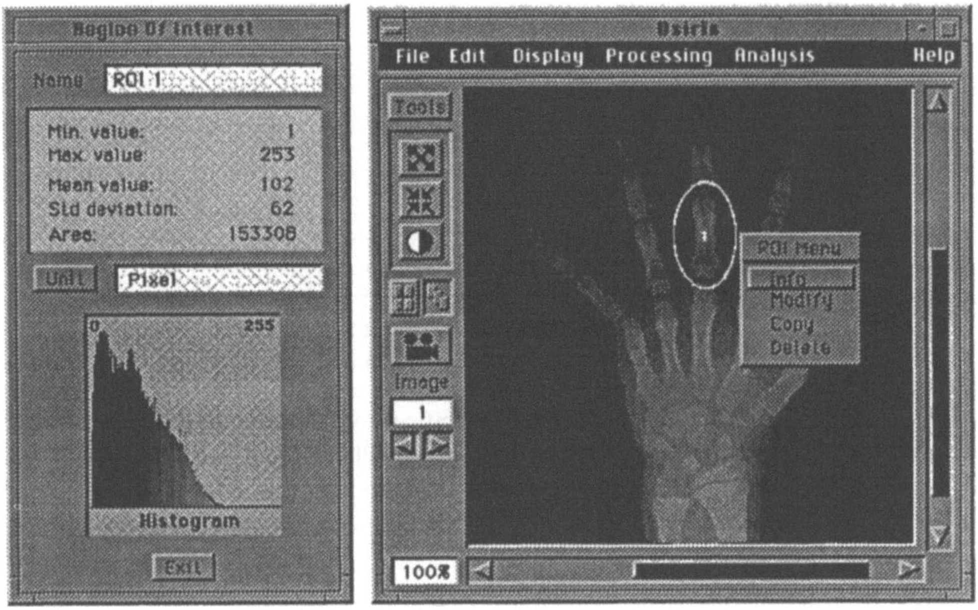

Fig. 1: OSIRIS main window (containing an image with a ROI: region of interest) and ROI window associated.

Functionalities

OSIRIS provides the traditional modes of visualisation (zoom in, zoom out, invert, etc). Multiple images can be displayed in two different modes (Stack and Tile modes). Several image manipulation tools are offered such as filters, analyses, measure of distances, modification of the color lookup table [2].

Medical images can be accompanied with different information. OSIRIS allows to define ROI's (Regions Of Interest) as well as annotations. A ROI is defined by a contour on the image and a label. The contour can be determined either by a geometrical shape, by hand or by an image processing tool (isocontour for example). A second window is associated to a ROI and can be displayed at request (fig. 1). It contains some information about the ROI (histogram and statistics). An annotation allows to emphasize some points or regions in the image. It is defined by a label accompanied by an arrow pointing on the interesting region. A pop-up window, called Post-it, can also be associated to the annotation and can be displayed at request in a specific window.

Stack and Tile Modes

Medical images often come in sets, OSIRIS allows to display them in two modes: Stack and Tile modes (fig. 2). In the Stack mode, the set of images can be displayed sequentially but only one image at a time. The operations applied to the selected image can be automatically applied to other images of the stack. OSIRIS provides also a dynamic Stack mode called Movie mode in which images are visualised sequentially in time, to reproduce real time movements. In the Tile mode, the images are displayed side by side in a grid that is optimized to be as square as possible based on the number and the size of images to be displayed.

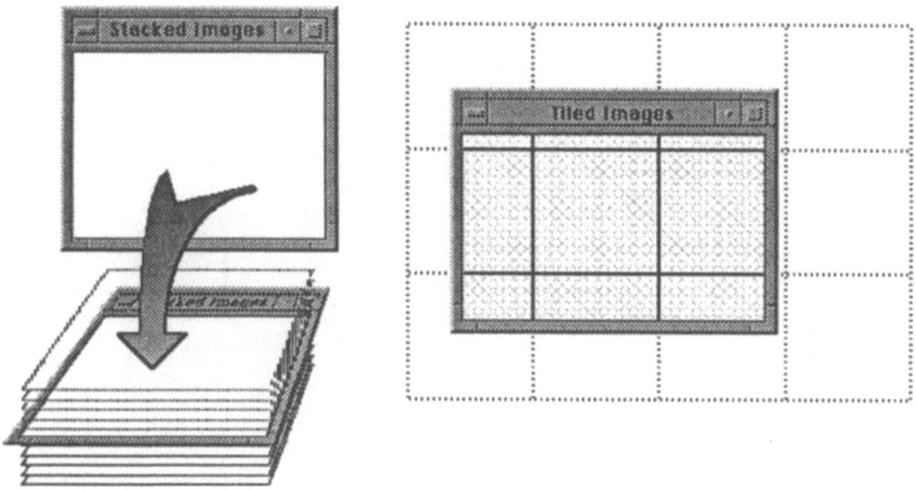

Fig. 2: Different visualisation modes of a set of images: Stack and Tile modes.

Software Architecture

OSIRIS is being developed with the object oriented programming language C++ in order to obtain a well-structured and adaptable software. A first prototype developed on a Macintosh [3] has shown that a strong and flexible software is needed to serve as a platform for the future extensions required by physicians. The figure 3 gives an overview of the class architecture of OSIRIS.

Conclusions

A software is being developed to display medical images. This software is designed to be portable on different Unix and Macintosh systems. The present version works on a SUN under X11 and OSF-Motif window system and on any Macintosh II. An important effort has been made on the design of the interface and the modular software architecture in order to be able to serve OSIRIS as a platform for future developments in image processing and image analysis applications.

References

1. Apple R., Ratib O., Scherrer J.R.: The Use of a File Format Description Language for Easy Handling of Images from Multiple Sources. Proceedings of EuroPACS 90, 8th International Meeting. Trieste, Italy, May 1990.

2. Ratib O., Ligier Y., Funk M., Girard C., Perrier R., Hochstrasser D.: PACS Workstation: User Interface Design. Proceedings of NATO ASI meeting on "Picture Archiving and Communication System (PACS) in Medicine". Evian, 1990.

3. Ratib O., Huang HK.: CALIPSO, an Interactive Software Package for Multimodality Medical Image Analysis on a Personal Computer. J. of Medical Imaging, 3: 205-216, 1989.

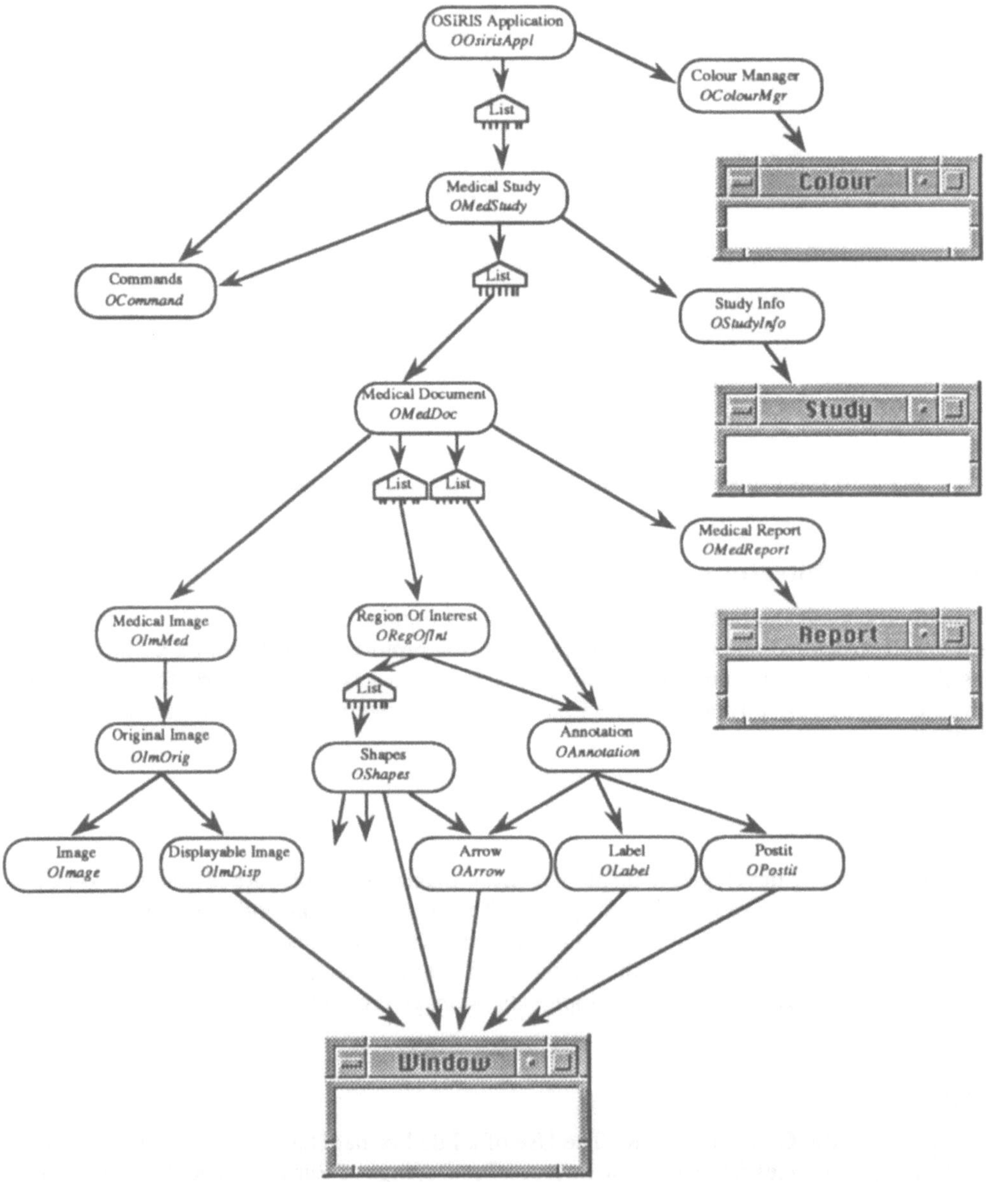

Fig. 3: OSIRIS class architecture: main ownership links.

Dedicated software packages to support PACS research

Fenno P. Ottes, Elizabeth M.S.J. van Gennip, Luuk Steenbergen and Bas M. van Poppel

BAZIS Central Development and Support Group Hospital Information System,
P.O. Box 901, 2300 AX Leiden, Netherlands.

Abstract This paper is concerned with two software packages that have been developed. Each of these supports a research activity of BAZIS in the field of PACS, i.e. Diagnostic Image Quality and Technology Assessment. The diagnostic quality of (digital) imaging systems can be measured in an objective way by means of the Receiver Operating Characteristic (ROC) technique. The performance of ROC studies involves many practical difficulties. It was recognized that the use of dedicated software could bring relief here. The financial consequences of the implementation of PACS have been studied by various groups. Our comparison of such studies in the literature revealed many discrepancies. An international dialogue is needed to elicit the reasons behind these differences. We have developed a software package that calculates the costs and savings of the PACS system in comparison with the film-based system. The international use of the package will facilitate analysis and support the required dialogue.

Keywords: diagnostic image quality, ROC, technology assessment, costs and savings.

The FEASIBLE package

To support the evaluation of the diagnostic quality of imaging systems we have developed the software package FEASIBLE, which is an acronym for Feature Evaluation And System Inspection By Logged Experiments.
The diagnostic quality of images can be measured by means of an Observer Performance study. Such a study is based on psychophysics and includes the Receiver Operating Characteristics (ROC) technique for the statistical analysis of the data. An ROC study is usually set-up to measure and compare the diagnostic quality of two (or more) series of images. When the question (often occuring in the PACS field) is whether the diagnosis made using a digital image displayed on a workstation is at least as good as that using the conventional film viewed on a light box, a series of radiographs on film will be compared with digital images that are produced after digitization of the films.

An ROC study consists of the following phases. First the set-up of the study as a whole is defined. This includes the careful preparation of the required image series. Then a number of psychophysical sessions are performed. Because of their essential role in making the diagnosis, radiologists act here as observers. Images must be presented in a randomized sequence to avoid that the observer will be able to predict the nature of the image. For each image one or more diagnostic questions on eg. the presence of pathologies should be answered using a certainty scale. When all data

NATO ASI Series, Vol. F 74
Picture Archiving and Communication Systems (PACS) in Medicine. Edited by H. K. Huang et al.

have been collected the data analysis will generate an ROC curve. The parameters of the curve, eg. the area under the curve, are a measure for the diagnostic quality.

The ROC methodology has many practical difficulties, such as:

- It must be conducted according to a strict protocol. If the protocol is not obeyed some or even all data could become useless.
- Images must be displayed in a random sequence, but data analysis can only take place when the observer responses are reordered.
- The data analysis involves arithmetics, which is too complicated to be performed by hand.

To support an ROC study we have developed the FEASIBLE package. Parts of the package have been based on previous work at the universities of Chicago (1) and of Arizona (2). It is written in Fortran and Pascal, and runs on an any kind of PC with MS-DOS. The human interface of the package is based on hierarchically ordered text menus, so that the not-yet-experienced user will be guided through the package. The program can be interrupted at any time during a session, which is important because medical professionals act as observers.

Until now FEASIBLE has been distributed internationally to more than 30 interested research groups. Its has been applied in various ROC studies including those within the scope of the Dutch PACS project and at the Leiden University Hospital (3,4). The results of one of these studies can be found in Figure 1. An improved version has been completed recently (5).

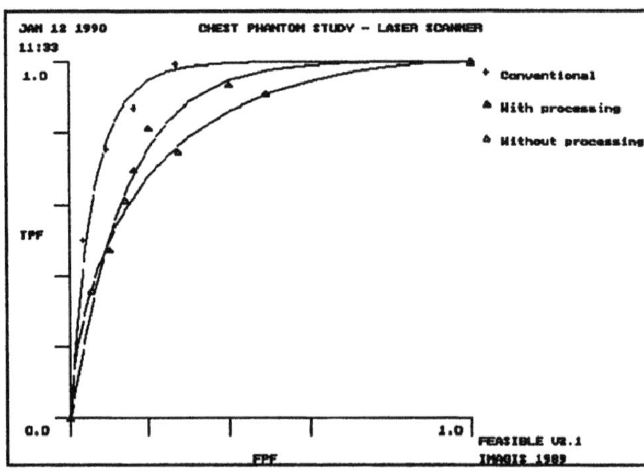

Figure 1. ROC curves plotted with the latest FEASIBLE version (2.1). In these studies chest phantom images were used. The data was pooled over all observers. Two ROC curve obtained with laser scanned images, with and without postprocessing, are shown in comparison with a curve of the conventional radiographs. The symbols indicate the raw data points directly derived from the rating scale histograms. The continued line indicates the curve obtained after the maximum likelihood estimation.

The FEASIBLE package has proven to be very useful in ROC studies, it guides the experimenter during the design process and provides automated data collection. Studies performed with FEASIBLE by different groups are because of the structured set-up better comparable.

The CAPACITY package

For the decision whether to implement PACS or not, it is crucial to estimate the costs and savings of such an implementation. Technology Assessment is concerned the financial consequences of all advantages and disadvantages of PACS. For practical reasons only the costs of equipment, consumables, floor space and personnel are considered here. It is expected that costs will change, eg. cost of PACS components will decrease, and costs for personnel will rise. The results of various existing studies comparing the cost of a film-based system and a PACS system appear to differ largely (6). To be able to quantify the effects of these changes and to support an international dialogue on this issue the sofware package CAPACITY has been developed (7).

It is based on a description of the current film-based situation in the hospital, a model of the PACS architecture and an algorithm for the trends in pricing. The data, that the user enters, are first compared by the Critique module with expert values using rules of thumb. CAPACITY calculates the annual costs of PACS and of the film-based system. It does so for a number of years in which the system could be purchased (3, Figure 2).

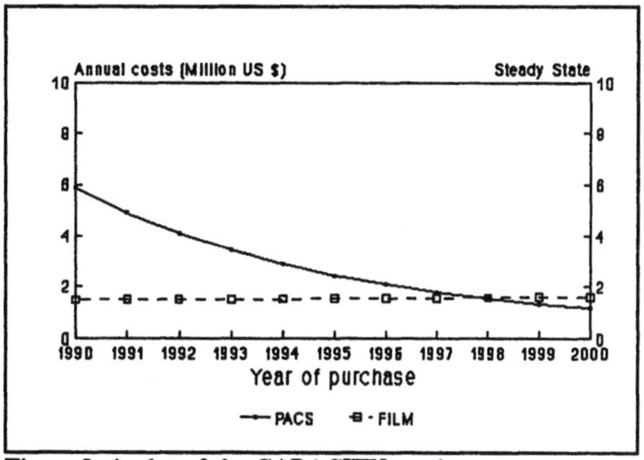

Figure 2. A plot of the CAPACITY results.

CAPACITY is written in TURBO Pascal, runs on any PC with MS-DOS, and is available for interested groups. The 500 input parameters required for one case are entered through menu's.

CONCLUSIONS

The development of these software packages has given us a better knowledge of the design of the kind of studies they support. This knowledge has led and will lead to further improvement of newer versions. We have been able to support other research groups by making the packages available to them, but also by assisting them in the setup of their studies and in the application of the package. The further development and dissemination of the packages shall continue to contribute to mutually fruitful research cooperations.

References

1. C.E. Metz, H.B. Kronman, P.-L. Wang, and J.-H. Shen, ROCFIT: A modified Maximum Likelihood Algorithm for Estimating a Binormal ROC Curve from Confidence-rating Data, Univ. Chicago, 1985.
2. G.W. Seeley, M. Borgstrom, J. Mazzeo, A general interactive computer program for running signal detection experiments. Behaviour Research Methods and Instrumentation 14, 6, 555, 1982.
3. BAZIS, University Hospital Utrecht and Philips Medical Systems, Concluding report of the Dutch PACS project, phase 1 (1986-1989). March 1990, available through BAZIS.
4. F.P. Ottes, J.P.J. de Valk, H.M. Kroon, F.H. Barneveld Binkhuysen, G.W. Seeley, L.H.L. Winter, L.J. Schultze Kool, Diagnostic image quality evaluation chain: applications of the FEASIBLE package in practice, Proc. Medical Imaging III, Newport Beach, SPIE Vol. 1093, pp. 2-9, 1989.
5. F.P. Ottes, L. Steenbergen, F.H. Barneveld Binkhuysen, G.W. Seeley, FEASIBLE software supports the ROC evaluation of digital images. Proc. Medical Imaging IV, SPIE Vol 1234, 1990, in press.
6. E.M.S.J. van Gennip, F.P. Ottes, B.M. van Poppel, J.H.T.H. Andriessen, Why do cost-benefit studies of PACS disagree?, Proc. Medical Imaging IV, SPIE Vol 1234, 1990, in press.
7. B.M. van Poppel, A.R. Bakker, J.B.M. Wilmink, A package for cost and critical analysis of picture archiving and communication indicating its true yield (CAPACITY), Med. Inform. 1990, 15, No 1, 67-75.

Modelling and simulation: indispensable tools for PACS development

W.J.J. Stut jr.[1], M.R. van Steen[2], L.P.J. Groenewegen[3], A.R. Bakker[1]

[1] BAZIS, Central Development and Support Group Hospital Information System, P.O. Box 901, 2300 AX Leiden, Netherlands
[2] TNO Institute of Applied Computer Science, Delft, Netherlands
[3] Department of Computer Science, Leiden University, Netherlands

Abstract: This paper illustrates the use of PACS simulation. We briefly describe the modelling methodology used. By reporting on a concrete PACS network simulation we show that modelling and simulation are indispensable tools for PACS development.

Keywords: PACS design, modelling, simulation, performance prediction.

Introduction

BAZIS decided to support PACS development by modelling and simulation. The use of simulation models addressing hardware, software and image-related working methods, has the following advantages [2]. First, simulation can be used to study the performance of future systems. Although the results of small-scale PACS experiments are promising, it remains to be seen whether current solutions will be also satisfactory for hospital-wide systems. Second, simulation can be used to actively support the design of software. The applicability of this form of software development follows from the observation that modelling software for future PACS systems in fact is the equivalent of designing this software. Finally, simulation can be used to answer so-called what-if-questions in the operational phase.

Modelling method

In order to capture the full complexity of PACS in a simulation model, and to take full advantage of simulation as a design tool, we have started the development of a new modelling method [3]. This method is based on semantic data models and parallel decision processes, and can be used for both system analysis and design. A semantic data model specifies objects and their structural relationships. A decision process models the behaviour (i.e. dynamics) of an object by using states and transitions between states.

The advantage of this method is that, due to its reflecting reality directly, it elegantly combines simulation modelling with software engineering. Furthermore, the modular approach facilitates model maintenance. The first systems modelled with the new method were imaging procedures in a hospital, a PACS network and an optical disk juke box.

NATO ASI Series, Vol. F 74
Picture Archiving and Communication Systems (PACS) in Medicine. Edited by H. K. Huang et al.

Case Study: ImNet

Within the scope of the HIPACS project (one of the projects within the AIM program of the EEC), BAZIS simulated the ImNet network as conceived by the Aachen University of Technology [1]. There were three main reasons to build an ImNet simulation model. First, although a prototype network has been installed, there is no experience concerning its behaviour under PACS production load. Second, simulation can be used to optimize configurations in existing hospitals. Finally, the modelling and simulation also lead to better insight into the behaviour of the network.
The characteristics of ImNet are the following:
- fibre-optic transmission channels that, due to the lower transfer rate of the connected PACS stations, offer an effective transfer rate of 1 MB/s. (The next version of ImNet is expected to operate with a transfer rate of ca. 12 MB/s.)
- a hyperstar topology (in which multiple data transfers can take place at the same time at full network speed), with line-switching (i.e. for two communicating stations the network looks like a point-to-point link).
- the network control is decentralized. When establishing links, the routing of commands is performed by independent switch units.
- when requesting an image, at least two connections have to be established: a connection to send a GetRequest command, and a connection in the opposite direction to send the requested image. When requesting a case (i.e. the images of a new examination together with some relevant historical images) this procedure is repeated for each image.
- a network channel between a switch unit and a PACS station can be used for only one data transfer; a network channel between two switch units can be simultaneously used for opposite data transfers.

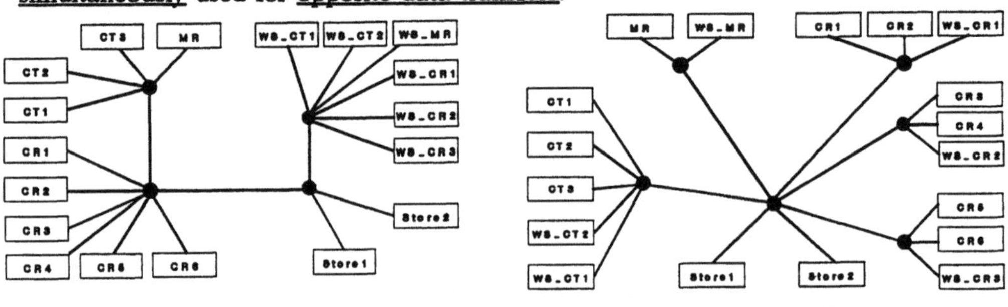

configuration A configuration B

The above figure shows two of the simulated configurations. In configuration A the acquisition stations (CT, MR, CR) and the workstations (WS_CT, WS_MR, WS_CR) are put into different groups. In configuration B the stations are grouped per modality. In order to focus on the performance of the network itself, no layered storage hierarchy was included in the simulations. The workload was based on an information analysis at the Aachen Klinikum, and on general data regarding image production and image use as available at BAZIS. The image size was 512x512x12 for CT and MR, and 2048x2048x12 for CR. When making a report, a CT case consisted of 10-30 images and a MR case of 20-40 images. A CR case had either 2 or 4 images.

Simulation results

The simulation has shown that the waiting times strongly depend on the occurrence of interfering image requests. After all, if workstations are connected to the same switch unit, they share the same network channel to the storage units. Since this network channel can be used by only one workstation at the time, the image requests coming from this cluster of workstations are served one after the other. With an 1.0 MB/s transfer rate, the time needed to send a 6 MB CR image is at least 6.0 s; for a CT or MR image the transfer time is at least 0.375 s. Consequently, the waiting time for a CT or MR case increases significantly if a CT case interferes.

Table 1 shows the average waiting time per case for all users for both configurations. Since a network channel between two switch units can be simultaneously used for opposite image transfers, in configuration B an image transfer to a workstation may overlap with an image transfer from an acquisition station of the same cluster. In general, the waiting times in configuration B are shorter; however, due to the coincidental occurrence of interference, the waiting time for the user at the MR workstation has increased.

Table 1. The average waiting time [s] for the various users. In configuration B* the segment size is 1.0 MB.

Conf	WS_CT1	WS_CT2	WS_MR	WS_CR1	WS_CR2	WS_CR3
A	17.7	19.2	20.9	23.2	23.8	24.3
B	15.6	14.7	25.4	20.3	22.1	21.7
B*	14.3	12.8	15.5	22.1	22.7	22.9

Apart from modifications in the configuration, we also studied the effect of parameters such as transfer rate and segment size. As an example we will discuss the segment size used.

Instead of sending an image in a single step, an image may be split into segments, that are sent one after the other. For each segment a new connection has to be established. After it has been sent, the connection is closed and a next segment is handled. However, this next segment may belong to another image that has been requested by another user.

The use of segments decreases the waiting time for individual CT and MR images significantly. However, Table 1 shows that the eventual effect on the waiting time for a CT or MR case is less significant. This is due to the fact that e.g. a CT case consists of 10-30 images; if a CR image (split into 6 segments of 1 MB) interferes, the waiting time for the CT case may still be increased with 6 (= 6x1.0) seconds! The average waiting time for a CR user has increased. After all, splitting a CR image into 6 segments increases the expected number of interfering transfers.

Conclusions

We have shown that simulation is a powerful tool to study the effect of design decisions before they are actually implemented. We thereby prevent that future hospital-wide PACS systems will be based only on current experiences and solutions.

References

1 Fasel B., Vossebürger F., Meyer-Ebrecht D.: High Speed Networks for PACS - Introduction and Comparison. In Proc International Symposium CAR'89, Lemke e.a.(eds), June 1989, pp. 577-580.

2 Stut jr. W.J.J., van Steen M.R., Groenewegen L.P.J., Bakker A.R.: PACS design issues: the importance of modelling and simulation. Accepted for publication in Journal of Digital Imaging.

3 Stut jr. W.J.J., van Steen M.R., Groenewegen L.P.J., Bakker A.R.: An integrated approach for simulation-based system development based on transformations of system specifications. In Proc European Simulation Multiconference, B.Schmidt e.a.(eds), June 1990, pp. 655-660.

Image Registration and Communication Standards in Medical Imaging

A. Todd-Pokropek and M. Oghabian
Dept of Medical Physics
University College London
U.K.

ABSTRACT

A major potential benefit in digital medical imaging is the possibility of extracting quantitative information, for example, of tissue character. However, there is an additional need to be able to register very different types of images, so as to be able to deduce physiological information within a precise anatomical identification. This is a 3-D problem, and after appropriate preprocessing, critical points can be established on diverse images such that they may be iteratively translated and rotated in 3-D. Results using a VAX linked to a Kontron image processor are encouraging. A more general problem is associated with defining communication standards so as to be able to transfer images from a variety of different medical imaging systems. A variety of standards have been proposed (AAPM, ACR/NEMA etc) and there is a requirement for an appropriate (international) data dictionary to be established to define the logical format, for which a proposal is presented.

1. INTRODUCTION

The basic requirement of a Picture Archiving and Communication System (PACS) network is that all images should be handled digitally, eliminating the need for film and hardcopy. The benefits would be that all images would be immediately available throughout a hospital complex, no films would be lost, and a great saving would be achieved both in eliminating film, and reducing delays between referral and reporting. This is of course the specification of an ideal system after all the intermediate problems have been solved. Based on this we can estimate the technical requirements in terms of Gigabytes of storage per day, number of displays, capacity of the communication network, and maximum acceptable delays. However, such a system will considerably change the working practices of a hospital, and then, rather like the Heisenberg uncertainty principle (-once you observe it, so it changes) these estimates can only be very approximate.

NATO ASI Series, Vol. F 74
Picture Archiving and Communication Systems (PACS) in Medicine. Edited by H. K. Huang et al.

It has been suggested that all present estimates of the economic value of PACS are both very uncertain, and in reality indicate poor performance in terms of their economic viability. However, a possible solution to this is available, in terms of what has been called 'value added PACS' that is making the PACS system deliver performance, in terms of clinical utility, which was previously unavailable, by means of 'image processing'. The ability to be able to perform image processing in PACS may be the critical requirement in economic viability. Examples of such image processing are in the area of quantitation and image registration.

2. IMAGE PROCESSING METHODS AND QUANTITATION

Image processing has been extensively developed in certain areas of medical imaging, in particular nuclear medicine, as well as in non medical areas, such as remote sensing (satellite images). Much of the work in this area is devoted to techniques of segmentation, classification and quantitation.

Certain methods of quantitation are essentially straightforward, and in common use. For example, measures of size, for example volumes and sizes of organs, as in estimating foetal head diameter in Ultrasound (US), tumour size in MRI, CT, NM (nuclear medicine) etc, can be performed without difficulty. Slightly more indirect measures such as ejection fraction and cardiac wall motion in NM, US and DSA, are derived from the same methodology, but are more strictly 3-D rather than 2-D (image based) operations. Tissue characterization, as employed in various methodologies, is even more indirect. An example of this is in the measurement of T1 and T2 relaxation times in NMR, estimating tissue character based on a knowledge of T1/T2 values is known tissue types, and then preparing a map of tissue types based on this and anatomical information.

Segmentation involves the separation of the image or image series into distinct identifiable regions. The most common method employed is that of 'edge detection' based on the 'image gradient'. Many better methods are available.

Two distinct methods (in image processing) must be taken into account: model based (fitting) and model forming. A model based method is one which, based on information (such as T1/T2 values) fits the observed data, taking into account statistical effects, to produce additional diagnostic information (hopefully). Edge detection is model based. An alternative, more clinical example, as often employed in NM, is to use a compartmental model to establish the physiological basis of some environment, and then by fitting the observations (radioactivity as a function of time) deriving and indirect measurement such as blood flow or regional oxygenation. Model forming methods are those where, based on the statistical properties of the image data, a self forming hypothesis is made, again based on which clinical observations can be derived. An example would be using cluster analysis, where certain regions of the image are observed to have similar statistical properties (means and standard deviations etc) and those some segmentation - that is: division of the image into distinct identifiable regions, can be performed. When a series of images exists, for example in time, such methods can be very powerful in terms of feature extraction, both segmentation and classification.

With model forming methods, a fundamental problem exists. The initial 'models' determined are usually unrelated to any physiological, physical or clinical phenomena. The model must be 'rotated' so as to identify effects which relate to some meaningful set of properties. That is, the statistically derived model must be constrained by the imposition of external knowledge. In other words, model forming methods must be constrained by some external (real) model to produce meaningful results. Model fitting methods often involve an inverse approach, allowing unwanted variability to be absorbed within statistical 'side-effects'.

While this might appear to be obscure, it is clear that, given a series of images, for example images with different acquisition parameters in NMR, at different times (in NM etc) or in the classification of a series of images from different patients, the human observer may not be an ideal classifier. Considerable improvements can be obtained using such well established image processing methods to guide the observed. Image subtraction is a trivial (that is easy to implement) but very powerful example of such a technique. Thus a prerequisite of quantitation is often segmentation, that is the identification of regions on which the measurements are to be made. These quantitative measurements may be direct and simple (such as volume) or much more obscure and difficult to validate (such as rCBF).

3. REGISTRATION

In many circumstances the clinical interpretation (and quantitation) of an image from a single image modality is inadequate. There are potentially many situations where different imaging methods (parameters), or the use of time information may add clinical value. This requires the alignment of anatomical or functional information from several image sets, which **must** be performed in **3-D**. The handling, display and interpretation of image in 3-D, as in CT, MRI, NM and soon, US is of considerable importance and many systems for surface and volume rendered display of medical data exist.

The registration and superimposition of images from different sources (multi-modality registration) is therefore an interesting and important area associated with image processing. Within a given system, however, data from probably quite different applications areas need to be handled simultaneously and presents communication problems since orientation, physical matrix size and position, need to be defined generally. Position with respect to anatomical landmarks, or some reference frame, may be especially problematic. Spatial distortion is likely to be even worse, and needs to be corrected.

We have been working on systems for registering images from MRI together with data from nuclear medicine (PET and SPECT), as have many other workers, for example C.T. Chen, and, in other areas, D.I. Barnea. The problem has been constrained to one of translation, scale change and rotation in 3-d, and the question of distortion (warping) has been ignored. Thus (at least) two approaches are possible: accurately defining position in some (portable) global reference system, or using some registration procedure to perform the alignment. In the latter case, a reasonably good starting position is still required.

For similar images, an iterative procedure using the Stochastic Sign Criteria (SSC) has been found to be helpful. In this study, external reference marks were not used, and since the images were very dissimilar, sets of critical points have been defined or detected automatically, which are then registered using an iterative algorithm. These points do not need to be paired. In general it has been found that such methods do not have a very sharp (global) minimum, and also suffer from the presence of many local minima. Thus the iterative procedures tend to be rather slow.

All methods, based on some distance measure, suffer from the presence of multiple local minima, when minimization algorithms are used to reduce the distance between two edges, or surfaces in 3-D (surface fitting).

A surface fitting algorithm evaluated by a least square distance function (LSD) which is minimized by an appropriate fast local minimization algorithm (for example, the Powell method) has been implemented, and, to improve the method, extended to multi-resolution (scale) space). The fitting starts using low resolution data, stepping to higher resolution as the desired fit passed an appropriate threshold. This largely eliminates the human operator interaction, often required by the presence of local minima and leads to a faster convergence in a global search space. In addition we attempt to eliminate any local minima during the search by appropriately setting threshold values for each resolution level. The choice of threshold level directly influences the number of search steps (transformation) required for the next higher resolution level, and in turn the accuracy and computation cost of the algorithm. A sequence of lower resolution surface images of sizes 128*128, 64*64, 32*32 are generated. The inverse transformation was applied on each surface data of varying size, and individual distance errors and the cumulative error was measured. Sets of MRI and PET images were used primarily to test the procedure, implemented on a Kontron image processing system connected to a microVAX.

Surface detection on the serial slices of each scan type was performed. For the surface to be fitted a solid model, and for the other scans, a series of 3-D points are produced. Automatically, the border of each binary surface is found, and a 3-D box (VOI) is introduced around each target surface to be fitted. The centroid of each data set, inside the VOI box, is then computed and moved to a selected point. A number of points are then selected in one surface data set, and their distance are measured with respect to the other surface by ray tracing, until the intersection with the second surface (objective data) is detected. The distances between corresponding points are then computed using the L2 norm distance function (LSD) for each parameter setting. The fitting program allows visual inspection of the two surfaces on the monitor, which allows for modification of the fitting parameters, or termination of the program, if the fit appear inappropriate (or it starts diverging).

The input to the multi-resolution image registration system is two surface images obtained by one of the surface extraction techniques. A set of images in scale space (decreasingly lower in resolution and smaller in size) are created from the original high resolution images. Since at a lower level of resolution the number of possible search locations decreases, the LSD function converge to an approximate minimum in a very short duration of time, even if the entire 3-D image is searched (global search). In lowest resolution level all possible combinations of transformation parameters were tested for the best LSD value between the two sets. The method uses the best transformation parameters, and the most promising search location found so far at a given level to pass them to a higher level for further investigation.

Since the density distribution of the error measures can be approximated by a normal distribution for large values of n (the number of critical points), the cumulated error has been taken into account instead of individual error measures, where the mean and SD of such data can be computed. The threshold for stopping the evaluation of the LSD can, then, be derived from the value of cumulative error, where this threshold should not eliminate those locations that are likely to be important as the true match location of the higher resolution level.

The choice of points, included in registration (position of selected points) should satisfy our assumption that the number of sampled points are a representative of all image points and they reflect all the misregistration information about the whole data set. Here this was accomplished by dividing the area of surface into m region having different distinct slopes (surface normal), and selecting one points from each region.

A series of experiments have indicated that the registration of data from different image modalities may be assisted by using such multi-resolution methods, tested on these different sets of clinical images. An example of our results is shown in Fig 1.

4. COMMUNICATIONS

Since a major problem (not restricted to the area of image registration) is the ability to transfer images from a variety of sources between different machines, an important task is defining and establishing the communications between such machines such that the data collected can be transferred. This implies not only defining the medium (Mag/tape, RS232 channel, ACR/NEMA port) but also the logical format and communication protocol. While ACR/NEMA may indeed be the protocol of choice and is widely recommended, this remains to be established since few implementations exist. Thus an investigation is required into the practicality of transferring large data sets between both routine (existing) PACS devices and future and research systems.

Communications, as always, needs to be separated into distinct levels. The physical levels are surely best defined by the standards community at large, and thus, ACR/NEMA seems to be an oddity and rather isolated. The ability to transfer data between systems should be totally independent of the medium. However, areas which specifically concern PACS, and which must be defined by that community are associated with the logical format of the data, and the high level (e.g. at the level of the user interface) mechanisms for study transfer. Note that the requirement is for the transfer of studies, and not files as such, since a study may comprise many files. The file structure itself is an issue within the definition of the logical format.

Imaging studies may be very complex, and involve many different types of data. This seems to be been inadequately treated by the ACR/NEMA definitions. A study is often NOT just a collection of slices, or even an n-D data set. In fact the transfer of multidimensional blocks of image data presents almost no difficulty. Many of the problems start with the definitions involved within the 'patient header blocks', that is, patient and medical data related to the study. Some of these can be handled easily (patient name, age, sex..), some by defining appropriate conventions (orientation in 3-D ...), but many definitions and parameters remain complex, idiosyncratic and obscure. Examples seem to include: pulse sequence definition in MRI, centre of rotation information in SPECT, specification of contrast injection method in DSA, cardiac/ respiratory gating information in general and 'model parameters' in PET. All of these types of definition are likely to be applications specific, and in many cases may be (imaging) device specific.

A general question which must be answered is whether such transfer formats are intended to be reversible. Is it a requirement that transmitting a study back to the system it came from should result in a data set which is indistinguishable from the original, or should only the 'meaningful' parameters be transmitted, and then restored? Part of the answer lies in the definition of the PURPOSE of the transfer format, is it for interpretation on a possibly alien system, or is it (also) for archiving?

The AAPM format [AAPM10 1980] as developed by various workers [C. Maguire et al 1989] is an alterative logical format for data transfer which has been adopted as a standard for nuclear medicine under the COST B2 initiative. Data comprises a header block containing, in pure ASCII, keywords and their associated values, followed by data in binary form. An example of a keyword pair would be patient sex:= male;. Many such keyword pairs have already been defined for a variety of applications.

Data as such comprises not just images, but also time activity curves, and potentially, lists of constants. However, neither the ARC/NEMA nor AAPM definitions address the problems of such a collection of sets of different types of data. For example, a clinical study might comprise a set of images, plus curves, constants, regions of interest, etc. A suitable method for grouping such data into (to the user) a single object, must be defined.

The ACR/NEMA standard is, in terms of logical format not so different from the AAPM standard. However, it does not use the keyword pair concept, preferring integer tags, which reduces 'clarity'. The AAPM standard on the other hand says nothing about packets, or file transfer protocols. There is may be an advantage in terms of 'readability', portability, and maintainability of using ASCII keywords. However there is a speed and compactness advantage in not using ASCII keywords. Any such a system in order to become widely used, requires a formal method for maintaining such a data dictionary, which would need to be properly defined. A proposal (which could also apply to the ACR/NEMA standard, or rather a joint ACR/NEMA/AAPM standard) would be as follows.

1) Establish a bulletin board and mailer.

2) On this system define
   a) general terms (keywords) for all applications
   b) applications specific terms (application dictionaries)
   c) manufacturer specific terms (user group dictionaries)

3) When a new term is required, check for uniqueness and add to a user group dictionary

4) Establish a mechanism for transfer of terms between different dictionaries

Thus 'user groups' could exist with their own application or device specific codes, such that a reversible transfer protocol could be defined. Any term existing within such a local group would not need to be known by other groups, and such terms would therefore would be ignored (or turned into comments) in any conversion program. Thus terms which are found to be clinically significant would need to be transferred from the users dictionaries upwards, firstly to the application specific sets of terms, and perhaps eventually to the general dictionary.

As an example, to be specific, orientation needs to be defined generally, pulse sequence in MRI needs to be defined to some extent for all manufacturers (application specific) but with the ability for manufacturers to have their specific terms defined for special purposes only. Centre of rotation correction might end up existing only within a user group dictionary. The ability to maintain such a system requires an internationally accessible communications network.

The study transfer interface, and protocols for the transfer of studies is partly a problem to be handled within a given standards definition, and partly in terms of the Human Computer Interface (HCI). There is also a strong desire for portable HCIs such that a clinician trained on one system can readily use a different system.

5. SUMMARY AND CONCLUSIONS

While there is considerable activity in the area of image networks in Medicine, the ability to improve the value of such system by use of image processing techniques seems to be quite important. An example of an image processing of potential value is in the area of quantitation. However, further clinical benefit can be obtained by handling medical images derived from different sources, superimposing the information, and enabling the comparison of anatomic and various types of physiological information. Such image registration needs to be performed in 3-D, and is far from straightforward. A variety of different methods have been described for handling this kind of fitting operation.

A further problem which needs to be solved in general in PACS, and in particular for the image registration application is that of handling the communications required between many different types of imaging devices. While several standards have been developed for these purposes, these need to be extended and rendered more flexible. For this purpose, it is suggested that the definition of an appropriate data dictionary must be defined, together with appropriate methods for its maintenance.

6. ACKNOWLEDGEMENT

The text of this paper is based on several other reports prepared for ISCAMI, EuroPACS, and the European Radiology Training Program. The first author would like to thank colleagues from UCL, St. Mary's Hospital and Great Ormond St. Hospital for Sick Children for their help and advice.

REFERENCES

1. Pelizzari C.A., Chen G.T. Y., Accurate Three-Dimensional Registration of CT, PET ,and NMR Images of the brain: J. of Comp. Assis. Tomog. Jan 1989.
2. Barnea D.I., Silverman H.F., A class of algorithm for fast digital registration,:IEEE Trans. Compu., 1972, 21, 179-186
3. Baxter BS, Hitchner LE, Maguire GQ, A standard format for digital image exchange, AAPM Report 10. New York, Amer Inst. Physics 1982.
4. The ACR/NEMA format for digital image transmission. NEMA, Washington, 1988.
5. Maguire GQ, Noz ME, Image formats: five years after the AAPM standard for digital image interchange. Technical report #CUCS-369-88, Columbia University, New York.

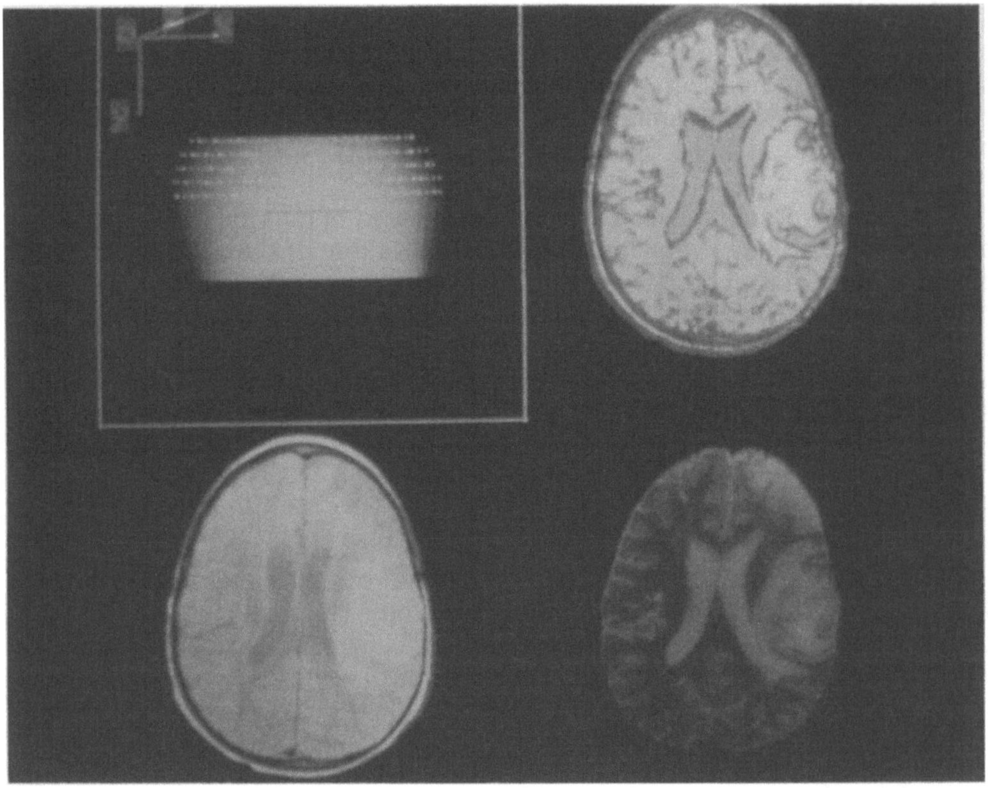

Fig 1. An example of the registration of MRI imgaes obtained with different pulse suquences on different days from the same patient. Top left shows a 3-D impression of the 2 data sets, the second being indicated by dots, top right shows two slices superimposed, with a grey image from the second data sets superimposed by contours from the first, while the two corresponding slices are shown below bottom left and right. The registration was performed in 3-D.

Filmless Nuclear Medicine Department

Douglas J. Wagenaar, J. Anthony Parker, Kevin Donohoe, and Gerald M. Kolodny*

Division of Nuclear Medicine, Department of Radiology, Beth Israel Hospital, Boston, Massachusetts, USA
*Dr. Kolodny is a stockholder in Sudbury Systems, Inc., Sudbury, MA USA

Abstract: The implementation of an all-digital department of nuclear medicine is described. Image display, archival, and retrieval have been performed without the use of film for three years. This PACS system is based on a network of imaging workstations with on-line access to an optical disc archive. Studies as old as three years can be retrieved within seconds for comparison with repeated scans of the same patient. The digital format allows the operator to use gray scale manipulation and cine features for full flexibility in display of any archived study. The filmless nuclear medicine department is shown to be practical and cost-effective.

Keywords: PACS, Network, Nuclear Medicine, Digital Medical Imaging

Introduction

The advance of computer technology and the concurrent development of medical imaging modalities which utilize this increased computational capacity have made the issue of implementing an effective PACS a major concern in every modern radiology department. Although screen/film remains dominant in projection radiography, efforts to supplant film with electronic image receptors are underway in mammography [1,2] and chest radiography [3]. The high resolution of a digital mammogram and the patient volume involved with a screening program would place the most severe burden on a PACS. Since nuclear medicine images require far fewer pixels to display the available information and since it is a mature digital technology, it is logical to begin to implement a complete PACS system within a department of nuclear medicine.

The present work describes the design and implementation of an all-digital department of nuclear medicine. The advantages of this PACS include an on-line, immediate access archive, the ability to perform computer manipulations and analysis, and the prevention of lost studies. Images dating back more than three years (nine years for cardiac studies) can be retrieved in seconds from an optical disc and be displayed and manipulated alongside a current study of the same patient.

System Design

The principal requirements emphasized in the design of our PACS were modularity, free data exchange, sufficient on-line storage to facilitate retrieval of previous studies for comparison, and a convenient backup procedure. Modularity is necessary to avoid the situation in which the operation of the department is crippled by the failure of one major component. A network of four image workstations (Sudbury Systems, Inc., Sudbury, MA) was chosen to meet this requirement. Each image workstation has a corresponding 512 x 512 x 8-bit display monitor, 8-inch , 5.25-inch and 3.5-inch floppy disc drives , a 40 Mbyte hard disc drive, joysticks for gray-scale manipulation, and image processing and analysis software. Free data exchange is the ability of the PACS to communicate with any imaging device within the department with minimal effort. Fast commercial hard wire networks are available from many

NATO ASI Series, Vol. F 74
Picture Archiving and Communication Systems (PACS) in Medicine. Edited by H. K. Huang et al.

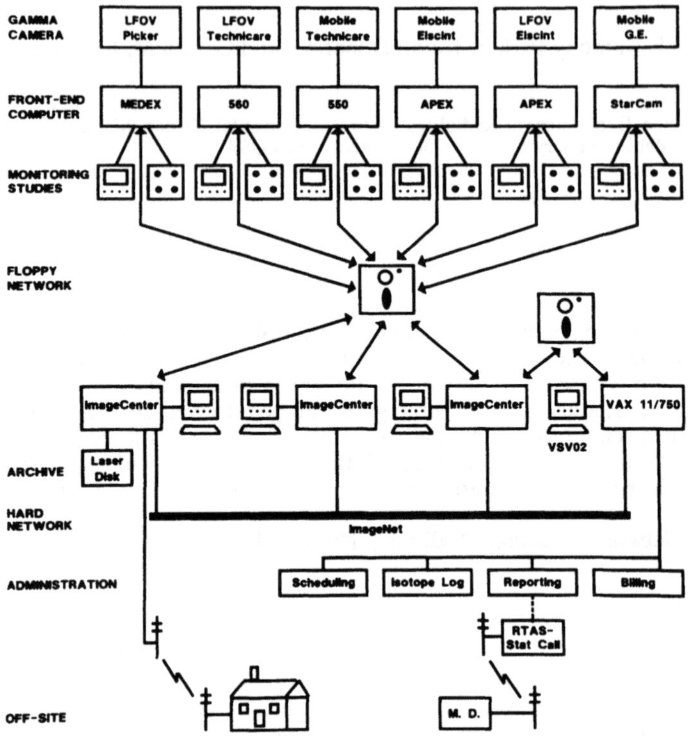

Figure 1. Schematic diagram of an all-digital nuclear medicine department. No photography or film is used. Gamma cameras from four manufacturers are interfaced to the network by means of floppy disc readers. Remote image workstations located within the homes of the physicians are linked to the network via modem.

vendors, but they only work with that one vendor's equipment. This is not acceptable in the usual mult-vendor environment. The ACR-NEMA Digital Imaging and Communications standard [4] has not been implemented on a widespread basis for a variety of reasons [5,6]. In our case, the free exchange requirement was fulfilled through the use of floppy discs. The image data from each gamma camera are written to a floppy disc in the format of the manufacturer, and the network software is capable of converting the data from any floppy disc to the network format. The on-line storage in the PACS is provided by two write once, read many (WORM) removable optical discs. Each of these 12-inch discs is capable of storing 2.2 gigabytes of data. One 2.2 gigabyte optical disc therefore has the capacity to hold about 12,000 studies, or approximately 2 years' worth of data . Finally, backup and recovery functions are available to the image workstation network. Studies from one optical disc are automatically transferred to the other disc during the backup procedure. One image station is occupied, but the remainder of the system operates normally except for a modest loss in speed on network operations.

Figure 1 is a functional diagram of the all-digital department. Six gamma cameras (four manufacturers) communicate with the ImageNet through the floppy disc network. The figure also shows a VAX 11/750 computer (Digital Equipment Corporation, Maynard, MA) which is used for physicians' reports, billing, and scheduling. After a study has been acquired, transferred to the image network, and interpreted by the physician, the study report is dictated into a digitized voice system (RTAS; Sudbury Systems). As soon as the report is dictated, it is available for review by the referring physician over any telephone. Transcribed reports are stored with the corresponding image data on the optical disc archive. All three staff physicians have image workstations in their homes which are connected to the PACS via modem to facilitate readings after hours and on weekends. The all-digital department includes a portable image station based on a laptop PC-compatible computer. With this transportable image station, the full range of digital analysis is available at conferences; the gray scale window and level can be adjusted, and cine views can be shown at locations far from the reading room.

Discussion

The overall response to our system has been very favorable. Personnel who use our networked system include staff physicians and physicists, resident physicians, technologists, and medical students. Newcomers begin to use the system productively after only brief instruction, in large part due to on-line help software. Referring physicians are impressed with the speed with which a current study can be called up, as well as the instant access available to previous scans for comparison. The all-digital department can be shown to be cost-effective when compared with film-based departments over a 5-year period. The initial costs for the digital setup do exceed those for film ($17,000 vs. $8,400), but this difference is more than accounted for by differences in consumable supplies. Substantial savings of $1.69 per study (or $13,520 per year) are realized because the floppy discs are reused as opposed to film, chemicals, and folders necessary for each study. Technologist labor savings are figured to be about $1.00 per study due to the elimination of film processing.

Our experience with the all-digital department has suggested the need for certain enhancements. Obviously the longer the PACS is in use, the greater the storage requirements of the archive will be, so a larger capacity optical disc (or other device) is always desirable. The image workstation network can reproduce printed hard–copies of patient studies, but it would be preferable to have the ability to produce a report in which the printed image is combined with the physician's report in a single document. Since the filmless department relies exclusively on CRT monitors for presentation of digital images, we recently initiated a study of the monitor settings (brightness and contrast) and gray scale windows preferred by our three staff physicians for four common nuclear medicine studies. In conclusion, we are operating an all-digital nuclear medicine department based on modular image workstations which has proven to be cost-effective, highly reliable and easy to use, and able to integrate nuclear medicine studies from four different manufacturers. Current patient studies can be compared with studies taken many months earlier with the ability to digitally manipulate and analyze both studies. Patient billing and scheduling, physicians reports, and modem connections to physicians home-based workstations are integrated into our PACS.

1. Wagenaar DJ, DiBianca FA, and Vance JE 1990 (abstract): Digital mammography using a kinestatic charge detector, Med Phys 17(4): 740-1

2. Bjorkholm P and Annis M 1986 (abstract): Digital mammography. Med Phys 13(4): 614

3. Sonada M, Takano M, Miyahara J, Kato H 1983: Computed radiography utilizing scanning laser stimulated luminescence. Radiology 148: 833-8

4. American College of Radiology, National Electrical Manufacturers Association. ACR-NEMA Digital Imaging and Communications Standard: ACR-NEMA Standards Publication No. 300-1985. Washington, D.C.: ACR-NEMA, 1985

5. Wang Y, Lodwick GS, Zielonka JS 1985: Overview of ACR/NEMA digital imaging and communication standard, in Schneider RH, Dwyer SJ (eds): 3rd International Conference of PACS (PACIII) for Medical Applications. Proc SPIE 536: 132-8

6. McNeill KM, Osada M, Martinex R, et al. 1990: Evaluation of the ACR-NEMA standard ;for communications in digital radiology. IEEE Trans Med Imag 9(3): 281-9

Factors Affecting Detection of Pulmonary Nodules in Digital Images

W. Wiesmann; P. Vassallo; P.E. Peters

Institute of Clinical Radiology, University of Muenster; 4400 Muenster, Albert-Schweitzer-Strasse 33
Germany

Abstract: 120 chest films with and without nodules were digitized with matrices of 1024 and 2048 pixel and analyzed on a 1024-CRT by 5 radiologists. A significant deterioration in diagnostic performance was observed for the digitized images compared with the original films. Detection rate was influenced by contrast of the lesion against the surroundings and unaffected by lesion size. A strong relationship was noted between detection rate, reporting time and the location of the lesion.

Keywords: Radiography, digital; observer performance; images, display

Introduction

Of all radiographic examinations chest radiography places particularly high demands on contrast and spatial resolution of an imaging system. Chest radiography is therefore the ideal test for assessing and comparing different digital systems and the pulmonary nodule has been shown to be a good test signal for such experiments. Presence or absence of such lesions can be relatively easily confirmed and is of great significance to the patient.

A pulmonary nodule can be characterized by its size, the delineation of its borders, its relative and absolute densities, and its contrast to the background noise. Our aim was to assess the influence of these factors on perception of pulmonary nodules on digital chest radiographs with different matrix sizes.

Materials and Methods

A total of 120 chest films with and without pulmonary lesions were digitized with matrices of 1024 and 2048 pixel and a depth of 8 bits. In 60 films intrapulmonary lesions were simulated by a new method using a silver nitrate solvent (potassium ferricyanide) [14]. This solution was applied in different concentrations and for different durations within the lung on one side of the original film to produce nodules of varying density, edge gradient and size. The films were then refixed and dried. The intrapulmonary nodules produced were similar to true pulmonary lesions, and a trial with three experienced radiologists showed that true and artificial lesions were not reliably differentiated.

NATO ASI Series, Vol. F 74
Picture Archiving and Communication Systems (PACS)
in Medicine. Edited by H.K. Huang et al.

The optical density was measured in a defined area within the pulmonary nodule using a microdensitometer. The averaged background optical density within 5 mm from the edge of the nodule were also calculated. The standard deviation of background optical density represents the degree of background noise. The above results were correlated with those of ROC studies as well as the reporting time required for every image. At least 3 reporting sessions were performed by 5 board certified radiologists; one with the original films and two with the digitized images viewed on a 1024 matrix CRT.

Results

Reporting sessions using conventional chest films showed an area under the ROC curve of 0.91. Digitization with a matrix of 1024 by 1024 pixels resulted in a significant deterioration in diagnostic performance. The average area under the ROC curve for all 5 observers in the second session was only 0.76. The difference between these two values was statistically significant ($p < 0.01$). No improvement in diagnostic performance was obtained with a matrix size of 2048 x 2048 pixels in the third session. Here the area under the ROC curve (0.76) was identical to that of the second session (Fig. 1).

Both the conventional film reading sessions and the reporting sessions at the workstation with a matrix of 1024 and 2048 pixels showed no correlation between nodular diameter and detection rate. Even other factors such as optical density of the lesion, optical density of the surroundings, standard deviation of optical density within the surroundings and edge gradient of the nodules influenced detection rate to a small extent. The correlation factors for these parameters ranged between 0.4 and 0.5.

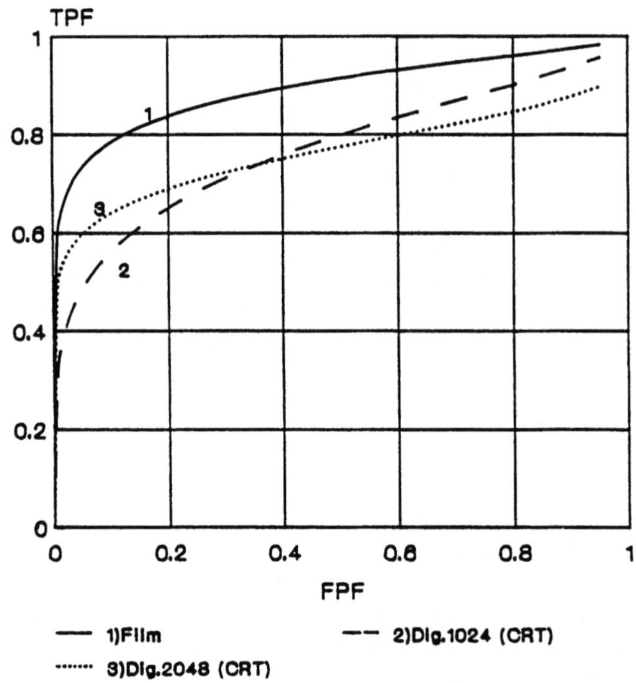

Fig. 1: ROC - curves of the session 1 - 3 for all probands
(Area under curves for session 1 = 0.91; for session 2 + 3 = 0.76)

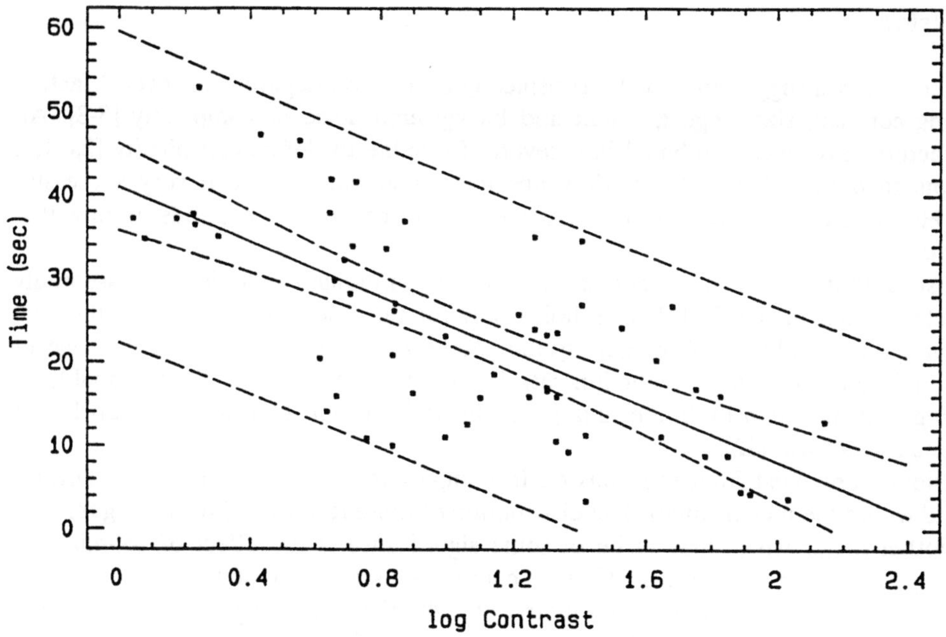

Fig. 2: Regression of reporting time on nodule's contrast (correlation factor = 0,7)

The best correlation was noted between reporting time and the logarithm of the nodule's contrast (Fig. 2).

Correlation factors between 0.6 and 0.7 were achieved. A strong relationship was observed between detection rate, reporting time and the location of the lesions. Particularly when reporting films at the CRT, difficulties were observed with lesions located in the periphery of the lung adjacent to the thoracic wall, in the mediastinum and behind the diaphragm. These diagnostic difficulties resulted in a prolonged reporting time and a reduced diagnostic performance. (Fig. 3)

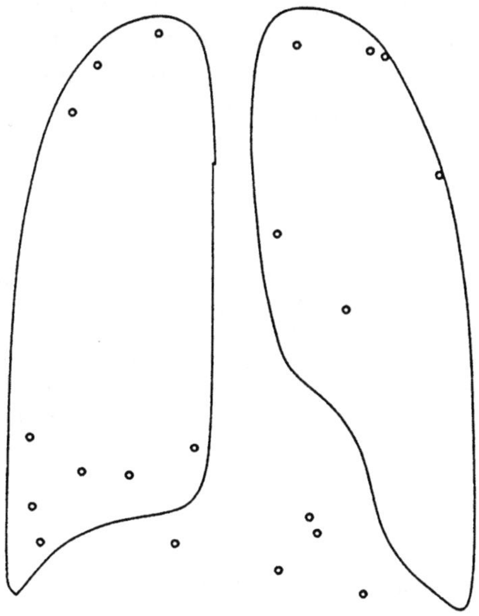

Fig. 3: Distribution of nodules with a confidence level ≥ 3 (conf. level 1 = definite lesion; conf. level 5 = no lesion)

Discussion

The literature has suggested that the conspicuity of a lesion depends on several factors including contrast, size, edge gradient and background noise or complexity [5-8]. All these factors have been combined into several formulae to define conspicuity [2,6,13]. Attempts to define these interrelationships have, however, not been very successful [6,13]. Hallberg et al. [4] have shown that the level of conspicuity depends on how it is measured.

Most authors agree that detection rate of pulmonary nodules is essentially unaffected by lesion size [9-11]. This holds only for those lesions which lie within the limits of spatial resolution of imaging system. Revesz et al. have suggested that one of the main factors influencing detection rate is contrast of the lesion. "The probably of detecting a nodular density in the lung parenchyma is proportional to the logarithm of the nodule's contrast" [12].

Lesions were most frequently missed in complex regions adjacent to the thoracic wall and within the mediastinum. The distribution of these lesions is shown in figure 3.

Another factor which may to be of some significance is the effect of zooming on detecting rate. With a matrix of 1024, a continous assessment of the hole image is possible, whereas in order to utilize a matrix of 2048 pixels zooming is required, so that only part of the image can be analyzed at any one time. This piecewise analysis of digitized images causes deterioration in diagnostic performance, since the ability to compare different parts of an image by rapid jumps with the eyes is lost [6-8]. Analysis of the whole image with a matrix of 2048 pixels may improve detection rate of pulmonary nodules [1,3].

Through the development of improved fluorescent screens, an increased luminance and contrast of CRT's may be expected in the future. Besides, the choice of an appropriate look-up table in combination with histographic equalization may help to improve the diagnostic performance of film reading on the CRT [1].

Literature

1) Arenson, R.L.; D.P. Chakraborty; S.B. Seshadri; H.L. Kundel:
The Digital Imaging Workstation.
Radiology 176,303-315 (1990)
2) Brogdon, B.G.; C.A, Kelsey; R.D. Moseley: Factors Affecting Perception of Pulmonary Lesions
Radiol.Clin. of North America 21,633-653 (1983)
3) Cox, G.G.; L.T. Cook; J.H. McMillan; St.J. Rosenthal; S.J Dwyer,III:
High Resolution 2560 x 2048 x12 Bit Digital Displays for Chest Radiography - A Comparison with Conventional Film and Digital Hardcopy
RSNA Chicago SE (1989)
4) Hallberg, J.R.; C.A. Kelsey; D. Briscoe:
Some effects of method on the measured conspicuity of chest lesions
Invest.Radiol. 13,439-443 (1978)
5) Hendee, W.R.: The Perception of Visual Information
RadioGraphics 7,1213-1219 (1987).

6) Kundel, H.L.: Images, Image Quality and Observer Performance
Radiology 132,265-271 (1979)
7) Kundel, H.L.; C.F. Nodine: A Visual Concept Shapes Image Perception
Radiology 146,363-368 (1983).
8) Kundel, Harold L.: Visual Perception and Image Display
Radiol.Clin. of North America 24,69 - 78 (1986)
9) MacMahon, H.; C.J. Vyborny; G. Powell; K. Doi; Ch.E. Metz:
The Effect of Pixel Size on the Detection Rate of Early Pulmonary Sarcoidosis
in Digital Chest Radiographic Systems
SPIE 486,14-20 (1984).
10) MacMahon, H.; C.J. Vyborny; C.E. Metz; K.Doi; V. Sabeti; St.L. Solomon :
Digital Radiography of Subtle Pulmonary Abnormalities:
An ROC Study of the Effect of Pixel Size on Observer Performance
Radiology 158,21-26 (1988).
11) Oestmann,J.W.; R.Greene; D.C.Kushner, P.M.Bourgouin; L.Linetsky; H.J.Llewellyn :
Lung Lesions: Correlation between Viewing Time and Detection
Radiology 166,451-453 (1988).
12) Revesz, G.; C. Haas: Television display of radiographic images with superimposed simulated lesions
Radiology 102,197-199 (1972)
13) Revesz, G.: Conspicuity and Uncertainty in the Radiographic Detection of Lesions
Radiology 154,625-628 (1985)
14) Vassallo, P.; W. Wiesmann; N. Roos; P.E. Peters
A Simply Method for the Simulation of Intrapulmonary Nodules
RSNA Chicago SE (1989)

Distributed Acquisition of Digital Images in a Rural Setting

Charles E. Willis[1] and Don F. Schomer[2]

[1] Teleradiology Networks Company, 8914 Rustling Branches, San Antonio, TX 78250
[2] Department of Radiology, University of Texas Health Science Center, 7703 Floyd Curl Drive, San Antonio, TX 78284

Abstract: Hospitals of less than 100 beds constitute over half of hospitals in the West South Central US. These hospitals can seldom afford to maintain a full-time radiologist on staff, and thus depend largely on intermittent interpretation by itinerant radiologists. Acquisition of images from these hospitals in digital form and transmission to a central location for remote diagnosis would provide continuous and emergent coverage with response times similar to large hospitals. Routine interpretation of digitized images from a group of small hospitals will generate a library of images that exceeds the number found in a large hospital. We are engaged in development and installation of a distributed network to collect and interpret images from 6 rural hospitals and 1 mobile CT facility. Design of the network requires assumptions regarding the volume of images produced by each hospital. Design decisions include spatial resolution necessary for diagnosis, data compression, transmission rate, mode of transmission, features of the radiologist's workstation, and method of archival storage. A distributed acquisition system can be installed and operated in a cost-effective manner using existing technology. System design and cost are affected dramatically by transmission rate. The required transmission rate can be determined empirically with minimum capital expenditure.

Keywords: Teleradiology, Rural Radiology, Telemedicine

Introduction

A time-honored "standard of care" is that every medical image will be interpreted by a radiologist qualified to assess subtle findings. The timely return of expert knowledge from the radiologist to the primary care physician is the single greatest problem in the practice of Radiology today. This problem is most acute in rural health care settings [1].

The ability to deliver quality health care in a rural setting is complicated by limited resources and staffing problems. Even hospitals that have sufficient resources to maintain a full-time radiologist on staff may not be able to hire one because of a shortage of radiologists and individual preferences for large hospitals, large patient loads, and metropolitan settings. Without a radiologist on staff, small hospitals are exposed to significant risk of medical/legal liability.

To mitigate this risk, small hospitals retain the services of itinerant radiologists who visit the hospital several days per week to interpret accumulated studies. This arrangement is less than satisfactory for all parties. There is a substantial delay in feedback of results to staff physicians for routine x-ray studies. There is no on-call Radiology support for emergency room physicians. Without a radiologist in-house, rural hospitals make limited use of mobile CT and MRI services. Hospital capabilities are underutilized because patients who could be easily managed locally with on-hand Radiology support, are evacuated instead to avoid further risk [2]. The itinerant radiologist has large blocks of nonproductive time during his transit between hospitals.

NATO ASI Series, Vol. F 74
Picture Archiving and Communication Systems (PACS) in Medicine. Edited by H. K. Huang et al.

Teleradiology is the practice of Radiology at a distance. Since the early 1970's [3,4], physicians, scientists, and engineers have worked to apply telecommunications technology to improve the quality of health care delivery. Today, technology has matured such that a variety of commercial products are available for Teleradiology. Selection of specific products requires assumptions about the caseload, definition of performance requirements, and analysis of profitability.

Materials and Methods

The observations noted here are based on a review of the literature, an evaluation of commercial products, and a survey of Radiology operations at six rural hospitals within 120 miles of San Antonio for a total of 319 beds. Our study included a mobile CT operation that serves several of these hospitals.

Results and Discussion

The caseload from rural hospitals can be estimated from the number of beds. We plan operations based on 15 radiologic studies per bed-month. On weekdays, about half these studies are generated between 6AM and 9AM, 40% are produced from 9AM to 5PM, and 10% occur during the night. We expect that studies other than plain film will comprise 10% of the total caseload. While interpretation time varies widely among individual radiologists and exams [5], a reasonable interpretation rate is 15 studies per hour. At this rate, two radiologists could adequately serve all six hospitals.

The functions of the Teleradiology Network are as follows: (1) acquire digital images from ordinary x-ray, CT, MR, Nuclear Medicine, and US modalities at rural hospitals; (2) transmit the digital images and associated patient information to a central node; (3) receive images and temporarily store, schedule for interpretation, and ultimately archive them; (4) transfer digital images to a radiologist; (5) display and manipulate images for interpretation; (6) record the interpretation; and (7) report the radiologist's findings to the hospital. The technological approach to accomplish each of these seven critical functions can be selected by comparison to performance requirements. Final decisions that determine the system design are based on functional requirements and profitability.

Profitability is the greatest obstacle to establishing a Teleradiology network. One vendor offers a turnkey point-to-point satellite transmission system featuring near real-time interpretation of images at a cost of $270K for a central node and $156K per hospital. Satellite transmission charges alone constitute 40% of the radiologist's interpretation fee for a typical two-view chest study. A recent report [6] describes a PC-based Teleradiology system that costs $344K for a central interpretation facility and $196K per hospital. We have identified a PC-based system that will support our requirements costing only about $64K for the central node and $44K for each hospital.

The inexpensive PC-based system is not designed to support 512 kbps satellite transmission, but can operate at 56 kbps via the switched-56 digital network offered by commercial telephone companies. While slower than satellite, switched-56 transmission costs for a chest study are only 4% of the interpretation fee. The size of image files (2K X 2K X 12 bits) has led some investigators [6] to expect typical transmission times per study to exceed 1 hour at 56 kbps. With compression of image data, a practical necessity for any Teleradiology or PACS system, and an average of two views per plain film study, transmission time drops to about 10 minutes per study. Public domain software is available for lossless compression ratios up to 4:1. Operations developed around a switched-56 network should adapt readily to the 64 kbps data channels of ISDN when it becomes available [7]. A rural hospital's cost for the PC-based system including hardware lease, transmission costs, and fees for emergent interpretation would total about $5K per month, or about $10 per study.

This cost does not seem excessive until one realizes that interpretation services are provided by itinerant radiologists essentially at no cost to the hospital. The Radiology Department may be the only profitable operation in a rural hospital, and may subsidize other critical services. Hospitals with less than 50 beds may not be able to absorb this cost. Two conditions could make the cost acceptable: if the hospital perceives that better resource management will offset the expense, or if neither a staff nor visiting radiologist can be recruited.

Rural hospitals may be resistant to changes in routine operations to accommodate Teleradiology. A Teleradiology system cannot support interventional studies such as angiography. Although ultrasonography can be performed by a trained technician, fluoroscopy requires that a radiologist visit the hospital periodically to perform the exams. While advanced scheduling of screening exams is not a problem, fluoroscopy visitation compromises the Teleradiology concept. Fluoroscopy exams might otherwise be performed by a Physician Assistant or Technologist with remote supervision, however, this is not yet accepted practice. In any event, contrast-enhanced studies would require in-house supervision by a physician trained to recognize and manage allergic drug reactions.

The diagnosis of plain film studies from a CRT is a significant deviation from the standard of care that enjoys lively discussion in the literature [5,8,9,10]. Certainly some details will be lost in digitizing the film and displaying on a device with limited contrast. The extent that losses will affect the radiologist's detection of disease is not yet clear. We feel that remote diagnosis of a digitized image within 24 hours represents a reasonable risk when compared to the alternative of waiting 72 hours for interpretation of the film.

Literature

1. Bryan, E.L., et al.: Communications' Role in Rural Health Care. unpublished manuscript available from Vortech Data, Inc., Reston, VA.

2. Barton, E., et al.: Influence on patient management of general practitioner direct access to radiological services. British Journal of Radiology. 60, 893-896 (1987)

3. Gershon-Cohen, J. and Cooley, A.G.: Telognosis. Radiology. 55, 582-587 (1950)

4. Murphy, R.L.H, et al.: Microwave Transmission of Chest Roentgenograms. American Review of Respiratory Disease. 102, 771-777 (1970)

5. Simsky, B.S., et al.: Receiver Operator Characteristic Analysis of Chest Image Interpretation with Conventional, Laser-printed, and High Resolution Workstation Images. Radiology. 174, 775-780 (1990)

6. Batnitzky, S., et al.: Teleradiology: An Assessment. Radiology 177, 11-17 (1990)

7. Lear, J.L., et al.: Ultra-High-Speed Teleradiology with ISDN Technology. Radiology. 171, 862-863 (1989)

8. Seely, G.W., et al.: Total Digital Radiology Department: Spatial Resolution Requirements. American Journal of Radiology. 148, 421-426 (1987)

9. Arenson, R.L. et al.: The Digital Imaging Workstation. Radiology. 176, 303-315 (1990)

10. Cox, G.G., et al.: Chest Radiography: Comparison of High Resolution Digital Displays with Conventional and Digital Film. Radiology. 176, 771-776 (1990)

Application of a PACS in daily routine

Marco Wiltgen[1], Günther Gell[1] and Gerhard H. Schneider[2]

[1] Institut für Medizinische Informatik, Statistik und Dokumentation Universität Graz A-8036 Graz, Austria
[2] Universitätsklinik für Radiologie Universität Graz A-8036 Graz, Austria

Abstract: An overview about the different PACS-projects in Austria is presented and the experiences with a PACS in routine are discussed.

Keywords: Picture Archiving and Communication Systems, PACS

Overview about the PACS projects in Austria

The main activities in PACS research in Austria are in Graz and Vienna. Only a few PAC-Systems are realized and in use. The most systems are still in a planning stage.
Graz: In collaboration with Siemens Erlangen a PACS was installed step by step at the Department of Radiology of the University of Graz. This PACS is used in routine. A second PACS is planned in the Department of Pediatric Surgery. In a later state of the project the two PACS will be connected.
Vienna: At the Hanusch hospital a Philips PACS is in use since one year.This PACS includes an Angiography and a Cardiography. Three other projects are in a planning stage. For the Social Medical Center (SMZ Ost) a Siemens PACS which includes a lot of modalities is planed. This PACS will be based on 3 FDDI-rings. In the new general hospital (AKH) a PACS, which connects several CT-scanner , several NMR-scanner and an Angiography, will be installed. At the Lorenz-Böhler hospital a Philips AT&T PACS will be used.

The PACS configuration at the Department of Radiology of the University of Graz

In the year 1985 there was the begin of a pilot project of the Department of Radiology of the University of Graz and Siemens Erlangen. The PACS was

NATO ASI Series, Vol. F 74
Picture Archiving and Communication Systems (PACS) in Medicine. Edited by H. K. Huang et al.

installed step by step and includes at the moment two networks based on Ethernet: an hospital network wich connects several departments and a local network. The local network joins 3 Siemens SOMATOM DR CT-scanner, an evaluation console, an archive with two drives for optical disks and a PACS diagnostic console. The diagnostic console is the gateway between the two networks. The hospital network connects the SOMATOM DR CT-scanner from the Radiotherapy, which is about 1 km away from the Department of Radiology, a Philips GYROSCAN NMR-scanner and several terminals of a RIS. This RIS was developed in the Department of Radiology and is named AURA [1]. The PACS is linked to the RIS in that way, that the optical image archive is managed by the RIS [2].

PACS in routine use

All the examinations from the 4 CT-scanners are archived on optical disk [3]. Only a few manual manipulations are needed to archive and retrieve images. Most operations are done automatically by software processes in the background.

Archiving: The archiving is initiated by entering patient data, the examination number, the archive identification and the number of images to be transferred at the console of the CT-scanner. These are the only manual operations needed, all the following operations are executed by the software processes in the background. First the images are converted into the ACR-Nema standard format [4]. Then the patient-, examination- and image data are submitted into the IMS (Image Managment System) database. After the examination is submitted completly into the IMS, the images are copied on optical disk. If the optical disk is full, a corresponding message is sent back to the CT-scanner and a new disk must be inserted into the disk drive. The disks are identified by running numbers. After the report text is inserted into the RIS, the number of the disk where the images are stored and the examination identification are added automatically to the report text. The CT-scanners are older models with an inflexible PACS interface and they are not able to accept patient identification data from the RIS. Therefore this data must be entered again and errors may occur. If some patient data in the PACS and the RIS are not corresponding, the PACS-RIS compare the other data (for example the examination number) and if they are corresponding the disk

number and the examination identification are added to the report text.
The managment of the optical disks and the retrieving is performed by the RIS.
Retrieving: The retrieving is initiated by selecting a patient with patient´s name and birthdate at a RIS terminal. Then a list of the examinations is displayed. After the selection of an examination the corresponding images are retrieved. If the images are not on the disk which is actually in the disk drive, the number of the desired disk is displayed at the archive console and a message that the archive needs operating is sended to the console of the CT-scanner. After the disk is loaded the images are copied to the diagnostic console.
The diagnostic console has three high resolution monitors. This console makes it possible to display 4 overview images (with 16 images) simultaniously on the overview screen. On the other two monitors 16 images with 256x256 pixels and 4 images with 512x512 pixels can be displayed simultaniously. All the operations on the console are activated by a mouse. If images are transferred from the CT-scanner to the diagnostic console the RIS is inspected automatically for previous examinations. If there are previous examinations the report texts are copied to the diagnostic console.

Performance data

At the moment we can store 26000 images on one optical disk. The images are stored with a matrix of 256x256 pixels and a compression of a factor two.
If an examination with 30 images is transferred to the archive it takes 5 minutes until the whole examination is submitted into the image data base and another 5 minutes to copy the examination on optical disk.
After the initiation of the retrieving, an examination with 30 images is available at the diagnostic console after 7 minutes.

Experiences and problems

In summary we have made good experiences with the archiving on optical disks. Due to the problemless handling of the optical disks there really exists economization of labour and time.
Because of it´s central function in the department, failures which occur in some parts of the PACS may spread out and hamper the routine work.

Examples of such failures are software crashes, the interruption of the connection between 2 processes or a hardware crash. Especially during the night and weekend shifts when no systemmanager is present, the routine work is then critically influenced.
The reporting on the diagnostic console is still not accepted in routine. The main reason is that not all the images of an examination and the images of the previous examinations can be displayed simultaniously on the monitors of the console.

Planned developments

We want to connect peripherical departments (e.g. neurosurgery) to the PACS. Then image consols are needed which are mainly or exclusively used for image viewing. Such consols (based on PC) will be installed in the surgery and allows the surgeon a direct access to the images.
There is also an increasing need for software which allows the user easily to detect malfunctions of the PACS during the routine work and to eliminate the troubles himselves.

References

1. Gell G.:AURA: A Clinical Data Bank Based on Free Text. In: R. R. O'Moore, B. Barber, P. L. Reichertz, F. Roger (eds): Lecture Notes in Medical Informatics, Vol. 16, Springer, Berlin-Heidelberg-New York (1982) 850-856.
2. Gell G., Schneider G. H., Wiltgen M.: PACS-RIS Interfacing: Experiences and Problems. In: Lemke H. U., Rhodes M. L:, Jaffe C. C., Felix R. (eds.): Computer Assisted Radiology, Springer, Berlin-Heidelberg-New York (1989) 623-627.
3. Gell G., Schneider G. H.,Wiltgen M., Porsch G.: PACS - Praktische Erfahrungen: Archivierung, Kommunikation, Befundung. In Schneider G. H., Vogler E., Kocevar K. (eds.): Digitale Bildgebung Interventionelle Radiologie Integrierte digitale Radiologie, 6. Grazer Radiologisches Symposium (1989) Blackwell Ueberreuter Wissenschaft Berlin 536-540
4. Digital Imaging and Communications.
ACR-NEMA publication no 300, 1985, Washington D. C.

List of Authors

Subject Index

NATO ASI Series F

Including Special Programmes on Sensory Systems for Robotic Control (ROB) and on Advanced Educational Technology (AET)

Vol. 1: Issues in Acoustic Signal – Image Processing and Recognition. Edited by C. H. Chen. VIII, 333 pages. 1983.

Vol. 2: Image Sequence Processing and Dynamic Scene Analysis. Edited by T. S. Huang. IX, 749 pages. 1983.

Vol. 3: Electronic Systems Effectiveness and Life Cycle Costing. Edited by J. K. Skwirzynski. XVII, 732 pages. 1983.

Vol. 4: Pictorial Data Analysis. Edited by R. M. Haralick. VIII, 468 pages. 1983.

Vol. 5: International Calibration Study of Traffic Conflict Techniques. Edited by E. Asmussen. VII, 229 pages. 1984.

Vol. 6: Information Technology and the Computer Network. Edited by K. G. Beauchamp. VIII, 271 pages. 1984.

Vol. 7: High-Speed Computation. Edited by J. S. Kowalik. IX, 441 pages. 1984.

Vol. 8: Program Transformation and Programming Environments. Report on a Workshop directed by F. L. Bauer and H. Remus. Edited by P. Pepper. XIV, 378 pages. 1984.

Vol. 9: Computer Aided Analysis and Optimization of Mechanical System Dynamics. Edited by E. J. Haug. XXII, 700 pages. 1984.

Vol. 10: Simulation and Model-Based Methodologies: An Integrative View. Edited by T. I. Ören, B. P. Zeigler, M. S. Elzas. XIII, 651 pages. 1984.

Vol. 11: Robotics and Artificial Intelligence. Edited by M. Brady, L. A. Gerhardt, H. F. Davidson. XVII, 693 pages. 1984.

Vol. 12: Combinatorial Algorithms on Words. Edited by A. Apostolico, Z. Galil. VIII, 361 pages. 1985.

Vol. 13: Logics and Models of Concurrent Systems. Edited by K. R. Apt. VIII, 498 pages. 1985.

Vol. 14: Control Flow and Data Flow: Concepts of Distributed Programming. Edited by M. Broy. VIII, 525 pages. 1985.

Vol. 15: Computational Mathematical Programming. Edited by K. Schittkowski. VIII, 451 pages. 1985.

Vol. 16: New Systems and Architectures for Automatic Speech Recognition and Synthesis. Edited by R. De Mori, C.Y. Suen. XIII, 630 pages. 1985.

Vol. 17: Fundamental Algorithms for Computer Graphics. Edited by R.A. Earnshaw. XVI, 1042 pages. 1985.

Vol. 18: Computer Architectures for Spatially Distributed Data. Edited by H. Freeman and G.G. Pieroni. VIII, 391 pages. 1985.

Vol. 19: Pictorial Information Systems in Medicine. Edited by K. H. Höhne. XII, 525 pages. 1986.

Vol. 20: Disordered Systems and Biological Organization. Edited by E. Bienenstock, F. Fogelman Soulié, G. Weisbuch. XXI, 405 pages.1986.

Vol. 21: Intelligent Decision Support in Process Environments. Edited by E. Hollnagel, G. Mancini, D.D. Woods. XV, 524 pages. 1986.

NATO ASI Series F

Vol. 22: Software System Design Methods. The Challenge of Advanced Computing Technology. Edited by J.K. Skwirzynski. XIII, 747 pages. 1986.

Vol. 23: Designing Computer-Based Learning Materials. Edited by H. Weinstock and A. Bork. IX, 285 pages. 1986.

Vol. 24: Database Machines. Modern Trends and Applications. Edited by A.K. Sood and A.H. Qureshi. VIII, 570 pages. 1986.

Vol. 25: Pyramidal Systems for Computer Vision. Edited by V. Cantoni and S. Levialdi. VIII, 392 pages. 1986. *(ROB)*

Vol. 26: Modelling and Analysis in Arms Control. Edited by R. Avenhaus, R.K. Huber and J.D. Kettelle. VIII, 488 pages. 1986.

Vol. 27: Computer Aided Optimal Design: Structural and Mechanical Systems. Edited by C.A. Mota Soares. XIII, 1029 pages. 1987.

Vol. 28: Distributed Operating Systems. Theory und Practice. Edited by Y. Paker, J.-P. Banatre and M. Bozyiğit. X, 379 pages. 1987.

Vol. 29: Languages for Sensor-Based Control in Robotics. Edited by U. Rembold and K. Hörmann. IX, 625 pages. 1987. *(ROB)*

Vol. 30: Pattern Recognition Theory and Applications. Edited by P.A. Devijver and J. Kittler. XI, 543 pages. 1987.

Vol. 31: Decision Support Systems: Theory and Application. Edited by C.W. Holsapple and A.B. Whinston. X, 500 pages. 1987.

Vol. 32: Information Systems: Failure Analysis. Edited by J.A. Wise and A. Debons. XV, 338 pages. 1987.

Vol. 33: Machine Intelligence and Knowledge Engineering for Robotic Applications. Edited by A.K.C. Wong and A. Pugh. XIV, 486 pages. 1987. *(ROB)*

Vol. 34: Modelling, Robustness and Sensitivity Reduction in Control Systems. Edited by R.F. Curtain. IX, 492 pages. 1987.

Vol. 35: Expert Judgment and Expert Systems. Edited by J.L. Mumpower, L.D. Phillips, O. Renn and V.R.R. Uppuluri. VIII, 361 pages. 1987.

Vol. 36: Logic of Programming and Calculi of Discrete Design. Edited by M. Broy. VII, 415 pages. 1987.

Vol. 37: Dynamics of Infinite Dimensional Systems. Edited by S.-N. Chow and J.K. Hale. IX, 514 pages. 1987.

Vol. 38: Flow Control of Congested Networks. Edited by A.R. Odoni, L. Bianco and G. Szegö. XII, 355 pages. 1987.

Vol. 39: Mathematics and Computer Science in Medical Imaging. Edited by M.A. Viergever and A. Todd-Pokropek. VIII, 546 pages. 1988.

Vol. 40: Theoretical Foundations of Computer Graphics and CAD. Edited by R.A. Earnshaw. XX, 1246 pages. 1988.

Vol. 41: Neural Computers. Edited by R. Eckmiller and Ch. v. d. Malsburg. XIII, 566 pages. 1988.

NATO ASI Series F

Vol. 42: Real-Time Object Measurement and Classification. Edited by A. K. Jain. VIII, 407 pages. 1988. *(ROB)*

Vol. 43: Sensors and Sensory Systems for Advanced Robots. Edited by P. Dario. XI, 597 pages. 1988. *(ROB)*

Vol. 44: Signal Processing and Pattern Recognition in Nondestructive Evaluation of Materials. Edited by C. H. Chen. VIII, 344 pages. 1988. *(ROB)*

Vol. 45: Syntactic and Structural Pattern Recognition. Edited by G. Ferraté, T. Pavlidis, A. Sanfeliu and H. Bunke. XVI, 467 pages. 1988. *(ROB)*

Vol. 46: Recent Advances in Speech Understanding and Dialog Systems. Edited by H. Niemann, M. Lang and G. Sagerer. X, 521 pages. 1988.

Vol. 47: Advanced Compuling Concepts and Techniques in Control Engineering. Edited by M. J. Denham and A. J. Laub. XI, 518 pages. 1988.

Vol. 48: Mathematical Models for Decision Support. Edited by G. Mitra. IX, 762 pages. 1988.

Vol. 49: Computer Integrated Manufacturing. Edited by I. B. Turksen. VIII, 568 pages. 1988.

Vol. 50: CAD Based Programming for Sensory Robots. Edited by B. Ravani. IX, 565 pages. 1988. *(ROB)*

Vol. 51: Algorithms and Model Formulations in Mathematical Programming. Edited by S. W. Wallace. IX, 190 pages. 1989.

Vol. 52: Sensor Devices and Systems for Robotics. Edited by A. Casals. IX, 362 pages. 1989. *(ROB)*

Vol. 53: Advanced Information Technologies for Industrial Material Flow Systems. Edited by S. Y. Nof and C. L. Moodie. IX, 710 pages. 1989.

Vol. 54: A Reappraisal of the Efficiency of Financial Markets. Edited by R. M. C. Guimarães, B. G. Kingsman and S. J. Taylor. X, 804 pages. 1989.

Vol. 55: Constructive Methods in Computing Science. Edited by M. Broy. VII, 478 pages. 1989.

Vol. 56: Multiple Criteria Decision Making and Risk Analysis Using Microcomputers. Edited by B. Karpak and S. Zionts. VII, 399 pages. 1989.

Vol. 57: Kinematics and Dynamic Issues in Sensor Based Control. Edited by G. E. Taylor. XI, 456 pages. 1990. *(ROB)*

Vol. 58: Highly Redundant Sensing in Robotic Systems. Edited by J. T. Tou and J. G. Balchen. X, 322 pages. 1990. *(ROB)*

Vol. 59: Superconducting Electronics. Edited by H. Weinstock and M. Nisenoff. X, 441 pages. 1989.

Vol. 60: 3D Imaging in Medicine. Algorithms, Systems, Applications. Edited by K. H. Höhne, H. Fuchs and S. M. Pizer. IX, 460 pages. 1990.

Vol. 61: Knowledge, Data and Computer-Assisted Decisions. Edited by M. Schader and W. Gaul. VIII, 421 pages. 1990.

Vol. 62: Supercomputing. Edited by J. S. Kowalik. X, 425 pages. 1990.

Vol. 63: Traditional and Non-Traditional Robotic Sensors. Edited by T. C. Henderson. VIII, 468 pages. 1990. *(ROB)*

Vol. 64: Sensory Robotics for the Handling of Limp Materials. Edited by P. M. Taylor. IX, 343 pages. 1990. *(ROB)*

Vol. 65: Mapping and Spatial Modelling for Navigation. Edited by L. F. Pau. VIII, 357 pages. 1990. *(ROB)*

NATO ASI Series F

Vol. 66: Sensor-Based Robots: Algorithms and Architectures. Edited by C. S. G. Lee. X, 285 pages. 1991. *(ROB)*

Vol. 67: Designing Hypermedia for Learning. Edited by D. H. Jonassen and H. Mandl. XXV, 457 pages. 1990. *(AET)*

Vol. 68: Neuro-Computing. Algorithms, Architectures and Applications. Edited by F. Fogelman Soulié and J. Hérault. XI, 455 pages. 1990.

Vol. 69: Real-Time Integration Methods for Mechanical System Simulation. Edited by E. J. Haug and R. C. Deyo. VIII, 352 pages. 1991.

Vol. 70: Numerical Linear Algebra, Digital Signal Processing and Parallel Algorithms. Edited by G. H. Golub and P. Van Dooren. XIII, 729 pages. 1991.

Vol. 71: Expert Systems and Robotics. Edited by T. Jordanides and B. Torby. XII, 744 pages. 1991.

Vol. 72: High-Capacity Local and Metropolitan Area Networks. Architecture and Performance Issues. Edited by G. Pujolle. X, 536 pages. 1991.

Vol. 73: Automation and Systems Issues in Air Traffic Control. Edited by J. A. Wise, V. D. Hopkin and M. L. Smith.

Vol. 74: Picture Archiving and Communication Systems (PACS) in Medicine. Edited by H. K. Huang, O. Ratib, A. R. Bakker and G. Witte. XII, 442 pages. 1991.

MIX
Papier aus verantwortungsvollen Quellen
Paper from responsible sources
FSC® C105338

If you have any concerns about our products,
you can contact us on
ProductSafety@springernature.com

In case Publisher is established outside the EU,
the EU authorized representative is:
Springer Nature Customer Service Center GmbH
Europaplatz 3, 69115 Heidelberg, Germany

Printed by Libri Plureos GmbH
in Hamburg, Germany